# Forest Hydrology and Watershed

# Forest Hydrology and Watershed

Special Issue Editors

**Fan-Rui Meng**
**Qiang Li**
**M. Altaf Arain**
**Michael Pisaric**

MDPI • Basel • Beijing • Wuhan • Barcelona • Belgrade

MDPI

*Special Issue Editors*

Fan-Rui Meng
University of New Brunswick
Canada

Qiang Li
University of Victoria
Canada

M. Altaf Arain
McMaster University
Canada

Michael Pisaric
Brock University
Canada

*Editorial Office*
MDPI
St. Alban-Anlage 66
4052 Basel, Switzerland

This is a reprint of articles from the Special Issue published online in the open access journal *Forests* (ISSN 1999-4907) from 2018 to 2019 (available at: https://www.mdpi.com/journal/forests/special_issues/Hydrology_Watershed).

For citation purposes, cite each article independently as indicated on the article page online and as indicated below:

LastName, A.A.; LastName, B.B.; LastName, C.C. Article Title. *Journal Name* **Year**, *Article Number*, Page Range.

ISBN 978-3-03921-385-6 (Pbk)
ISBN 978-3-03921-386-3 (PDF)

# Contents

About the Special Issue Editors . . . . . . . . . . . . . . . . . . . . . . . . . . . . . . . . . . vii

Preface to "Forest Hydrology and Watershed" . . . . . . . . . . . . . . . . . . . . . . . . . . ix

Tayierjiang Aishan, Florian Betz, Ümüt Halik, Bernd Cyffka and Aihemaitijiang Rouzi
Biomass Carbon Sequestration Potential by Riparian Forest in the Tarim River Watershed,
Northwest China: Implication for the Mitigation of Climate Change Impact
Reprinted from: *Forests* **2018**, *9*, 196, doi:10.3390/f10020196 . . . . . . . . . . . . . . . . . . . . . 1

Xu-dong Huang, Ling Wang, Pei-pei Han and Wen-chuan Wang
Spatial and Temporal Patterns in Nonstationary Flood Frequency across a Forest Watershed:
Linkage with Rainfall and Land Use Types
Reprinted from: *Forests* **2018**, *9*, 339, doi:10.3390/f9060339 . . . . . . . . . . . . . . . . . . . . . 16

Liangliang Duan and Tijiu Cai
Quantifying Impacts of Forest Recovery on Water Yield in Two Large Watersheds in the Cold
Region of Northeast China
Reprinted from: *Forests* **2018**, *9*, 392, doi:10.3390/f9070392 . . . . . . . . . . . . . . . . . . . . . 36

Yoshitaka Oishi
Evaluation of the Water-Storage Capacity of Bryophytes along an Altitudinal Gradient from
Temperate Forests to the Alpine Zone
Reprinted from: *Forests* **2018**, *9*, 433, doi:10.3390/f9070433 . . . . . . . . . . . . . . . . . . . . . 54

Yiping Hou, Mingfang Zhang, Shirong Liu, Pengsen Sun, Lihe Yin, Taoli Yang, Yide Li,
Qiang Li and Xiaohua Wei
The Hydrological Impact of Extreme Weather-Induced Forest Disturbances in a Tropical
Experimental Watershed in South China
Reprinted from: *Forests* **2018**, *9*, 734, doi:10.3390/f9120734 . . . . . . . . . . . . . . . . . . . . . 68

Qinli Yang, Shasha Luo, Hongcai Wu, Guoqing Wang, Dawei Han, Haishen Lü and
Junming Shao
Attribution Analysis for Runoff Change on Multiple Scales in a Humid Subtropical Basin
Dominated by Forest, East China
Reprinted from: *Forests* **2019**, *10*, 184, doi:10.3390/f10020184 . . . . . . . . . . . . . . . . . . . . 89

Krysta Giles-Hansen, Qiang Li and Xiaohua Wei
The Cumulative Effects of Forest Disturbance and Climate Variability on Streamflow in the
Deadman River Watershed
Reprinted from: *Forests* **2019**, *10*, 196, doi:10.3390/f9040196 . . . . . . . . . . . . . . . . . . . . . 107

Na Liu, Guang Bao, Yu Liu and Hans W. Linderholm
Two Centuries-Long Streamflow Reconstruction Inferred from Tree Rings for the Middle
Reaches of the Weihe River in Central China
Reprinted from: *Forests* **2019**, *10*, 208, doi:10.3390/f10030208 . . . . . . . . . . . . . . . . . . . . 123

Zhipeng Xu, Wenfei Liu, Xiaohua Wei, Houbao Fan, Yizao Ge, Guanpeng Chen and Jin Xu
Contrasting Differences in Responses of Streamflow Regimes between Reforestation and Fruit
Tree Planting in a Subtropical Watershed of China
Reprinted from: *Forests* **2019**, *10*, 212, doi:10.3390/f10030212 . . . . . . . . . . . . . . . . . . . . 137

**Ruibo Zhang, Bakytbek Ermenbaev, Tongwen Zhang, Mamtimin Ali, Li Qin and Rysbek Satylkanov**
The Radial Growth of Schrenk Spruce (*Picea schrenkiana* Fisch.  et Mey.)  Records the Hydroclimatic Changes in the Chu River Basin over the Past 175 Years
Reprinted from: *Forests* **2019**, *10*, 223, doi:10.3390/f10030223 . . . . . . . . . . . . . . . . . . . . . . . **152**

**Siyi Tan, Hairong Zhao, Wanqin Yang, Bo Tan, Kai Yue, Yu Zhang, Fuzhong Wu and Xiangyin Ni**
Forest Canopy Can Efficiently Filter Trace Metals in Deposited Precipitation in a Subalpine Spruce Plantation
Reprinted from: *Forests* **2019**, *10*, 318, doi:10.3390/f10040318 . . . . . . . . . . . . . . . . . . . . . . . **163**

**Ziyi Liang, Fuzhong Wu, Xiangyin Ni, Bo Tan, Li Zhang, Zhenfeng Xu, Junyi Hu and Kai Yue**
Woody Litter Increases Headwater Stream Metal Export Ratio in an Alpine Forest
Reprinted from: *Forests* **2019**, *10*, 379, doi:10.3390/f10050379 . . . . . . . . . . . . . . . . . . . . . . . **176**

# About the Special Issue Editors

**Fan-Rui Meng**, Dr., started his career with a Bachelor and Master Degrees in Forest Engineering from Northeast Forestry University in Harbin, China. He then completed his Ph.D. in forest ecology from the University of New Brunswick, Canada. In 1997, Dr. Meng joined the Faculty of Forestry and Environmental Management as Industrial Research Chair, and served as the Research Director for the Noranda/Avenor Forest Watershed Management and Conservation Research Centre and later on Bowater/Nexfor Forest Watershed Research Centre.

**Qiang Li**, Dr., is a postdoctoral fellow in the Department of Civil Engineering at the University of Victoria, Canada. He obtained his Ph.D. degree at the University of British Columbia, Okanagan, Canada in 2018. He completed his Master's studies at the University of New Brunswick, Canada in 2013. His finished his BSE in electrical engineering at Xi'an Jiaotong University City College in 2010. His research interests include forest change and climate change impacts on hydrological processes, surface water and groundwater interactions, and global water sustainability.

**M. Altaf Arain**, Dr., is a professor in the School of Geography and Earth Science at McMaster University, Hamilton, Ontario, Canada. He is also the Director of the McMaster Centre for Climate Change. He is also associate member of the McMaster Department of Civil Engineering and United Nations University. He was the president of the Canadian Geophysical Union (CGU)—Biogeosciences Section—from 2009 to 2012. He graduated from the Department of Hydrology and Water Resources, at the University of Arizona, Tucson, Arizona, USA. He has expertise in both micrometeorological measurements and land-surface atmosphere interaction modeling.

**Michael Pisaric**, Dr., is a Professor and current Chair of the Department of Geography and Tourism Studies at Brock University, Ontario, Canada. He received his Ph.D. from Queen's University in Kingston (Ontario) in 2001. Michael employs a number of paleoecological techniques to carry out his research, including lake-sediment analysis (fossil pollen, stomata and charcoal) and dendrochronology. He has carried out research in a number of regions throughout the world, including Siberia, western Canada and Montana (USA). He currently has ongoing research projects in Northwest Territories, Yukon, and Southern Ontario.

# Preface to "Forest Hydrology and Watershed"

Qiang Li [1], Fan-Rui Meng [2], M. Altaf Arain [3] and Michael F.J. Pisaric [4]

[1]   Department of Civil Engineering, University of Victoria, 3800 Finnerty Road, Victoria, BC V8W 2Y2, Canada

[2]   Faculty of Forestry and Environmental Management, University of New Brunswick, 28 Dineen Drive. Fredericton, NB E3B 5A3, Canada

[3]   School of Geography and Earth Sciences, McMaster University, 1280 Main Street West, Hamilton, ON L8S 4K1, Canada

[4]   Department of Geography and Tourism Studies Brock University 1812 Sir Isaac Brock Way, St. Catharines, ON L2S 3A1, Canada

Hydrological processes in forested watersheds are influenced by environmental, physiological, and biometric factors such as precipitation, radiation, temperature, species type, leaf area, and extent and structure of forest ecosystems. Over the past two centuries, forest coverage and forest structures have been impacted globally by anthropogenic activities, for example, forest harvesting, and conversion of forested landscapes for plantations and urbanization [1–3]. In addition, climate change since the industrial revolution has resulted in profound impacts on forest ecosystems due to higher carbon dioxide ($CO_2$) concentration or $CO_2$ fertilization, warmer temperatures, changes in frequency and intensity of extreme weather events and natural disturbances. As a result, hydrological processes in forested watersheds have been altered by these natural and anthropogenic factors and these changes are expected to accelerate due to future changing climatic conditions. Hence, understanding how various environmental, physiological, and physical drivers interactively influence hydrological and biogeochemical processes in forest ecosystems is critical for sustainable water supply in forested watersheds. About 21% of the global population depends on water sources that originate in forested catchments, where forest coverage larger than 30% [4]. Furthermore, there are knowledge gaps in our understanding of mechanism of hydrological and hydrochemical cycles in forested watersheds. This special issue addresses these gaps in our knowledge and includes twelve papers in the following three major research themes in forest watershed areas.

## 1. Effects of forest cover change on hydrological regimes

The effects of forest cover change on hydrological processes were evaluated through statistical and hydrological modelling approaches with watershed size ranging from 250 km$^2$ to 1000 km$^2$ across several forests biomes [5–7]. In boreal forests, Duan et al. [8] identified that forest recovery and climate variability combined caused a decrease in water yield in two large watersheds (>2500 km$^2$) in northeast China. In addition, forest recovery was identified as a more important driver for the water yield variation, which accounted for as much as 87.4% with the rest due to climate variability. In sub-boreal forests, Giles-Hansen et al. [9] found that forest disturbance (e.g., logging, mountain pine beetle infestation, and wildfire) and climate variability simultaneously increased the water yield in the Deadman River watershed (878 km$^2$) in British Columbia, Canada and revealed that forest disturbance played a more important role than climate variability in annual water yield variations. In sub-tropical regions, Xu et al. [10] determined that forest planting has reduced high flow and increased low flow significantly. In contrast, fruit tree plantation augmented high flow while had no significant impacts on low flow in a mid-sized watershed (261 km$^2$) due to the different forest structure and management practices adopted in orchards. In a similar region, Yang et al. [11] adopted the Soil and Water Assessment Tool in Qingliu River basin (1070 km$^2$) to explore climate variability, land-use change, and human activity in a large watershed at different temporal scales. At the annual scale, climate variability had the highest impact on annual runoff, while human activities had higher impact on seasonal and monthly runoff than other factors. In tropical forests, Hou et al. [12] revealed that weather-induced forest disturbance is the major culprit for water availability and low flow in a mid-sized watershed (450 km$^2$) in south China.

The overall conclusion of these studies was that reforestation decreases total water yields, while decrease in forest cover increase it at the watershed level [6,13] mainly because evapotranspiration from forest ecosystem is much higher as compared to other land-use types. However, due to the differences in watershed properties (i.e. topography, landforms, water storage), hydrological responses (i.e., water yield, high, and low flows) to forest cover change are not consistent across different sizes or spatial extents of watersheds. Therefore, we suggest that conclusions gained from one spatial scale cannot simply be extrapolated to others and more case studies are needed to advance our knowledge in this research theme.

## 2. Mechanism of hydrological and hydrochemical cycles in forest watersheds

Forests play an important role in hydrological and hydrochemical cycles. A better understanding of the mechanisms of these processes could enhance our knowledge of forest ecosystems. Huang et al. [14] quantified the impacts of rainfall and forest landscape on flood frequency, which suggested that rainfall and landscape are critical factors causing alterations in low-flow flooding events, while water infrastructure (e.g., dams) played a more important role in higher return period flood events. Aishan et al. [15] estimated the total biomass and carbon storage in the lower reaches of the Tarim River. In addition, they showed that total biomass decreased with increasing distance to the river with the threshold distance being 800 m. Oishi [16] examined the water storage capacity of bryophytes in mountainous forest regions in Japan and found that water storage capacity was affected by the forms of bryophyte communities (life forms) and their interactions. Liang et al. [17] examined how non-woody and woody litter affects water quality (e.g., metals in headwater streams) and identified that input of woody litter could significantly increase flow discharge and metal export ratio in the stream. Tan et al. [18] found that closed canopy forests have a greater capacity to filter trace metals from precipitation than a gap-edge canopy in a subalpine forest.

## 3. Long-term hydroclimatic changes using dendrochronology

Understanding long-term hydroclimatic dynamics has important implications for regional water resource management and provides important information concerning emerging regional climate change. Liu et al. [19] reconstructed summer streamflow for the past 186 years in the middle reaches of the Weihe River, China. This first-ever study in the Weihe River captured historic hydroclimate changes and observed that more extreme drought and flood events occurred in the 20th century as comparted to past. Similarly, Zhang et al. [20] reconstructed drought indices (i.e., standardized precipitation-evapotranspiration index (SPEI)) in the Chu River basin over the past 175 years. They identified that the changes in the reconstructed SPEI index were aligned with historical regional climate dynamics. These two studies provided an insight into the usefulness of dendrochronology for understanding long-term regional hydroclimate dynamics.

In summary, this special issue is a reflection of current trends in forest hydrology studies at the watershed level. More importantly, these studies stressed the role of forests or forest changes in hydrological and hydrochemical cycles and addressed several emerging research/environment issues at different spatial and temporal scales. We hope general readers, forest managers, environmental planners, and policy makers will find this information useful and productive.

**References**

1. Evaristo, J.; McDonnell, J.J. Global analysis of streamflow response to forest management. *Nature* **2019**, *570*, 455–461.

2. Foley, J.A.; DeFries, R.; Asner, G.P.; Barford, C.; Bonan, G.; Carpenter, S.R.; Chapin, F.S.; Coe, M.T.; Daily, G.C.; Gibbs, H.K.; et al. Global consequences of land use. *Science* **2005**, *309*, 570–574.

3. Vorosmarty, C.J.; Green, P.; Salisbury, J.; Lammers, R.B. Global water resources: Vulnerability from climate change and population growth. *Science* **2000**, *289*, 284–288.

4. Wei, X.; Li, Q.; Zhang, M.; Giles-Hansen, K.; Liu, W.; Fan, H.; Wang, Y.; Zhou, G.; Piao, S.; Liu, S. Vegetation cover - another dominant factor in determining global water resources in forested regions. *Global Change Biology* **2017**.

5. Andreassian, V. Waters and forests: from historical controversy to scientific debate. *Journal of Hydrology* **2004**, *291*, 1–27.

6. Zhang, M.; Liu, N.; Harper, R.; Li, Q.; Liu, K.; Wei, X.; Ning, D.; Hou, Y.; Liu, S. A global review on hydrological responses to forest change across multiple spatial scales: Importance of scale, climate, forest type and hydrological regime. *Journal of Hydrology* **2017**, *546*.

7. Brown, A.E.; Zhang, L.; McMahon, T.A.; Western, A.W.; Vertessy, R.A. A review of paired catchment studies for determining changes in water yield resulting from alterations in vegetation. *Journal of Hydrology* **2005**, *310*, 28–61.

8. Duan, L.; Cai, T. Quantifying impacts of forest recovery on water yield in two large watersheds in the cold region of northeast China. *Forests* **2018**.

9. Giles-Hansen, K.; Li, Q.; Wei, X. The Cumulative Effects of Forest Disturbance and Climate Variability on Streamflow in the Deadman River Watershed. *Forests* **2019**.

10. Xu, Z.; Liu, W.; Wei, X.; Fan, H.; Ge, Y.; Chen, G.; Xu, J. Contrasting differences in responses of streamflow regimes between reforestation and fruit tree planting in a subtropical watershed of China. *Forests* **2019**.

11. Yang, Q.; Luo, S.; Wu, H.; Wang, G.; Han, D.; Lü, H.; Shao, J. Attribution Analysis for Runoff Change on Multiple Scales in a Humid Subtropical Basin Dominated by Forest, East China. *Forests* **2019**.

12. Hou, Y.; Zhang, M.; Liu, S.; Sun, P.; Yin, L.; Yang, T.; Li, Y.; Li, Q.; Wei, X. The Hydrological Impact of Extreme Weather-Induced Forest Disturbances in a Tropical Experimental Watershed in South China. *Forests* **2018**.

13. Li, Q.; Wei, X.; Zhang, M.; Liu, W.; Fan, H.; Zhou, G.; Giles-Hansen, K.; Liu, S.; Wang, Y. Forest cover change and water yield in large forested watersheds: A global synthetic assessment. *Ecohydrology* **2017**, *10*.

14. Huang, X.D.; Wang, L.; Han, P.P.; Wang, W.C. Spatial and temporal patterns in nonstationary flood frequency across a forest watershed: Linkage with rainfall and land use types. *Forests* **2018**.

15. Aishan, T.; Betz, F.; Halik, Ü.; Cyffka, B.; Rouzi, A. Biomass carbon sequestration potential by riparian forest in the Tarim RiverWatershed, Northwest China: Implication for the mitigation of climate change impact. *Forests* **2018**.

16. Oishi, Y. Evaluation of the Water-Storage Capacity of Bryophytes along an Altitudinal Gradient from Temperate Forests to the Alpine Zone. *Forests* **2018**.

17. Liang, Z.; Wu, F.; Ni, X.; Tan, B.; Zhang, L.; Xu, Z.; Hu, J.; Yue, K. Woody Litter Increases Headwater Stream Metal Export Ratio in an Alpine Forest. *Forests* **2019**.

18. Tan, S.; Zhao, H.; Yang, W.; Tan, B.; Yue, K.; Zhang, Y.; Wu, F.; Ni, X. Forest Canopy Can Efficiently Filter Trace Metals in Deposited Precipitation in a Subalpine Spruce Plantation. *Forests* **2019**.

19. Liu, N.; Bao, G.; Liu, Y.; Linderholm, H.W. Two Centuries-Long Streamflow Reconstruction Inferred from Tree Rings for the Middle Reaches of the Weihe River in Central China. *Forests* **2019**.

20. Zhang, R.; Ermenbaev, B.; Zhang, T.; Ali, M.; Qin, L.; Satylkanov, R. The Radial Growth of Schrenk Spruce (Picea schrenkiana Fisch. et Mey.) Records the Hydroclimatic Changes in the Chu River Basin over the Past 175 Years. *Forests* **2019**.

**Fan-Rui Meng, Qiang Li, M. Altaf Arain, Michael Pisaric**
*Special Issue Editors*

*forests*

*Article*

# Biomass Carbon Sequestration Potential by Riparian Forest in the Tarim River Watershed, Northwest China: Implication for the Mitigation of Climate Change Impact

Tayierjiang Aishan [1,2], Florian Betz [3], Ümüt Halik [2,*], Bernd Cyffka [3] and Aihemaitijiang Rouzi [2]

[1] Institute of Arid Ecology and Environment, Xinjiang University, Sheng Li Road 666, Urumqi 830046, Xinjiang, China; tayirjan@xju.edu.cn

[2] Key Laboratory of Oasis Ecology, College of Resources & Environmental Sciences, Xinjiang University, Sheng Li Road 666, Urumqi 830046, Xinjiang, China; ahmadjan_1983@yahoo.com

[3] Faculty of Mathematics and Geography, Catholic University of Eichstaett-Ingolstadt, Ostenstraße 14, 85071 Eichstatt, Germany; florian.betz@ku.de (F.B.); bernd.cyffka@ku.de (B.C.)

\* Correspondence: halik@xju.edu.cn; Tel.: +86-189-9911-1553

Received: 14 February 2018; Accepted: 3 April 2018; Published: 10 April 2018

**Abstract:** Carbon management in forests has become the most important agenda of the first half of the 21st century in China in the context of the mitigation of climate change impact. As the main producer of the inland river basin ecosystem in arid region of Northwest China, the desert riparian forest maintains the regional environment and also holds a great significance in regulating the regional/global carbon cycle. In this study, we estimated the total biomass, carbon storage, as well as monetary ecosystem service values of desert riparian *Populus euphratica* Oliv. in the lower reaches of the Tarim River based on terrestrial forest inventory data within an area of 100 ha (100 plots with sizes of 100 m × 100 m) and digitized tree data within 1000 ha (with 10 m × 10 m grid) using a statistical model of biomass estimation against tree height (TH) and diameter at breast height (DBH) data. Our results show that total estimated biomass and carbon storage of *P. euphratica* within the investigated area ranged from 3.00 to 4317.00 kg/ha and from 1.82 to 2158.73 kg/ha, respectively. There was a significant negative relationship ($p < 0.001$) between biomass productivity of these forests and distance to the river and groundwater level. Large proportions of biomass (64% of total biomass) are estimated within 200 m distance to the river where groundwater is relatively favorable for vegetation growth and biomass production. However, our data demonstrated that total biomass showed a sharp decreasing trend with increasing distance to the river; above 800 m distance, less biomass and carbon storage were estimated. The total monetary value of the ecosystem service "carbon storage" provided by *P. euphratica* was estimated to be $6.8 × 10^4$ USD within the investigated area, while the average monetary value was approximately $70 USD per ha, suggesting that the riparian forest ecosystem in the Tarim River Basin should be considered a relevant regional carbon sink. The findings of this study help to establish a better understanding of the spatial distribution pattern of *P. euphratica* forest under water scarcity and can also provide an alternative approach to local decision-makers for efficient and precise assessment of forest carbon resources for emission reduction programs.

**Keywords:** river discharge; groundwater level; riparian forest; climate change; watershed management

## 1. Introduction

Terrestrial ecosystem biomass is a basic quantitative characteristic of an ecosystem and also a major ecological parameter in determining carbon sequestration and carbon sink function. Among the numerous types of ecosystems, forest ecosystems play a crucial role in the global carbon budget [1–6]. In Central Asia, desert riparian forest ecosystems along large river systems, such as the Amu Darya, Syr Darya, and Tarim River, store a large amount of carbon in the aboveground (leaves, branches, and stems) and belowground (roots) biomass and provide the main biomass resources for human wellbeing in those continental arid regions [7–11]. Due to the extremely arid climate conditions of these regions, the productivity of riparian forest ecosystems is highly dependent on groundwater and soil moisture availability supplied by inland river water flow [12]. Therefore, it is crucial to understand the relationship between hydrological regime, ecological development, and resource management in order to apply a forest management that is able to sustain the biomass budget and carbon storage of floodplain forests in those regions.

The Tarim River, located in the arid region of the Northwest China, is one of the largest inland rivers in the world along with the Volga, Syr Darya, Amu Darya, and the Ural [13]. It is mainly supplied by glacier and snow melt water and precipitations from the Tian Shan Mountains. Desert riparian forests distributed along the river are extremely important natural barriers supporting the ecological stability of the region. *Populus euphratica* Oliv. is the dominant tree species of Tugai vegetation (composed of trees, shrubs, and herbals) in the extremely dry areas of Central Asia and forms the main biomass reservoir and natural carbon sink in the Tarim River Basin [10,11]. More than 90% of existing floodplain forests along the Tarim River are *P. euphratica* riparian forests [14,15].

Carbon storage is one of the major ecosystem services provided by the Tugai forests in addition to the reduction of sand and dust storms, moderation of desertification, and regulation of oasis climate [16,17]. Reduction of $CO_2$ in the atmosphere helps mitigate climate change through its various effects and, therefore, is an important contribution to peoples' well-being [1]. Tarim riparian forests, in contrast to China's other forests, are characterized by their patchy distribution, high carbon (C) density, and high anthropogenic disturbance and they function as an important carbon pool in China's northwest arid region.

Over the past 50 years, however, due to climate change and the rapid socio-economic development in Xinjiang (especially increasing water demands for cotton production), many tributaries of the watershed were disconnected from the Tarim River [18,19]. The water discharge in the main stream of the Tarim River has dramatically decreased. More seriously, more than 320 km of river channel along the lower reaches of the Tarim River were completely desiccated since the construction of Daxihaizi reservoir in 1973. Consequently, the groundwater level dropped to 10–12 m below the surface and became salinized [20–24]. As a result of ecosystem deteriorations, substantial amounts of carbon stored in this riparian vegetation have been lost due to the reduction in the amount of live biomass.

In response to the aforementioned environmental damage, the Chinese government invested $10.7 \times 10^8$ RMB (approx. \$1.8 billion USD) in the implementation of the "Integrated Water Resource Management of the Tarim River Basin" program to secure the ecological, economic, and social sustainability of the oases along the Tarim River [20]. The ecological water diversion project (EWDP) is one of the key sub-projects within this restoration program; it started in May 2000 and was implemented in the lower reaches of the Tarim River with multiple purposes ranging from regenerating degraded riparian forests to improving environmental conditions and maintaining the green corridor for economic development [20,25,26]. Accompanying the restoration project, the majority of the research mainly focused on the responses of the groundwater table and various eco-physiological and morphological parameters of riparian forests to the rehabilitation measures [20,21,26–30], but relative research about the quantification of the live biomass structure and carbon storage potential of *P. euphratica* desert riparian vegetation in the lower reaches of the Tarim River had not been reported. Understanding the biomass structure of *P. euphratica* forest and its spatial distribution under ongoing ecological restoration practices is needed for estimating the contributions of these forests to mitigate climate change impacts in the region. The main objectives of this study are to estimate the spatial variability of biomass productivity and capacity of carbon sequestration by desert riparian *P. euphratica* forest and to monetize its carbon storage ecosystem service. This study is expected to broaden our understandings on biomass carbon distribution patterns under the current restoration program in this region and to facilitate riparian restoration by increasing awareness of decision makers on the potentials of desert riparian forests to sequester carbon. Research findings would provide a scientific basis for evaluating the contribution of riparian ecosystem to climate change mitigation through sustainable management of water resources (allocation of substantial water for ecology) and riparian forests (facilitating rehabilitation of highly-degraded vegetation).

## 2. Materials and Methods

### 2.1. Study Area Description

The study area is situated at Arghan village (40°08′50″ N, 88°21′28″ E), between Taklimakan and Kuruk Tag desert, in the lower reaches of the Tarim River, Xinjiang Uyghur Autonomous Region, Northwest China (Figure 1). This area is located in an extremely arid climatic zone with an annual precipitation <15 mm (Figure 2) and potential annual evaporation of 2500–3000 mm [14,20,27,31]. Sparse vegetation is predominantly distributed on the river floodplain ecosystem. It comprises trees, shrubs, and herbs. *P. euphratica* is the dominant species. Nearly 70% of the existing species in our study area are of *P. euphratica* [32]. Shrubs include *Tamarix ramosissima* Ledeb., *Tamarix hispida* Willd., *Tamarix elongata* Ledeb., *Lycium ruthenicum* Murr., *Halimodendron halodendron* (Pall.) Voss., *Halostachys caspica* (M.B.) C.A. Mey., *Poacynum hendersonii* (Hook. F.) Woodson., *Alhagi sparsifolia* (B. Keller et Shap.) Shap., *Glycyrrhiza inflata* Bat., *Karelinia caspica* (Pall.) Less., *Inula salsoloides* (Turcz.) Ostrnf., and *Hexinia polydichotoma* (Ostent.) H.L. Yang [10,20,21]. Besides *Tamarix*, most shrubs and herbs are distributed within the range of 100 m from the river. In particular, due to the scarcity of precipitation, groundwater is the main source of water that is required to maintain the structure and functions of riparian ecosystem in this hyper-arid region. In addition, this ecosystem is highly vulnerable to climate change. The river flow in the lower reaches of the Tarim River is complex and is comprised of many intersections and meanders. The river bed divides into two branches at 4.2 km downstream from the Daxihaizi reservoir. The western branch is the old Tarim River and the eastern branch is the Qiwinkol River. The two branches are roughly parallel to one another and converge at Arghan, where vegetation coverage is relatively high and the anthropogenic impacts, such as grazing and fuel wood harvest on the sampling sites, are not intensive. Therefore, Arghan is an ideal location for the estimation of tree biomass and carbon storage and their spatial distribution under current ongoing water diversion practices. In addition, six groundwater gauges installed there by the Tarim River Basin Management Bureau enable long-term monitoring of groundwater level changes.

### 2.2. Data Collection and Processing

The assessments of biomass and carbon storage presented in this study were performed by three steps (Figure 3). In a first step, biomass was estimated for a total number of 4773 *P. euphratica* tree stands within 100 ha monitoring plots (Figure 4). For the biomass estimation, we used allometric formulas suggested by Chen and Li [33]. These formulas link diameter at breast height (DBH), tree height, and biomass and have already been successfully applied to similar study regions at the middle/lower reaches of the Tarim River [10,34] and at the Amu Darya in Turkmenistan [10,11]. The parameters have been measured for the permanent monitoring plots within the growing season in 2010 and 2011. The DBH of each tree was measured with a DBH meter and tree height was determined with a laser distance meter.

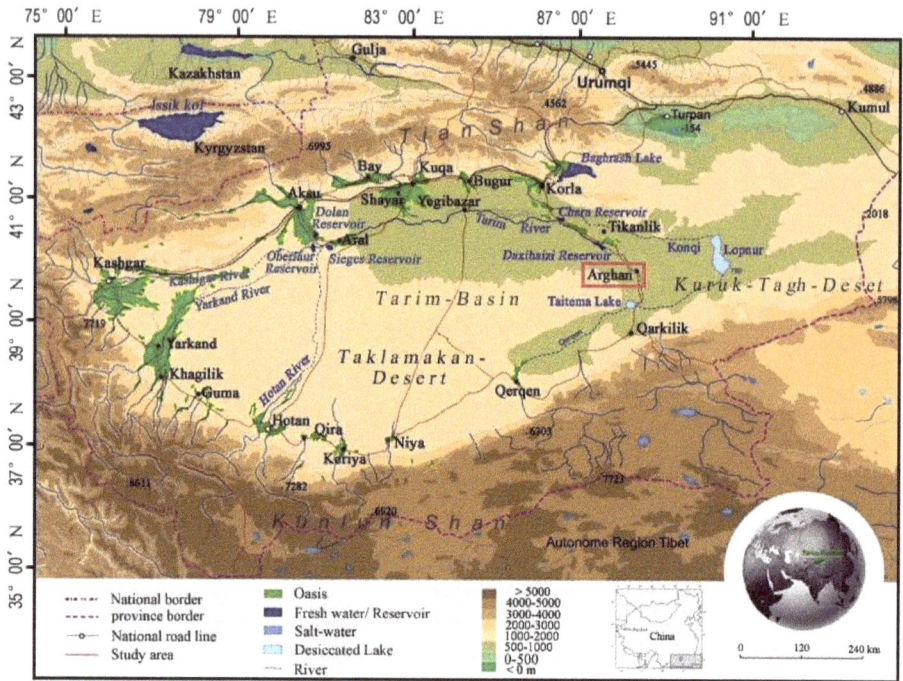

**Figure 1.** Locations of the lower reaches of the Tarim River and the Arghan transects.

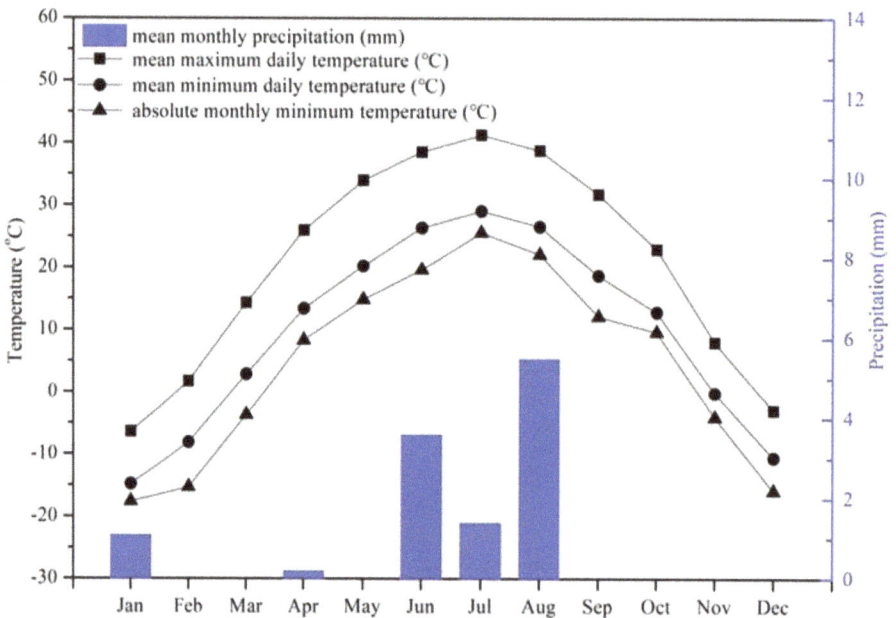

**Figure 2.** Climate diagram of the study area (based on data from Arghan for the period from 2013 to 2015).

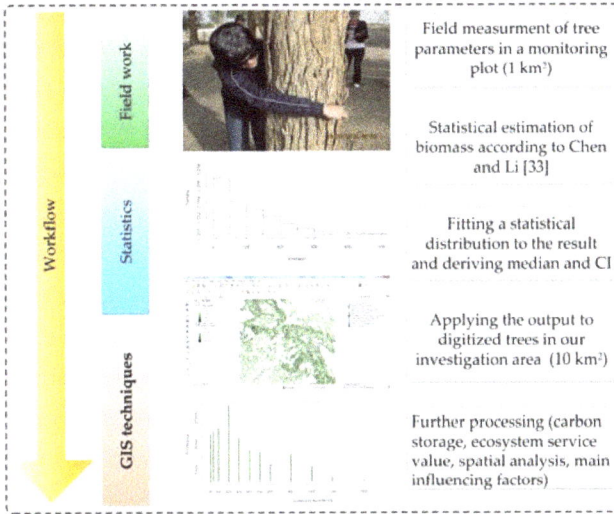

**Figure 3.** The workflow of this study (CI indicates confidence interval, GIS techniques refers to some of geographical information system tools such as digitization used in this study).

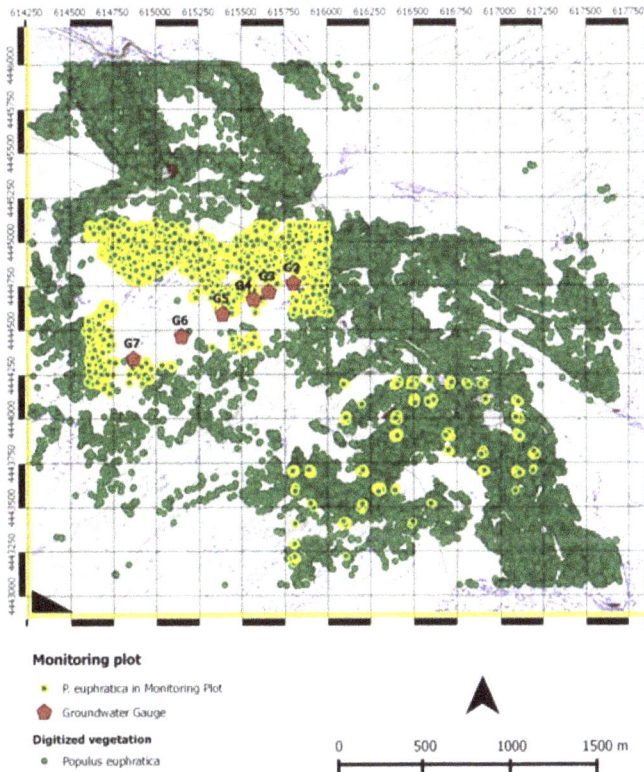

**Figure 4.** Data used for the biomass estimation of *P. euphratica*.

Once the allometric parameters were obtained for each tree in the monitoring plots, we calculated the biomass for each single tree using the formulas of Chen and Li [33] given in Table 1. Then, we fitted a statistical distribution to the tree biomasses. Due to the right-skewed nature of the empirical distribution, we chose the Weibull distribution, a widely-used theoretical distribution for analyzing the environmental data (Equation (1)):

$$F(x) = 1 - \exp^{-(\lambda x)^k} \tag{1}$$

where $\lambda$ is the scale factor and k the shape factor. For the fitting procedure, we used the maximum likelihood method, as suggested by Venables and Ripley [35]. This distribution was used to calculate the boundaries of the 95% confidence interval (CI), as well as the median of the biomass.

**Table 1.** Formulas of Chen and Li [33] for the relationship between the diameter at breast height (DBH) in centimeters, the tree height (H) in meters, and the biomass in kilograms.

| Biomass Fraction | Formula | Correlation Coefficient R |
|---|---|---|
| Trunk biomass (BT) | LogBT = log0.0382 + 0.8837 × logDBH²H | 0.99 |
| Branch and twig biomass (BB) | LogBB = log0.1072 + 0.6350 × logDBH²H | 0.89 |
| Leaf biomass (BL) | LogBL = log(1.41 × 10⁻³) + 0.8134 × logDBH²H | 0.71 |
| Root Biomass(BR) | LogBR = log0.1059 + 0.6185 × logDBH²H | 0.94 |
| Total Biomass (B) | B = BT + BB + BL + BR | |

Since the monitoring plots cover the most representative environmental situations for the study area, we assumed that the distribution could represent all possible tree biomass values occurring within the environmental setting. Thus, we used the 95% confidence intervals of the Weibull distribution to estimate the possible range of biomass for each *P. euphratica* tree in the study area. All euphratica poplars in the study area were digitized from a QuickBird satellite image with a resolution of 0.5 m. The trees could be clearly distinguished from other shrub-like species via the visible crown shadow by naked eye. Then, the minimum, maximum, and median of the biomass values obtained from the fitted Weibull distribution were assigned to the approximately 23000 points representing the poplars in the study area (Figure 4). Then, the point values for biomass were transferred into a 10 m × 10 m raster using the sum of all biomass values as cell value. In a subsequent step, the biomass values were transferred to carbon storage values by using a converting factor of 0.5, as suggested by the Intergovernmental Panel on Climate Change (IPCC) [1]. When the specific C content is unknown, different researchers have estimated C content as 50% of the absolutely dry mass of the stem, roots, and leafless branches [36]. For the calculation, we focused on the median of the biomass. This is, of course, a rough estimate only. Nevertheless, along with the 95% confidence interval, this approach gives a reliable range of biomass and associated carbon storage values of *P. euphratica* for the study area at the lower reaches of the Tarim River. However, further research is needed to reduce uncertainties.

To obtain the relationship with water availability, the main ecological driver of the study area, a buffer analysis was carried out for zones 100 m from the river. This reflects the assumption that the water supply is controlled by the river only, which is a rational assumption for a region with an annual precipitation of below 15 mm. For further analysis, the values of biomass and carbon storage of the raster dataset were extracted. Then, the sum of all cell values were calculated to obtain the biomass or carbon storage for a certain distance range from the river.

The final step was to valuate the carbon storage of *P. euphratica* as an ecosystem service. The approach chosen was to consider avoided economic damage assuming that the storage of carbon avoids the emission of $CO_2$ and, therefore, reduces the effect of climate change with its negative social and economic consequences. The monetary value of these consequences is also known as the social costs of carbon. Several well-known studies have been conducted for estimating these costs, for example, the Stern Report [37]. Tol [38] combined the results from 232 published estimates for the value of social costs of carbon and reported a median value of 87 USD/t C [38]. The value determined by Tol [38] was also used for the social costs of carbon in this paper. This avoided damage approach has been widely used in multiple settings, such as in Integrated valuation of ecosystem services and tradeoffs (InVEST) model [39]. However, this approach is not without its criticisms [40,41] and other methods, like using prices from carbon markets, are also available [42], with most methods suffering some flaws in their valuation due to market's inadequacy at capturing all the externalities and different accounting systems being used. Thus, the economic valuation presented in this study is one possible criterion to assess monetary values of carbon in order to provide references to policy-makers, even though it still has some drawbacks.

## 3. Results and Discussions

### 3.1. Spatial Distribution of Biomass, Carbon Storage, and Ecosystem Service Value

Estimation of biomass carbon storage of *P. euphratica* forest is complex due to both temporal and spatial variability, which result from the variation in individual growth patterns. These factors, along with water availability (groundwater and soil moisture), micro-topography, and salt placement, lead to a non-uniform/heterogeneity in biomass distribution in our study area. Currently accepted methods for biomass estimation of many other forest types assume a normal distribution. In this study, this assumption turned out not to be tenable. As such, in this study, Weibull distribution was fitted to a model of the distribution of *P. euphratica* biomass. The fitted Weibull distribution to the biomass of the monitoring plot is shown in Figure 5. This curve is clearly positively skewed, which indicates that trees with a low biomass have a much higher probability to occur than trees with a high biomass. From this, the 95% confidence intervals of 1.17 kg for the lower boundary and 544.62 kg for the upper boundary, as well as the median of 69.36 kg, were calculated.

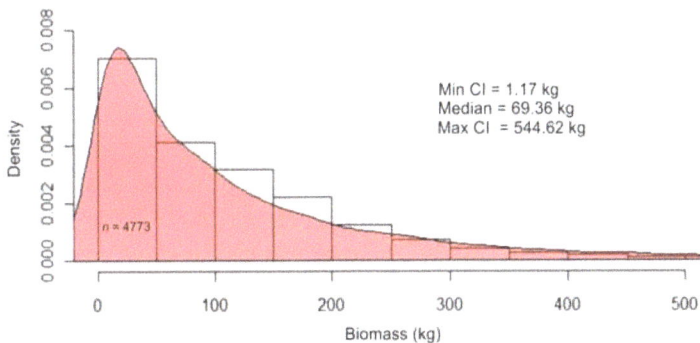

**Figure 5.** Weibull distribution fitted to the biomass values of *P. euphratica* within the monitoring plot.

These values (Table 2) have been assigned to the tree-points in Quantum GIS (QGIS) and summarized. Results show that the median value of biomass was 1,581,166 kg, which means 790,583.20 kg of stored carbon and a resulting ecosystem service of $68,780.74 USD. In addition to the absolute results, there was a clear spatial distribution pattern of biomass within the investigation area (Figure 6). Maximum values of biomass were found to concentrate close to the river channel or old river branches, while it values tended to decrease with increasing distance to the river. This phenomenon could be explained by the productivity and reproduction strategy of *P. euphratica*, which is heavily dependent on flooded areas and freshly-sedimented river banks [10,34,43]. Detailed analyses for this distribution were carried out by a buffer analysis with different distances to the river (Figures 7 and 8). The highest values for biomass, carbon storage, and ecosystem service values occurred within 200 m of the river. Within this zone, 20,915 kg/ha of biomass were stored, which accounted for 64% of the total value in the investigation area. In the distance of 200 m to 800 m only slight differences occurred.

Within this area, another 31% (6446 kg/ha) of the biomass was stored. Above a distance of 800 m the biomass decreased sharply and reached a level of zero greater than 2000 m from the river.

**Table 2.** Results from the estimation of biomass, carbon storage, and the value as an ecosystem service for the whole investigation area.

|  | Lower Boundary of CI | Median | Upper Boundary of CI |
|---|---|---|---|
| Biomass | 26,737.29 kg | 1,581,166 kg | 12,416,286 kg |
| Carbon Storage | 13,368.64 kg | 790,583.20 kg | 6,208,143 kg |
| Value | 1163.072 $ | 68,780.74 $ | 540,108.4 $ |

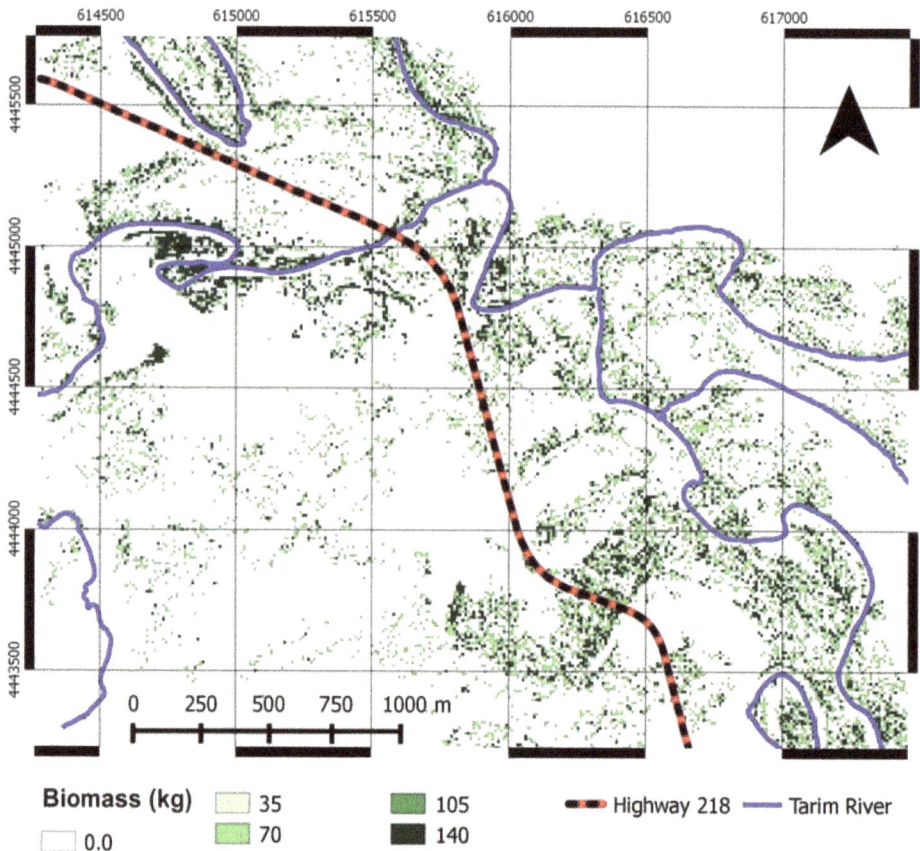

**Figure 6.** Spatial distribution of estimated potential biomass distribution for *P. euphratica* forests at Arghan in the lower reaches of the Tarim River.

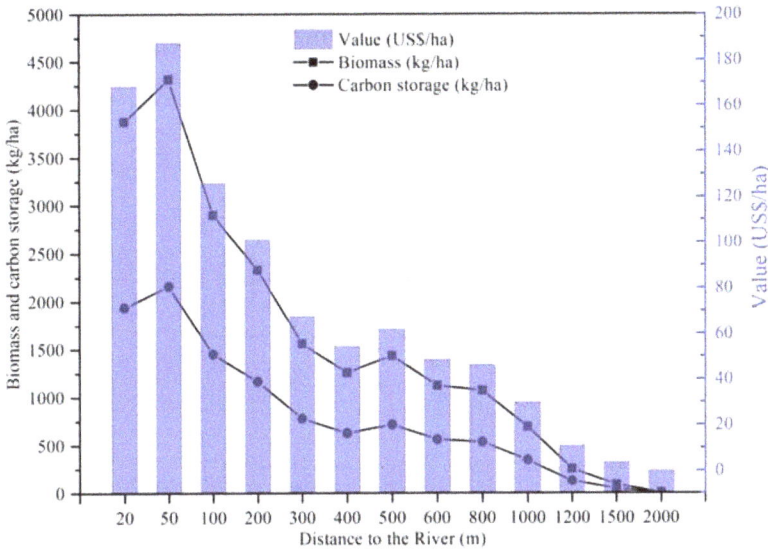

**Figure 7.** Biomass, carbon storage, and value of ecosystem service of *P. euphratica* forests at different distances from the river.

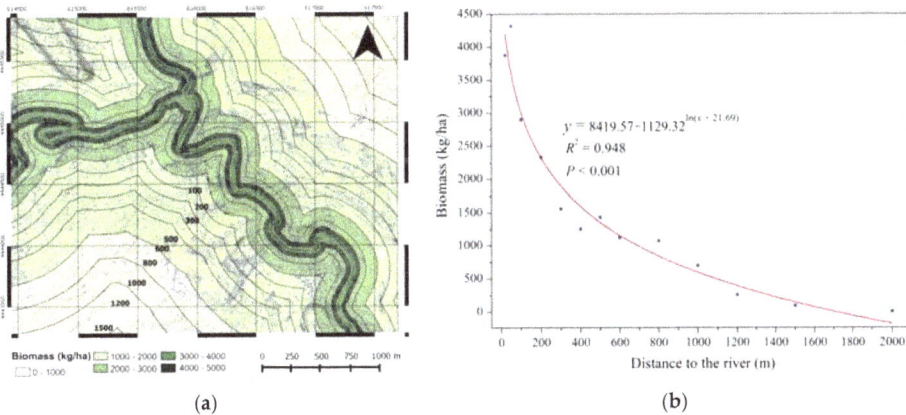

(a)                                    (b)

**Figure 8.** Biomass distribution pattern of *P. euphratica* along a gradient of water stress. (**a**) Biomass estimation in different buffer zones from the river; and (**b**) the relationship between biomass and distance from the river.

## 3.2. Dependency of Tree Biomass Production and Carbon Storage on Water Availability

There are many environmental factors influencing biomass production of riparian forest, such as climate, water availability, topographical conditions, and age structures of riparian vegetation [44–46]. As we pointed out above, nearly 30 years of river desiccation in the lower reaches of the Tarim River resulted in the decline of groundwater and massive destruction and degradation of natural vegetation along both sides of the river channel, which led to a lasting decrease in the carbon storage of riparian vegetation, with an increased accumulation of dead wood. Due to extremely harsh environmental conditions, riparian forests in the lower reaches of the Tarim River mainly depend on the limited groundwater to sustain their life cycle. As shown in Figure 8a, estimated tree biomass presented different values within different buffer zones from the river. Furthermore, we used regression analysis to analyze the correlation between biomass and distance to the river (Figure 8b) and found that there was a negative correlation between the biomass of *P. euphratica* forest and distance from the river; the determination coefficient was greater than 0.9 and the correlation is significant ($p < 0.001$). Within 20 m distance from the river, total tree biomass was estimated to be about 4000–5000 kg/ha, while 3000–4000 kg/ha of tree biomass was distributed within the area 20 m to 100 m from the river. Above 100 m distance from the river, less biomass was estimated and the amount of biomass decreased dramatically with increasing distance from the river channel. To the direction of desert margin, almost no biomass was found. Groundwater in arid regions of Central Asia is one of the decisive factors for promoting carbon sequestration capacities of the floodplain forests [10,11,34]. Figure 9 shows that there were obvious differences between groundwater levels at different distances to the river. Generally, groundwater levels were closely related to the distance from the river. With increasing distance to the river channel, groundwater levels also presented a decreasing trend. Trees growing under favorable groundwater conditions have healthy and large crowns with dense foliage, which contributes to an increase of above-ground biomass.

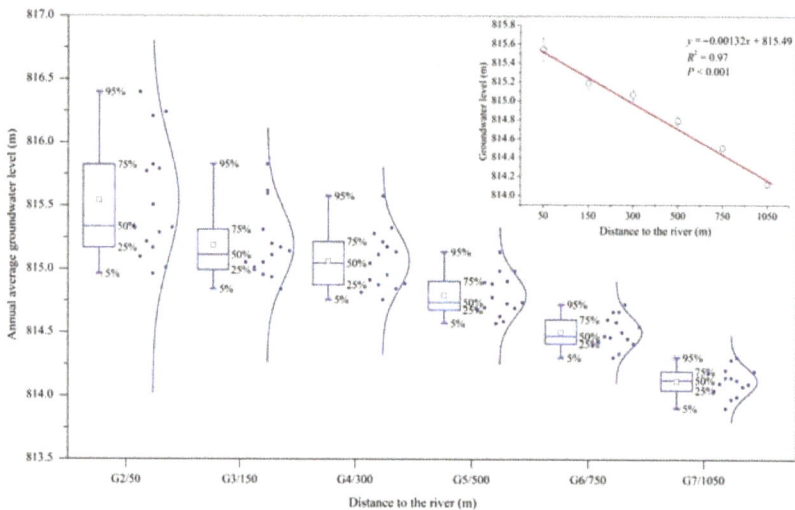

**Figure 9.** Groundwater levels at different distances to the river (G refers to the name of the groundwater monitoring well; see also Figure 4).

The zone far away from the river is characterized by a harsh environment due to highly limited water resources. Thus, normal growth of *P. euphratica* trees is prohibited, leading to less biomass production. At present, water diversion practices are the main environmental driver for eco-hydrological processes of the riparian ecosystem in the region. Seventeen water diversions in the lower reaches of the Tarim River have played a significant role in raising groundwater tables near to the main river course and in recovering floodplain forests to a certain degree [30,47,48]. On the other hand, these practices might be a major contributor for producing a higher biomass by means of improving water availability and leading to the sequestration of carbon. However, some studies also indicated that current water diversion practices are less effective for the large-scale establishment of new seedlings and juvenile trees [20,49]. It was observed that young forests have greater annual carbon sequestration rates while mature forests store more carbon [5]. Therefore, much effort should be carried out to create suitable habitat conditions for maintaining the current carbon storage and increasing the carbon sequestration capacities of riparian forests by changing their age.

### 3.3. Potential Effect of Water Diversion Project in Relation to Forest Carbon Storage

By the end of 2015, ecological water was diverted 17 times in irregular frequencies throughout 1714 days in 16 years with a total volume of $51.10 \times 10^8$ m³. As mentioned above, ecosystem services of desert riparian forests in the lower reaches of the Tarim River are highly controlled by water availability (particularly groundwater). Thus far, other important benefits of the water diversion project, such as increasing carbon sequestration potentials of desert riparian forests, have rarely been reported or mentioned. These benefits are mainly reflected in two aspects: (i) promoting the growth and recovery of the vegetation by satisfying its water requirement; and (ii) establishing more young seedlings and juveniles by increasing the overflow area of water delivery. Here, we mainly discuss the potential effects of water diversion projects on actual carbon storage and future carbon sequestration and how these particular ecosystem services could be improved by water management. Specifically, we examined how water availability (groundwater level and soil moisture) affects the distribution of biomass carbon storage of *P. euphratica* riparian forest.

The growth of desert riparian forest is dependent on the groundwater level, which is determined by the amount, duration, and magnitude of the current water diversion project in the region [32,34,50]. Furthermore, sufficient soil moisture, even though it has a limited effect on adult trees (e.g., *Populuseuphratica*.) and shrubs (e.g., *Tamarix* spp.), is also a key factor that must be considered, since it activates the soil seed bank, sustains herbs and shrubs with shallower root systems, and creates favorable habitat conditions for the establishment, survival, and growth of *P. euphratica* young seedlings. However, in our research area, high densities of *P. euphratica* seedlings mainly occurred in the vicinity of the zones that are highly affected by water diversions or around suitable micro-relief [20]. Hao and Li [50] found that changes in groundwater depth also led to corresponding changes of groundwater quality and soil moisture. Without a doubt, increased delivery of water in sufficient duration and magnitude into the downstream is highly desirable, since it is the only driver for rehabilitating degraded riparian ecosystems and producing more biomass. Hence, the amount and time of ecological water diversion need to be ensured according to the eco-hydrological requirements of riparian vegetation so that the capacity of carbon sequestration potentials in the desert riparian forest in the Tarim River Basin can be improved and maximized. However, water use conflicts between water demand and supply along the entire river basin and overexploitation of water in the upper and middle reaches of the river are still challenging [51].

In view of the above, increasing the carbon storage of *P. euphratica* forest through ecological water conveyance may be the most effective option to reduce the rise of atmospheric carbon dioxide concentration in the region. Numerous studies have quantified regulated ecosystem services of riparian forests along the lower reaches of the Tarim River by using an ecosystem service accounting method and the results of these studies highlighted the importance of water diversion effects on the carbon sequestration potentials of degraded riparian forests [16,52,53]. Currently, the impact of ecological water conveyance on the recharge of groundwater is very significant. In recent years, the suitable groundwater levels for the desert riparian forests have been achieved to some extent in the lower reaches of the Tarim River via the implementation of water diversions on 17 separate occasions [54]. From this point of view, our study implies that large-scale ecological restoration programs, such as rehabilitation and restoration of riparian forests by water diversion, could help to enhance regional/global carbon sinks, which may shed new light on the carbon sequestration benefits of such programs in the arid regions of Northwest China and also in other regions.

In the present study, we mainly focused on the quantification of *P. euphratica* forest biomass carbon, since it is a predominantly distributed species along the Tarim River. However, there are knowledge gaps regarding properties, processes, and practices affecting carbon sequestration in soil, deadwoods, and other vegetation (e.g., shrubs and herbs) that must be filled in order to develop a comprehensive sustainable forest management program along the lower reaches of the Tarim River.

*3.4. Implications of the Results for Climate Change Mitigation*

According to an IPCC report, various afforestation and reforestation efforts could reduce global atmospheric $CO_2$ by 25% by 2020 [55]. Reducing Emissions from Deforestation and Forest Degradation (REDD) and the Clean Development Mechanism (CDM) have been used in developing countries to promote low carbon paths to sustainable development [56]. China has developed its own pilot emission trading system (ETS) where Chinese Certified Emission reduction projects to trade carbon credits in the eastern provinces [57]. Although Xinjiang has not officially participated in the pilot project yet, a soon to be implemented nation-wide policy would provide the region more opportunities to trade its carbon credits with more developed regions on the coast. Water diversion-induced carbon accumulation would be an attractive alternative for emission trading, and achieving restoration of the degraded ecosystem would provide a multitude of ecosystem services.

## 4. Conclusions

In the present study, we estimated the total biomass, carbon storage, as well as the monetary ecosystem service value of the desert riparian *P. euphratica* in the lower reaches of the Tarim River and found that they were closely associated with water availability. Average total carbon stored by trees within 200 m distance to the river accounted for 64% of total estimated carbon, indicating that this area is, or might be, the main pool for carbon storage by riparian forests. This study demonstrated a first estimation of the carbon storage of *P. euphratica* in the lower reaches of the Tarim River. However, a static view on dynamic ecosystems presents only limited information about the system behavior. To assess the carbon storage dynamics, a chronosequence approach using tree-ring analysis or the analysis of remote sensing time series might be useful. Our study is rather a final statement about the carbon dynamics of the riparian ecosystems at the lower Tarim River, but a first attempt to deliver information for highlighting the relevance of this issue. The results of this study would provide the basis to evaluate the effect of the water diversion by calculating the stored carbon, which, in turn, could be traded as carbon offsets in China's emission trading system.

**Acknowledgments:** This research work was supported by National Natural Science Foundation of China (grant no: 31700386, U1703102), Doctoral Research Project of the Xinjiang University (grant no: BS160258), the Thousand Youth Talents Plan of China (Xinjiang Projects), the German Federal Ministry of Education and Research (BMBF) within the framework of the SuMaRiO project (01LL0918D), and the German VolkswagenStiftung within the framework of the EcoCAR project (Az.: 88497). We thank the Tarim River Basin Administration Bureau for providing hydrological data and the Forestry Department of Qarkilik (Ruoqiang) for their logistical support during our field work in Arghan. The authors are grateful to the anonymous reviewers for their constructive comments.

**Author Contributions:** All authors contributed to the design and development of this manuscript. Tayierjiang Aishan carried out the research and prepared the first draft of the manuscript; Florian Betz provided important advice and technical support on the methodology; Ümüt Halik and Bernd Cyffka conceived of and designed the overall concept for the research; data were collected and processed by Tayierjiang Aishan and colleagues from Xinjiang University; and Aihemaitijiang Rouzi was responsible for manuscript proofreading.

**Conflicts of Interest:** The authors declare no conflict of interest.

## References

1. Intergovernmental Panel on Climate Change (IPCC). *IPCC Guidelines for National Greenhouse Gas Inventories, Prepared by the National Greenhouse Gas Inventories Programme*; IPCC: Geneva, Switzerland, 2006; Volume 4, ISBN 4-88788-032-4.
2. Fang, J.Y.; Chen, A.P.; Peng, C.H.; Zhao, S.Q.; Ci, L.J. Changes in forest biomass carbon storage in China between 1949 and 1998. *Science* **2001**, *292*, 2320–2322. [CrossRef] [PubMed]
3. Piao, S.L.; Fang, J.Y.; Ciais, P.; Peylin, P.; Huang, Y.; Sitch, S.; Wang, T. The carbon balance of terrestrial ecosystems in China. *Nature* **2009**, *458*, 1009. [CrossRef] [PubMed]
4. Li, Y.; Wang, Y.G.; Houghton, R.A.; Tang, L.S. Hidden carbon sink beneath desert. *Geophys. Res. Lett.* **2015**, *42*, 5880–5887. [CrossRef]

5.  Fahey, T.J.; Woodburry, P.B.; Battles, J.J.; Goodale, J.C.L.; Hamburg, S.P.; Ollinger, S.V.; Wodall, C.W. Forest Carbon Storage: Ecology, Management and Policy. *Front. Ecol. Environ.* **2010**, *8*, 245–252. [CrossRef]
6.  Pan, Y.; Birdsey, R.S.; Fang, J.; Houghton, R.; Kauppi, P.E.; Kurz, W.A.; Phillips, O.L.; Shvidenko, A.; Lewis, S.L.; Canadell, J.G.; et al. A large and persistent Carbon Sink in the World's Forests. *Science* **2011**, *333*, 988–993. [CrossRef] [PubMed]
7.  Zerbe, S.; Thevs, N. Restoring Central Asian Floodplain Ecosystems as Natural Capital and Cultural Heritage in a continental Desert Environment. In *Landscape Ecology in Asian Cultures*; Hong, S.K., Kim, J.E., Wu, J., Nakagoshi, N., Eds.; Ecological Research Monographs; Springer: Tokyo, Japan, 2011; pp. 277–297. ISBN 978-4-431-87798-1.
8.  Grünzweig, J.M.; Lin, T.; Rotenberg, E.; Schwartz, A.; Yakir, D. Carbon sequestration in arid-land forest. *Glob. Chang. Biol.* **2010**, *9*, 791–799. [CrossRef]
9.  Chen, Y.L.; Luo, G.P.; Maisupova, B.; Chen, X.; Mukanov, B.M.; Wu, M.; Mambetov, B.T.; Huang, J.F.; Li, C.F. Carbon budget from forest land use and management in Central Asia during 1961–2010. *Agr. For. Meteorol.* **2016**, *221*, 131–141. [CrossRef]
10. Thevs, N.; Buras, A.; Zerbe, S.; Kühnel, E.; Abdusalih, N.; Ovezberdiyeva, A. Structure and wood biomass of near-natural floodplain forests along the Central Asian Rivers Tarim and Amu Darya. *Forestry* **2012**, *85*, 193–202. [CrossRef]
11. Buras, A.; Thevs, N.; Zerbe, S.; Wilmking, M. Productivity and carbon sequestration of *Populus euphratica* at the Amu River, Turkmenistan. *Forestry* **2013**, *86*, 429–439. [CrossRef]
12. Kuba, M.; Aishan, T.; Cyffka, B.; Halik, Ü. Analysis of connections between soil moisture, groundwater level and vegetation vitality along two transects at the Lower Reaches of the Tarim River, Northwest China. *Geo-Öko* **2013**, *34*, 103–127.
13. Hai, Y.; Wai, L.; Hoppe, T.; Thevs, N. Half a Century of Environmental Change in the Tarim River Valley—An Outline of Cause and Remedies. In *Watershed and Floodplain Management along the Tarim River in China's Arid Northwest*; Hoppe, T., Kleinschmit, B., Roberts, B., Thevs, N., Halik, U., Eds.; Shaker Press: Aachen, Germany, 2006; pp. 39–76.
14. Song, Y.D.; Fan, Z.L.; Lei, Z.D. *Research on Water Resources and Ecology of the Tarim River, China*; Peoples Press: Urumqi, China, 2000.
15. Huang, P. *Irrigation-Free Vegetation and It's Recovery in Arid Region*; Science Press: Beijing, China, 2002; pp. 15–50.
16. Huang, X.; Chen, Y.N.; Ma, J.X.; Chen, Y.P. Study on change in value of ecosystem service function of Tarim River. *Acta Ecol. Sin.* **2010**, *30*, 67–75. [CrossRef]
17. Cyffka, B.; Rumbaur, C.; Kuba, M.; Disse, M. Sustainable Management of River Oases along the Tarim River, P.R. China (SuMaRiO) and the Ecosystem Services Approach. *Geogr. Soc. Environ.* **2013**, *6*, 77–90. [CrossRef]
18. Giese, E.; Mamatkanov, D.M.; Wang, R. *Wasserressourcen und deren Nutzung im Flussbecken des Tarim (Autonome Region Xinjiang/VR China)*; Discussion Papers 25.; Zentrum für Internationale Entwicklungs- und Umweltforschung: Justus Liebig University Giessen, Germany, 2006; Volume 20, p. 63. Available online: http://geb.uni-giessen.de/geb/volltexte/2006/2661 (accessed on 13 January 2018).
19. Tao, H.; Gemmer, M.; Bai, Y.; Su, B.; Mao, W. Trends of stream flow in the Tarim River Basin through the past 50 years: Human Impact or Climate Change? *J. Hydrol.* **2011**, *400*, 1–9. [CrossRef]
20. Aishan, T.; Halik, Ü.; Cyffka, B.; Kuba, M.; Abliz, A.; Baidourela, A. Monitoring the hydrological and ecological response to water diversion in the lower reaches of the Tarim River, northwest China. *Quat. Int.* **2013**, *311*, 155–162. [CrossRef]
21. Halik, U.; Kurban, A.; Mijit, M.; Schulz, J.; Paproth, F.; Coenradie, B. The Potential Influence of Embankment Engineering and Ecological Water Transfer on the Riparian Vegetation along the Middle and Lower Reaches of the Tarim River. In *Watershed and Floodplain Management along the Tarim River in China's Arid Northwest*; Hoppe, T., Kleinschmit, B., Roberts, B., Thevs, N., Halik, U., Eds.; Shaker Press: Aachen, Germany, 2006; pp. 221–236.
22. Chen, Y.N.; Chen, Y.P.; Xu, C.C.; Ye, Z.X.; Li, Z.Q.; Zhu, C.G.; Ma, X.D. Effects of ecological water conveyance on groundwater dynamics and riparian vegetation in the lower reaches of Tarim River, China. *Hydrol. Process.* **2010**, *24*, 170–177. [CrossRef]
23. Betz, F.; Halik, Ü.; Kuba, M.; Aishan, T.; Cyffka, B. Controls on aeolian sediment dynamics by natural riparian vegetation in the Eastern Tarim Basin, NW China. *Aeolian Res.* **2015**, *18*, 23–34. [CrossRef]

24. Hoppe, T.; Kleinschmit, B.; Roberts, B.; Thevs, N.; Halik, Ü. *Watershed and Floodplain Management along the Tarim River in China's Arid Northwest*; Shaker Press: Aachen, Germany, 2006; ISBN 978-3-8322-5662-3.
25. Xu, H.L.; Ye, M.; Li, J.M. The ecological characteristics of the riparian vegetation affected by river overflowing disturbance in the lower Tarim River. *Environ. Geol.* **2009**, *58*, 1749–1755. [CrossRef]
26. Chen, Y.N.; Ye, Z.X.; Shen, Y.J. Desiccation of the Tarim River, Xinjiang, China, and mitigation strategy. *Quat. Int.* **2011**, *244*, 264–271. [CrossRef]
27. Xu, H.L.; Ye, M.; Li, J.M. Changes in groundwater levels and the response of natural vegetation to transfer of water to the lower reaches of the Tarim River. *J. Environ. Sci.* **2007**, *19*, 1199–1207. [CrossRef]
28. Halik, Ü.; Chai, Z.; Kurban, A.; Cyffka, B. The positive response of some ecological indices of Populus euphratica to the emergency water transfer in the lower reaches of the Tarim River. *Resour. Sci.* **2009**, *31*, 1309–1314.
29. Halik, Ü.; Aishan, T.; Kurban, A.; Cyffka, B.; Opp, C. Response of Crown Diameter of Populus euphratica to Ecological Water Transfer in the Lower Reaches of Tarim River. *J. Northeast For. Univ.* **2011**, *39*, 82–84.
30. Aishan, T.; Halik, Ü.; Kurban, A.; Cyffka, B.; Kuba, M.; Betz, F.; Keyimu, M. Eco-morphological response of floodplain forests (*Populus euphratica* Oliv.) to water diversion in the lower Tarim River, northwest China. *Environ. Earth Sci.* **2015**, *73*, 533–545. [CrossRef]
31. Wu, J.; Tang, D.S. The influence of water conveyances on restoration of vegetation to the lower reaches of Tarim River. *Environ. Earth Sci.* **2010**, *59*, 967–975. [CrossRef]
32. Aishan, T. Degraded Tugai Forests under Rehabilitation in the Tarim Riparian Ecosystem, Northwest China: Monitoring, Assessing and Modelling. Ph.D. Thesis, Katholische Universität Eichstätt-Ingolstadt, Eichstätt, Germany, 2016.
33. Chen, B.H.; Li, H.Q. Study on Biomass of natural Diversifolious Poplar Plantations in River Talimu, Xinjiang, Western China. *For. Sci. Technol. Xinjiang* **1984**, *3*, 8–16.
34. Thomas, F.M.; Jeschke, M.; Zhang, X.; Lang, P. Stand structure and productivity of *Populus euphratica* along a gradient of groundwater distances at the Tarim River (NW China). *J. Plant Ecol.* **2016**, *10*, 753–764. [CrossRef]
35. Venables, W.N.; Ripley, B.D. *Modern Applied Statistics with S*; Springer: New York, NY, USA, 2002.
36. Williams, R.A.; Schafer, S.E. Forest Carbon Sequestration and Storage of the Kargasoksky Leshoz of the Tomsk Oblast, Russia—Current Status and the Investment Potential. In *Climate Change and Terrestrial Carbon Sequestration in Central Asia*; Lal, R., Suleimenov, M., Stewart, B.A., Hansen, D.O., Doraiswamy, P., Eds.; CRC Press: Leiden, The Netherlands, 2007; pp. 363–370. ISBN 9788578110796.
37. Stern, N.H.; Peters, S.; Bakhski, V.; Bowen, A.; Cameron, C.; Catovsky, S.; Crane, D.; Cruickshank, S.; Dietz, S.; Edmondson, N.; et al. *Stern Review: The Economics of Climate Change*; Cambridge University Press: Cambridge, UK, 2006.
38. Tol, R.S.J. The Economic Effects of Climate Change. *J. Econ. Perspect.* **2009**, *23*, 29–51. [CrossRef]
39. Natural Capital Project. Available online: https://www.naturalcapitalproject.org/invest/ (accessed on 20 March 2018).
40. Nordhaus, W.D. A Review of the Stern Review on the Economics of Climate Change. *J. Econ. Lit.* **2007**, *45*, 686–702. [CrossRef]
41. Weitzman, M.L. A Review of the Stern Review on the Economics of Climate Change. *J. Econ. Lit.* **2007**, *45*, 703–724. [CrossRef]
42. Matzek, V.; Puleston, C.; Gunn, J. Can carbon credits fund riparian forest restoration? *Restor. Ecol.* **2015**, *23*, 7–14. [CrossRef]
43. Peng, S.H.; Chen, X.; Qian, J.; Liu, S.G. Spatial Pattern of *Populus euphratica* Forest Change as Affected by Water Conveyance in the Lower Tarim River. *Forests* **2014**, *5*, 134–152. [CrossRef]
44. Luo, Y.J.; Wang, X.K.; Zhang, X.Q.; Lu, F. *Biomass and Its Allocation of Forest Ecosystems in China*; China Forestry Publishing House: Beijing, China, 2013; pp. 44–45.
45. Schlesinger, W.H.; Belnap, J.; Marion, G. On carbon sequestration in desert ecosystems. *Glob. Chang. Biol.* **2009**, *15*, 1488–1490. [CrossRef]
46. Huxman, T.E.; Snyder, K.A.; Tissue, D.; Leffler, A.J.; Ogle, K.; Pockman, W.T.; Sandquist, D.R.; Potts, D.L.; Schwinning, S. Precipitation pulses and carbon fluxes in semiarid and arid ecosystems. *Oecologia* **2004**, *141*, 254–268. [CrossRef] [PubMed]
47. Keyimu, M.; Halik, Ü.; Rouzi, A. Relating Water Use to Tree Vitality of *Populus euphratica* Oliv. in the Lower Tarim River, NW China. *Water* **2017**, *9*, 622. [CrossRef]

48. Keyimu, M.; Halik, Ü.; Kurban, A. Estimation of water consumption of riparian forest in the lower reaches of Tarim River, northwest China. *Environ. Earth Sci.* **2017**, *76*, 547. [CrossRef]

49. Thomas, F.M. Ecology of Phreatophytes. In *Progress in Botany*; Springer: Berlin, Germany, 2014; Volume 75, pp. 335–375. ISBN 978-3-642-38796-8.

50. Hao, X.; Li, W. Impacts of ecological water conveyance on groundwater dynamics and vegetation recovery in the lower reaches of the Tarim River in northwest China. *Environ. Monit. Assess.* **2014**, *186*, 7605–7616. [CrossRef] [PubMed]

51. Rumbaur, C.; Thevs, N.; Disse, M.; Ahlheim, M.; Brieden, A.; Cyffka, B.; Duethmann, D.; Feike, T.; Frör, O.; Gärtner, P.; et al. Sustainable management of river oases along the Tarim River (SuMaRiO) in Northwest China under conditions of climate change. *Earth Syst. Dyn.* **2015**, *6*, 83–107. [CrossRef]

52. Mamat, Z.; Halik, Ü.; Keyimu, M.; Keram, A.; Nurmamat, K. Variation of the Floodplain Forest Ecosystem Service Value in the Lower Reaches of Tarim River, China. *Land Degrad. Dev.* **2017**, 1–11. [CrossRef]

53. Ma, X.; Feng, Q.; Yu, T.; Su, Y.; Deo, R.C. Carbon dioxide fluxes and their environmental controls in a riparian forest within the hyper-arid region of Northwest China. *Forests* **2017**, *8*, 379. [CrossRef]

54. Deng, M.J.; Yang, P.N.; Zhou, H.Y.; Xu, H.L. Water Conversion and Strategy of Ecological Water Conveyance in the Lower Reaches of the Tarim River. *Arid Zone Res.* **2017**, *34*, 717–726. [CrossRef]

55. Reyer, C.; Guericke, M.; Ibisch, P.L. Climate change mitigation via afforestation, reforestation and deforestation avoidance: And what about adaptation to environmental change? *New For.* **2009**, *38*, 15–34. [CrossRef]

56. Allen, M.R.; Barros, V.R.; Broome, J.; Cramer, W.; Christ, R.; Church, J.A.; Clarke, L.; Dahe, Q.; Dasgupta, P.; Dubash, N.K.; et al. *IPCC Fifth Assessment Synthesis Report—Climate Change 2014 Synthesis Report*; World Health Organization: Geneva, Switzerland, 2014; 167p.

57. Lo, A.Y.; Cong, R. After CDM: Domestic carbon offsetting in China. *J. Clean. Prod.* **2017**, *141*, 1391–1399. [CrossRef]

*forests*

MDPI

*Article*

# Spatial and Temporal Patterns in Nonstationary Flood Frequency across a Forest Watershed: Linkage with Rainfall and Land Use Types

Xu-dong Huang [1], Ling Wang [2,3,4], Pei-pei Han [5] and Wen-chuan Wang [1,6,*]

[1] School of Water Conservancy, North China University of Water Resources and Electric Power, Zhengzhou 450046, China; huangxudong@ncwu.edu.cn

[2] Jiangxi Provincial Key Laboratory of Soil Erosion and Prevention, Nanchang 330029, China; wangling_ln@126.com

[3] Jiangxi Institute of Soil and Water Conservation, Nanchang 330029, China

[4] College of Resources and Environment, Huazhong Agricultural University, Wuhan 430070, China

[5] Henan Yellow River Hydrological Survey and Design Institute, Zhenghzou 450002, China; hzauhanpeipei@163.com

[6] Collaborative Innovation Center of Water Resources Efficient Utilization and Support Engineering, Zhengzhou 450046, China

* Correspondence: wangwen1621@163.com; Tel.: +86-0371–6912-7211

Received: 10 April 2018; Accepted: 6 June 2018; Published: 8 June 2018

**Abstract:** Understanding the response of flood frequency to impact factors could help water resource managers make better decisions. This study applied an integrated approach of a hydrological model and partial least squares (PLS) regression to quantify the influences of rainfall and forest landscape on flood frequency dynamics in the Upper Honganjian watershed (981 km²) in China, the flood events of flood seasons in return periods from two to 100 years, wet seasons in return periods from two to 20 years, and dry seasons in return periods from two to five years show similar dynamics. Our study suggests that rainfall and the forest landscape are pivotal factors triggering flood event alterations in lower return periods, that flood event dynamics in higher return periods are attributed to hydrological regulations of water infrastructures, and that the influence of rainfall on flood events is much greater than that of land use in the dry season. This effective and simple approach could be applied to a variety of other watersheds for which a digital spatial database is available, hydrological data are lacking, and the hydroclimate context is variable.

**Keywords:** frequency analysis; flood; hydrological model; probability distribution; partial least squares (PLS) regression

## 1. Introduction

Flood frequency is the probability of a flood event in a certain period, which is relevant to planning and decision processes related to hydraulic works or flood alleviation programs [1]. Thorough knowledge of flood frequency dynamics is crucial in a watershed [2–4]. Variations in flood frequencies result from meteorological factors and underlying surface properties, including rainfall, flood control facilities, topography, soil and land use types [3,5]. Among these factors, flood control facilities represent passive defense structures against floods; topography and soil are relatively constant in short periods, whereas rainfall and land use are variable [3,5–8]. Therefore, determining the response of flood frequencies to rainfall and land use is crucial for water resource management in watersheds [5,9]. However, major challenges are associated with flood frequency response research, including failed estimation of flood frequencies, lack of data, and the high collinearity of rainfall and land use types [8,10].

To quantify the factors controlling flood frequency, estimation for flood frequencies is a prerequisite. Frequency analysis (FA) is a method used to fit the frequency of extreme hydrological events to their magnitudes by using probability distributions [6,10,11]. FA has played an important role in increasing the prediction accuracy of flood frequencies [2,10,11]. Other methods, such as empirical relationships and the fuzzy logic approach, have also been applied to a few watersheds [10]. Many FA methods have been developed and tested, including the index flood method, the rational method, and various regression-based methods [2], the disadvantages of conventional FA could be the significant uncertainties and bias, which are due to limited historical recorded data with sufficient spatial and temporal coverage and acceptable quality, sampling variability, model errors, and the errors in projections into the future [12,13]. In addition, this traditional technology is based on the assumption that the hydrological observations are independently and identically distributed and the conditions remain stationary [11,14]. Lastly, FA focuses on flood peak values; however, the severity of a flood is defined not only by the flood peak value but also by flood volume, duration, etc. [15]. In practical applications, the flood series are always not independent and exist in a nonstationary context, and the watershed always contains a large number of ungauged areas [16]. Overall, conventional FA provides a limited assessment of flood frequencies [14]. Using conventional FA on nonstationary flow series may lead to uncertain flood frequency predictions [11,17].

To overcome these limitations, a nonstationary FA framework combined with lognormal or generalized extreme value distribution models and hydrological models has been developed [5,18]. Nonstationary FA coupled probability distribution has been considered an effective improvement in flood frequency analysis, with observations that are not independent under nonstationary circumstances [2,11,14]. To address these challenges of data scarcity, hydrological models have the ability to complement available datasets from local gauging stations with a spatial extension. This modeling approach provides some advantages: (1) planned alterations in rainfall and land use can be considered, and hydrological modeling allows one to obtain the full hydrograph for the design, and (2) the estimation of design flows can be performed for ungauged watersheds if the parameters of the hydrological model are regionalized [4]. Recently, many studies have reported the use of hydrological models, such as the HECHMS, WRF/DHSVM, and SWAT models, to simulate floods [4,17,19], the SWAT model can divide a watershed into sub-watersheds and then discretize them into a series of hydrologic response units (HRUs), which are spatially identified as unique soil–land use combination areas. SWAT can also provide a wide range of flexibility for model formulation and calibration [20–22]. A nonstationary FA framework incorporated with SWAT is a relatively new modeling approach and has been shown to provide acceptable prediction results when the hydroclimate context is variable, data are lacking, or the spatiotemporal analysis is complex [20,21,23].

Multivariate regression approaches have great potential for quantifying the relative importance of rainfall and land use types in controlling flood frequencies. However, traditional multivariate approaches cannot easily overcome the limitations of rainfall and land use types, which are highly collinear predictors [8,22,24,25]. Therefore, non-independent data must be handled cautiously in quantitative analyses. Partial least squares (PLS) regression is an advanced method that combines the features of principal component analysis and multiple linear regressions. It has been widely used to overcome the issue of multicollinearity and noisy data in quantitative analyses by projecting variables on high-dimensional spaces [26], the importance of a predictor to variations in model fitting is given by the variable influence on projection (VIP) value. VIP values reflect the importance of terms in a model with respect to both Y, i.e., a variable's correlation to all responses, and X, i.e., the projection [27]. Variations with higher VIP values are considered more important [28]. PLS regression can be used to evaluate rainfall, forest, and other land use influences on flood events [8,22,26].

The objective of this paper is to study the influences of forest land use and precipitation on flood frequencies in the Upper Honganjian watershed in China. This investigation is separated into two parts: (1) revealing flood frequency destruction based on a nonstationary FA method and SWAT model, and (2) illuminating the response of flood frequency distribution to land use and precipitation based on PLS regression.

## 2. Materials and Methods

### 2.1. Study Area

The Upper Honganjian watershed, with a total area of 981 km², is located in the Yellow River Basin and lies between 36°2′ N to 36°34′ N and 110°50′ E to 112°10′ E, the average annual temperature is approximately 11.8 °C, and the average annual precipitation is approximately 558.5 mm. A large portion of precipitation occurs during the monsoon season from May to October. Floods occur primarily in July, August, and September [29], the topography of the watershed is undulating and characterized by mountain ranges, steep slopes, and deep valleys, the elevation varies from 572 m at the Dongzhuang gauging station to 2259 m at the highest point in the watershed (Figure 1), the main soil types are yellow loamy soil (50.3%) and brown soil (18.8%), which correspond to Alfisols and Entisols in the USA Soil Taxonomy [30], respectively. Most areas are covered by forest (43.4%) and farmland (34.3%).

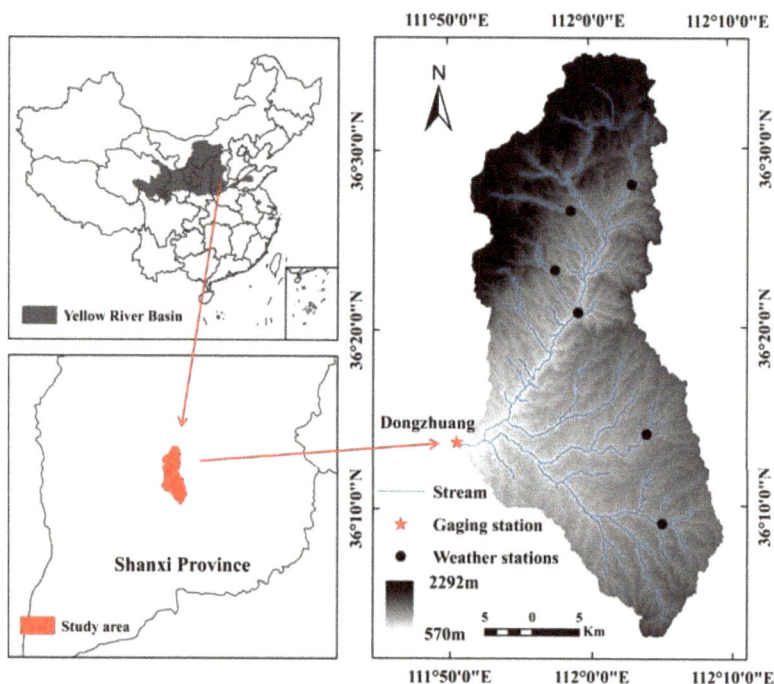

**Figure 1.** Location of Upper Honganjian watershed with observation stations.

### 2.2. Data Collection and Pretreatment

Hydrometeorological data: Forty-six years (1965–2010) of daily streamflow data were collected at the Dongzhuang station (the outlet of the Upper Honganjian watershed). Daily precipitation, solar radiation, wind speed, relative humidity, and max/min temperatures were attained from six weather stations (Figure 1). To account for seasonal variations, the streamflow data series were split into a wet season (i.e., May, June, and October), a flood season (i.e., July–September), and a dry season (i.e., November–April), the precipitation data were interpolated over the delineated sub-watersheds using a skewed normal distribution.

Topographical and soil coverage data for the model setup: The topographical data required by the SWAT model were derived from a digital elevation model (DEM) with a resolution of 25 × 25 m, which was obtained from the National Geomatics Center of China, the soil data, including a soil type map (1:100,000) and information on the related soil properties, were obtained from the Hydrological Bureau of Shanxi Province.

Land use data: The land use data for the 1980s were obtained from the Hydrological Bureau of Shanxi Province. Four land use domains were identified, namely, forest, farmland, urban, and grassland (Figure 2). Land use and soil data were extracted using ArcGIS Version 10.2. (Esri, Redlands, CA, USA).

Sub-watersheds and HRUs: In ArcSWAT 2012 (Esri, Redlands, CA, USA), the Upper Honganjian watershed was discretized into 33 sub-watersheds (Figure 3), which were then further subdivided into 207 HRUs based on land use, soil, and slopes.

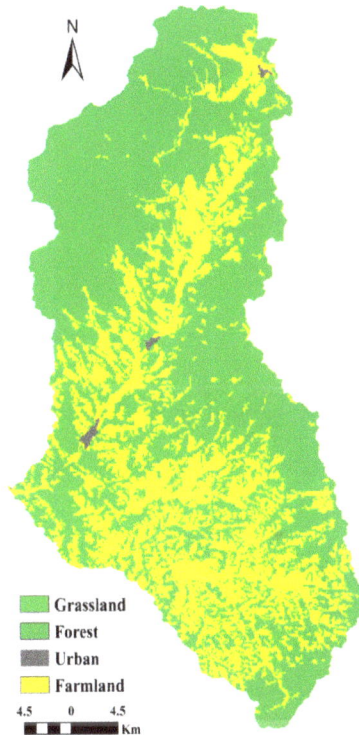

**Figure 2.** Land uses (farmland, urban, forest, and grassland) map for the 1980s in the Upper Honganjian watershed.

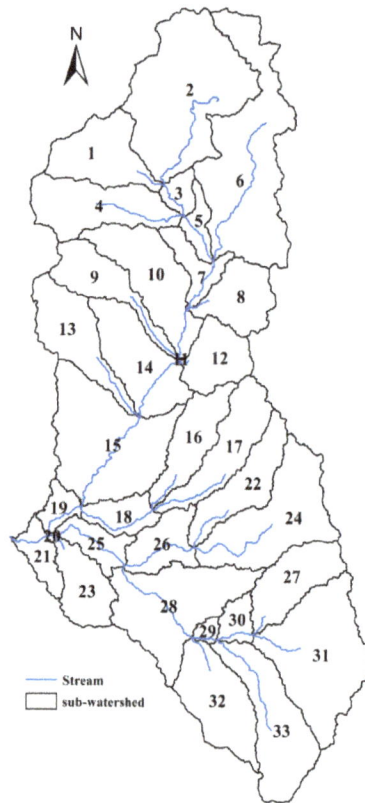

**Figure 3.** Map showing the sub-watersheds in the Upper Honganjian watershed.

*2.3. SWAT Model Calibration for Ungauged Sub-Watershed Streamflow*

Before using the SWAT model to predict the streamflow in ungauged sub-watersheds, calibration and validation of the model were performed. Calibration was performed with automated and manual techniques, the first step in the calibration process was determination of the most sensitive parameters for studying the watershed. Sensitivity analysis of the parameters in the SWAT model was performed using the LH–OAT analysis method, which combines the Latin hypercube (LH) sampling method and the one-factor-at-a-time (OAT) sensitivity analysis method. After 350 runs, the most sensitive parameters were detected. Autocalibration was the second step. This procedure was based on shuffled complex evolution (SCE–UA), which allows calibration of model parameters based on a single objective function [31]. In the last step, the SWAT model was manually calibrated against monthly and daily streamflow data, which were observed at the Dongzhuang gauge station, the calibration period was from January 1972 to December 1981. Manual calibration was performed to minimize total flow (minimized average annual percent bias), accompanied by visual inspection of the hydrographs, the parameters governing the surface runoff response were first calibrated, followed by those governing the fraction of streamflow that transform to baseflow. This preliminary calibration was followed by a fine-tuning at the daily time scale to ensure that the predicted versus measured peak flows and recession curves on a daily time step matched as closely as possible.

The validation period was from January 1982 to December 1991. Nash–Sutcliffe efficiency ($E_{NS}$), percent bias (*PBIAS*), and coefficient of determination ($R^2$) were used to evaluate the performance of the model. $E_{NS}$ was calculated as follows [8,32]:

$$E_{NS} = 1 - \frac{\sum_{i=1}^{n}(O_i - S_i)^2}{\sum_{i=1}^{n}(O_i - \overline{O})^2},\tag{1}$$

where $n$ is the discrete time step and $O_i$ and $S_i$ are the measured and simulated values, respectively. *PBIAS* is defined as follows [8,32]:

$$PBIAS = \frac{\sum_{i=1}^{n}(O_i - S_i)}{\sum_{i=1}^{n}O_i} \times 100,\tag{2}$$

where $O_i$ and $S_i$ are the measured and simulated values, respectively, and $n$ is the total number of paired values. $R^2$ was calculated as follows [8,32]:

$$R^2 = \left\{ \frac{\sum_{i=1}^{n}(O_i - \overline{O})(S_i - \overline{S})}{\left[\sum_{i=1}^{n}(O_i - \overline{O})^2\right]^{0.5}\left[\sum_{i=1}^{n}(S_i - \overline{S})^2\right]^{0.5}} \right\}^2,\tag{3}$$

where $n$ is the number of events, $O_i$ and $S_i$ are the measured and simulated streamflow values, respectively, and $O$ and $S$ are the mean observed and simulated values, respectively, the performance of the SWAT model (1) is considered acceptable when $R^2$ and $E_{NS}$ are greater than 0.5 and *PBIAS* ranges from ±15% to ±25%; (2) is good when $R^2$ is greater than 0.5, $E_{NS}$ is greater than 0.65, and *PBIAS* ranges from ±10% to ±15%; and (3) is very good when $R^2$ is greater than 0.5, $E_{NS}$ is greater than 0.75, and *PBIAS* is smaller than ±10% [8,32].

## 2.4. Evaluation of the Quantiles of Maximum Streamflow

Based on previous studies, the annual maximum (AM) series model was used in the multistep nonstationary FA in this study [2,18], the AM model is a framework that uses annual maximum values as appropriate estimators with a preferred distribution [10]. First, the AM series model was applied to identify the annual seasonal extreme streamflow. To identify the low and high outliers for the flow series, the Grubbs and Beck (1972) statistical test (see Appendix A.1 for more technical details) was used after the data were transformed to be normally distributed [33]. Second, to perform FA, several statistical tests were used, including the Wald–Wolfowitz test for randomness or autocorrelation and the Mann–Kendall test for stationarity (see Appendix A.2 and A.3 for more technical details), the Wald–Wolfowitz test and Mann–Kendall test were performed by SPSS 20 (IBM SPSS Inc., Chicago, IL, USA) and MATLAB 8.4 (The MathWorks Inc., Natick, MA, USA), respectively. Third, the appropriate probability distribution for the sub-watershed streamflow frequencies was identified. In this study, for the nonstationary modeling, a generalized extreme value (GEV) distribution model and a lognormal (LN2) distribution model (see Appendix A.4 and A.5 for more technical details) were chosen because the principles of the models are incorporated in the regional frequency analysis, although some of the parameters are allowed to change with time [18,34], the parameters were estimated using the maximum likelihood (ML) estimation method in MATLAB 8.4. To identify an appropriate probability distribution for fitting the observed hydrological data, goodness-of-fit tests were used in this study (similar methods were used by [10]), the Akaike information criterion (AIC), based on the principle of maximum entropy, and the Bayesian information criterion (BIC), proposed for use in the Bayesian framework, were used to assess the performance of the two models [35–38], the equations of the AIC and BIC are given as follows:

$$AIC = -2\log(L) + 2k\tag{4}$$

$$BIC = -2 \log (L) + 2 k \log (N), \tag{5}$$

where $L$ is the likelihood function, $k$ is the number of parameters of the distribution, and $N$ is the sample size. Finally, the quantiles of maximum streamflow and return periods were obtained and evaluated. Details on estimating the quantiles for the used distribution and return periods can be found in [18].

## 2.5. Determination of Flood Events and Return Periods

The flood events of the Upper Honganjian watershed were derived from the daily streamflow data collected at the Dongzhuang gauge station, the number of extreme streamflow events was calculated by counting the number of days in a year or season for which daily values exceed the quantiles of maximum streamflow. A flood event was identified when extreme streamflow events were continuous for six or more days in high streamflow periods [39], the notion of return period for extreme hydrological events is commonly used in hydrological nonstationary FA, the return period T is an event magnitude having a probability 1/T of being exceeded during any single event [10].

## 3. Results

### 3.1. SWAT Model Calibration and Validation

The calibrated SWAT parameters are listed in Table 1, the $E_{NS}$, $R^2$, and *PBIAS* values for the monthly and daily streamflow calibration and validation are listed in Table 2. All of the $E_{NS}$ and $R^2$ values for monthly streamflow were greater than 0.8, and the *PBIAS* values were in the range of ±10%, the statistical comparison between the measured daily streamflow and the simulation results showed good agreement, and the parameters that were calibrated for the daily streamflow of the model could be used to simulate every sub-watershed.

**Table 1.** Parameters for streamflow calibration of the SWAT in the Upper Honganjian watershed.

|  | **Parameter** | **Definition** | **Calibrated Value** |
|---|---|---|---|
| basin.bsn | ESCO | Soil evaporation compensation factor | 0.19 |
| basin.bsn | EPCO | Plant water uptake compensation factor | 1 |
| basin.bsn | SURLAG | Surface runoff lag time | 4 |
| .GW | GW_DELAY | Groundwater delay | 31 |
| .GW | GW_REVAP | Groundwater revap | 0.06 |
| .GW | ALPHA_BF | Baseflow alpha factor | 0.043 |
| .soil | SOL_AWC | Available water capacity of the soil layer | 0.2 |
| .sub | CH_N1 | Manning's '$n$' value for tributary channels | 0.014 |
| .rte | CH_N2 | Manning's '$n$' value for the main channel | 0.014 |
| .mgt | CN2 | SCS curve number | 62 (Forest) |
|  |  |  | 77 (Grassland) |
|  |  |  | 78 (Farmland) |
|  |  |  | 79 (Urban) |

**Table 2.** Accuracy of the SWAT model calibration and validation in the Upper Honganjian watershed.

| **Station** | **Period** | $E_{NS}$ [a] | *PBIAS* [b] | $R^2$ |
|---|---|---|---|---|
| *Monthly–Streamflow* | Calibration (1972–1981) | 0.80 | −0.10 | 0.81 |
|  | Validation (1982–1991) | 0.77 | −0.09 | 0.87 |
|  | Validation (1992–2001) | 0.75 | −0.13 | 0.80 |
|  | Overall (1972–2001) | 0.77 | −0.11 | 0.83 |
| *Daily–Streamflow* | Calibration (1971–1980) | 0.58 | −0.10 | 0.62 |
|  | Validation (1981–1990) | 0.56 | −0.09 | 0.67 |
|  | Validation (1991–2000) | 0.52 | −0.11 | 0.62 |
|  | Overall (1971–2000) | 0.55 | −0.10 | 0.64 |

[a] $E_{NS}$ = Nash–Sutcliffe efficiency; [b] *PBIAS* = Percent Bias.

## 3.2. Annual Extreme Streamflow and Appropriate Distribution

Figure 4 displays a box plot for the annual seasonal extreme streamflow data for each decade, the total annual number of days with extreme streamflow for the entire watershed for the 1970s, 1980s, 1990s, and 2000s were 306, 321, 291, and 196, respectively, the peak in the 1980s was well defined due to the frequency of the hydrometeorological days, which represented approximately 28.8% of all the data, while the low value in the 2000s represented approximately 17.6% of all the data.

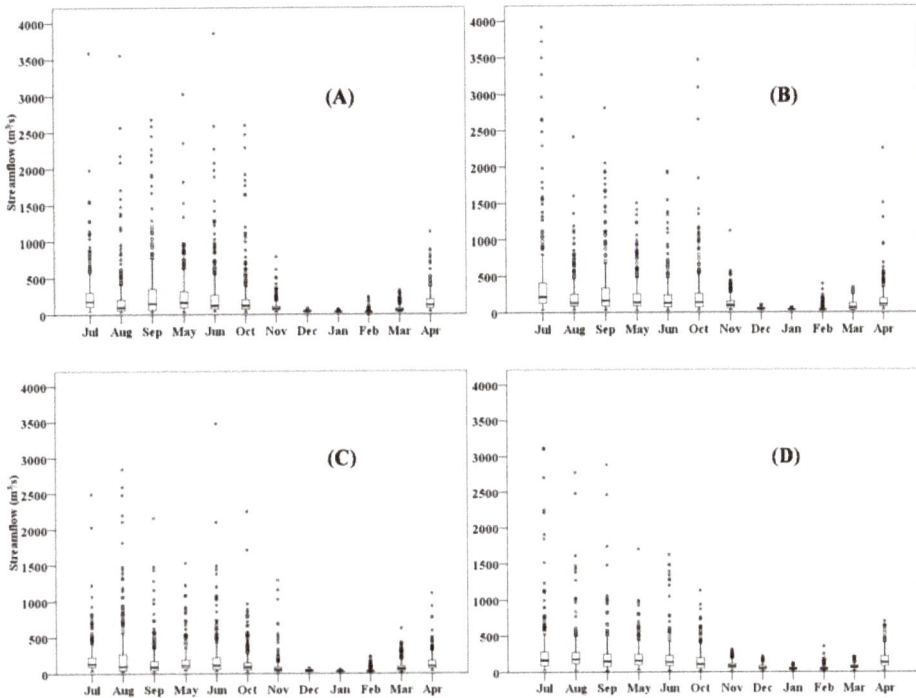

**Figure 4.** Box plot for the monthly extreme streamflow data in the flood season (July, August, and September), wet season (May, June, and October), and dry season (November–April) in the 1970s (**A**), 1980s (**B**), 1990s (**C**), and 2000s (**D**).

The Wald–Wolfowitz test results ($Z = -1.054$, $p$-value = 0.387) of the streamflow at the Dongzhuang station showed that the data were independent at $p < 0.05$, the Mann–Kendall test results ($K = -2.256$, $p$-value = 0.021) showed that the yearly maximum streamflow increased significantly after 1999 at $p < 0.05$, the corresponding parameters were estimated by the ML methods, the results of the goodness-of-fit tests for selecting an appropriate probability distribution for the sub-watershed streamflow frequency analysis using the AIC and BIC criteria are summarized in Table 3. Out of 66 cases (two model selection criteria × 33 sub-watersheds), the LN2 distribution model was preferred for 53 cases (i.e., 80.3%), whereas the GEV distribution model was favored in the remaining 13 cases (i.e., 19.7%). Both model selection criteria favored the LN2 distribution over the GEV distribution, as shown in Table 3. These results demonstrated that the LN2 distribution was preferable to the GEV distribution for modeling the partial time series data for the selected watersheds, the empirical probability curves as well as the confidence intervals of the LN2 distribution for the observed streamflow at the watershed scale are presented in Figure 5.

Table 3. Summary of the goodness–of–fit tests for the 33 sub-watersheds.

| Distribution | AIC | | BIC | |
|---|---|---|---|---|
| | LN2 [a] | GEV [b] | LN2 | GEV |
| No. of sub-watersheds being selected | 26 | 7 | 27 | 6 |
| Percentage of sub-watersheds (%) | 78.8 | 21.2 | 81.8 | 18.2 |

[a] Sampling LN2 = lognormal; [b] Sampling GEV = generalized extreme value.

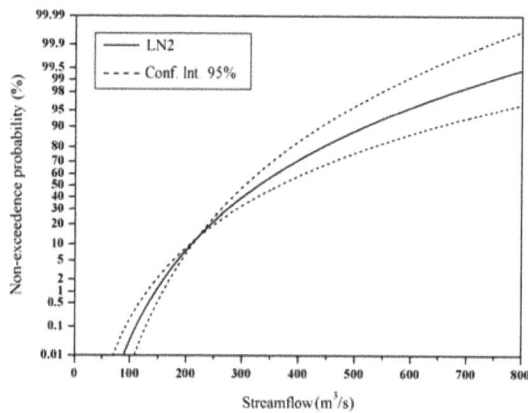

**Figure 5.** Empirical probability curves and confidence intervals (Conf. Int.) of the LN2 distributions for the observed stream flows at the watershed scale, the scale of y-axis is normal probability.

*3.3. Nonstationary Regional Frequency Analysis*

By fitting the LN2 distribution with the ML parameter estimation method to the sub-watershed streamflow data, the maximum streamflow quantiles at each sub-watershed were estimated for average recurrence intervals (ARIs) of two, five, 10, 20, 100, and 1000 years to analyze the flood events in each sub-watershed.

To analyze the seasonal variation in flood events at the watershed scale during different decades, the annual average number of flood events in different decades (1970s, 1980s, 1990s, and 2000s) were computed under different return levels (2–5, 5–20, 20–100, 100–1000, and >1000 years), the results are shown in Figure 6, the number of flood events during flood seasons for return periods covering two to 100 years, wet seasons for return periods covering two to 20 years, and dry seasons for the return periods of two to five years shows similar fluctuations. Similar variations can also be found for return periods that exceed 100, 20, and five years in flood seasons, wet seasons, and dry seasons, respectively. Compared with the lower and higher return periods, a significantly different flood event distribution is identified. As the seasons change (i.e., from the flood to dry season), the total number of flood events in each return periods change and the maximum and minimum numbers diminish and shift to shorter return periods.

For the return periods from two to five years, the flood events were most frequent during the 1980s in all seasons, and the flood events were least frequent in the 1990s for each season. For the return periods of five to 20 years, the maximum values occurred in the 1980s, whereas the minimum values occurred in the 1990s for both the flood and wet seasons. For the return periods of 20 to 100 years, the maximum values occurred in the 1980s for both the flood and dry seasons and in the 1970s for the wet season, and the minimum values were recorded in the 1990s for the flood season, the lowest values occurred in the 1980s, 1990s, and 2000s for the wet season, and the total number for the three decades was four. For the return periods of 100–1000 and >1000 years, the maximum values occurred in the 1970s for both the flood and wet seasons, whereas the minimum values were recorded in the 1990s for the flood season. Compared with the baseline return periods of two to five years, in the flood season, the mean annual number of flood events for the four decades increased for the return periods of five to 20 years and then showed a decreasing gradient for larger return periods. Similar variations occurred in each decade. In the wet season, compared with the baseline, the mean annual number of flood events over the four decades indicated a decreasing trend for return periods below 1000 years. Each decade exhibited similar dynamics. In the dry season, a strong gradient existed in the total number of events over the four decades.

Among the four decades, in the flood season, the highest number of events was found for the return periods of five to 20 years, and the lowest number was found for the return periods of >1000 years, the highest numbers were found for the return periods of two to five years in both the wet and dry seasons, while the lowest numbers were found for the return periods of 100 to 1000 years and the return periods of 20 to 100 years in the wet and dry seasons, respectively. Among the three seasons, the flood events in the flood season were more severe than in the other two seasons.

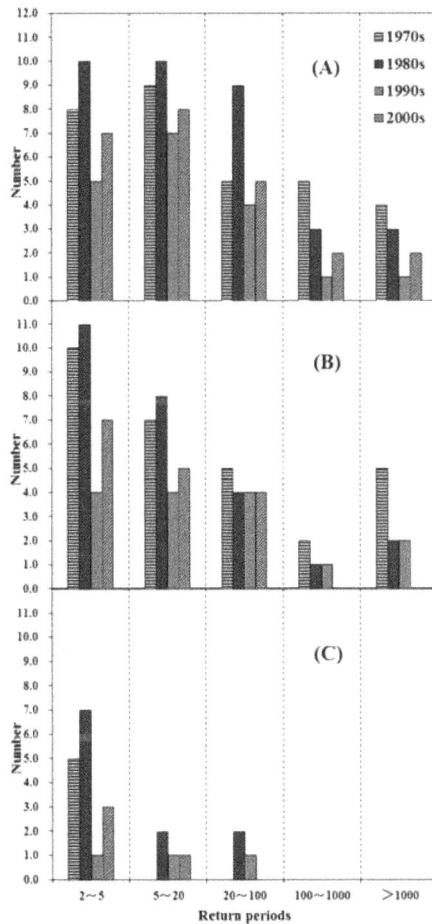

**Figure 6.** The average annual number of flood events under different return periods, i.e., 2 to 5, 5 to 20, 20 to 100, 100 to 1000, and >1000 years, for various decades (1970s, 1980s, 1990s, and 2000s) in the flood season (**A**), wet season (**B**), and dry season (**C**).

### 3.4. Contribution of Rainfall and Land Use to Flood Events

The performance of the PLS regression models is shown in Table 4. For the flood season, in the return periods of two to five years, the first component was dominated by forest on the negative side and explained 74.3% of the variation in flood events, the addition of the second component was dominated by rainfall, urban land, and farmland on the positive side and made the explanation approached 81.2% of the variation and generated the minimum mean square error of cross-validation (RMSECV) value, the addition of more components to the PLS regression models did not substantially improve the explanatory power but resulted in a higher RMSECV, indicating that the subsequent components were not strongly correlated with the residuals of the predicted variable according to [26]. For return periods of five to 20 years model, the first component was dominated by forest on the negative side, which explained 71.8% of the flood events variance in the dataset, the addition of the second component, dominated by rainfall, farmland, and urban land on the positive side, increased the model–explained variance to 78.0%. For the return periods of 20 to 100 and 100–1000 years, the first component was dominated by forest on the negative side, and the second component was dominated by urban land on the positive side. For the wet season, in return periods of two to five and five to 20 years, the first component was dominated by forest on the negative side, and the second component was dominated by rainfall, urban land, and farmland on the positive side. For the dry season, in the return periods of two to five years, the first component was dominated by forest on the negative side and farmland on the positive side, and the second component was dominated by rainfall and urban land on the positive side. In return periods > 5 years, the rainfall became to the dominate factor.

The relationship between the number of seasonal flood events (in the 1980s) and the proportion of land use types (1980s) was analyzed at the sub-watershed scale. Figure 7 shows the relative contributions of impact factors under different return periods, the influence of land use types on flood events decreased with increasing return periods. Farmland and urban areas had a positive effect on the number of events, while forest land had a negative effect. No statistically significant relationships were detected between grassland and flood events. For the flood and wet season, in the return periods of 2–5 and 5–20 years, the forest area had a significant negative effect on the number of flood events, while significant positive correlations were detected between both farmland and urban areas and flood events. VIP values of forest, farmland, and urban land (VIP > 1) were higher than those of grassland and rainfall. For the flood season, in the return periods of 20 to 100 and 100–1000 years, forest and urban land had VIP scores greater than 1, and rainfall, farmland, and grassland had VIP value less than 1, as shown in Figure 7. In the wet seasons, no statistically significant relationships between land use types and flood events were detected when the return period exceeded 100 years. For the dry season, in the return periods of two to five years, rainfall had the highest VIP score (1.337) and a larger positive regression (0.679), followed by the farmland (coefficient = 0.723; VIP = 1.301), forest (coefficient = −0.685; VIP = 1.246), and urban land (coefficient = 0.324; VIP = 1.129). These results show a significant negative correlation between the forest area and the number of flood events. Rainfall, farmland, and urban area had remarkable positive effects on the number of flood events, the VIP value (greater than 1) of rainfall was highest in return periods > 5 years in the dry season, followed by farmland, forest, urban land, and grassland with VIP values less than 1 (ranging from 0.428 to 0.087).

Table 4. Summary of PLS regression for floods in each season.

| Response Y | | | | | | |
|---|---|---|---|---|---|---|
| Seasons | Return Periods (Years) | $R^{2\,a}$ | Component | Explained in Y (%) | Cumulative Explained in Y (%) | RMSECV [b] |
| Flood | 2–5 | 0.81 | 1 | 74.3 | 74.3 | 0.89 |
| | | | 2 | 6.9 | 81.2 | 0.80 |
| | | | 3 | 0.1 | 81.3 | 0.81 |
| | 5–20 | 0.79 | 1 | 71.8 | 71.8 | 4.17 |
| | | | 2 | 6.2 | 78.0 | 3.97 |
| | | | 3 | 1.4 | 79.4 | 4.05 |
| | 20–100 | 0.70 | 1 | 67.9 | 67.9 | 0.75 |
| | | | 2 | 2.4 | 70.3 | 0.73 |
| | | | 3 | 0.1 | 70.4 | 0.74 |
| | 100–1000 | 0.64 | 1 | 60.2 | 60.2 | 5.55 |
| | | | 2 | 3.1 | 63.3 | 5.28 |
| | | | 3 | 0.8 | 64.1 | 5.47 |
| Wet | 2–5 | 0.82 | 1 | 75.1 | 75.1 | 0.71 |
| | | | 2 | 5.6 | 80.7 | 0.64 |
| | | | 3 | 1.3 | 82.0 | 0.61 |
| | | | 4 | 0.4 | 82.4 | 0.65 |
| | 5–20 | 0.79 | 1 | 71.3 | 71.3 | 0.85 |
| | | | 2 | 6.6 | 77.9 | 0.80 |
| | | | 3 | 1.2 | 79.1 | 0.81 |
| Wet | 20–100 | 0.34 | 1 | 28.3 | 28.3 | 0.79 |
| | | | 2 | 3.9 | 32.2 | 0.69 |
| | | | 3 | 1.6 | 33.8 | 0.71 |
| | 100–1000 | 0.27 | 1 | 22.2 | 22.2 | 0.65 |
| | | | 2 | 4.5 | 26.7 | 0.59 |
| | | | 3 | 0.3 | 27.0 | 0.63 |
| Dry | 2–5 | 0.84 | 1 | 78.5 | 78.5 | 0.91 |
| | | | 2 | 3.9 | 82.4 | 0.88 |
| | | | 3 | 1.6 | 84.0 | 0.84 |
| | | | 4 | 0.2 | 84.2 | 0.85 |
| | 5–20 | 0.61 | 1 | 60.0 | 60.0 | 0.74 |
| | | | 2 | 0.9 | 60.9 | 0.68 |
| | | | 3 | 0.1 | 61.0 | 0.69 |
| | 20–100 | 0.69 | 1 | 67.2 | 67.2 | 0.65 |
| | | | 2 | 1.6 | 68.8 | 0.54 |
| | | | 3 | 0.1 | 68.9 | 0.58 |
| | 100–1000 | 0.76 | 1 | 74.2 | 74.2 | 0.54 |
| | | | 2 | 1.6 | 75.8 | 0.49 |
| | | | 3 | 0.2 | 76.0 | 0.52 |

[a] Sampling $R^2$ = goodness of fit; [b] RMSECV = cross-validated root mean squared error.

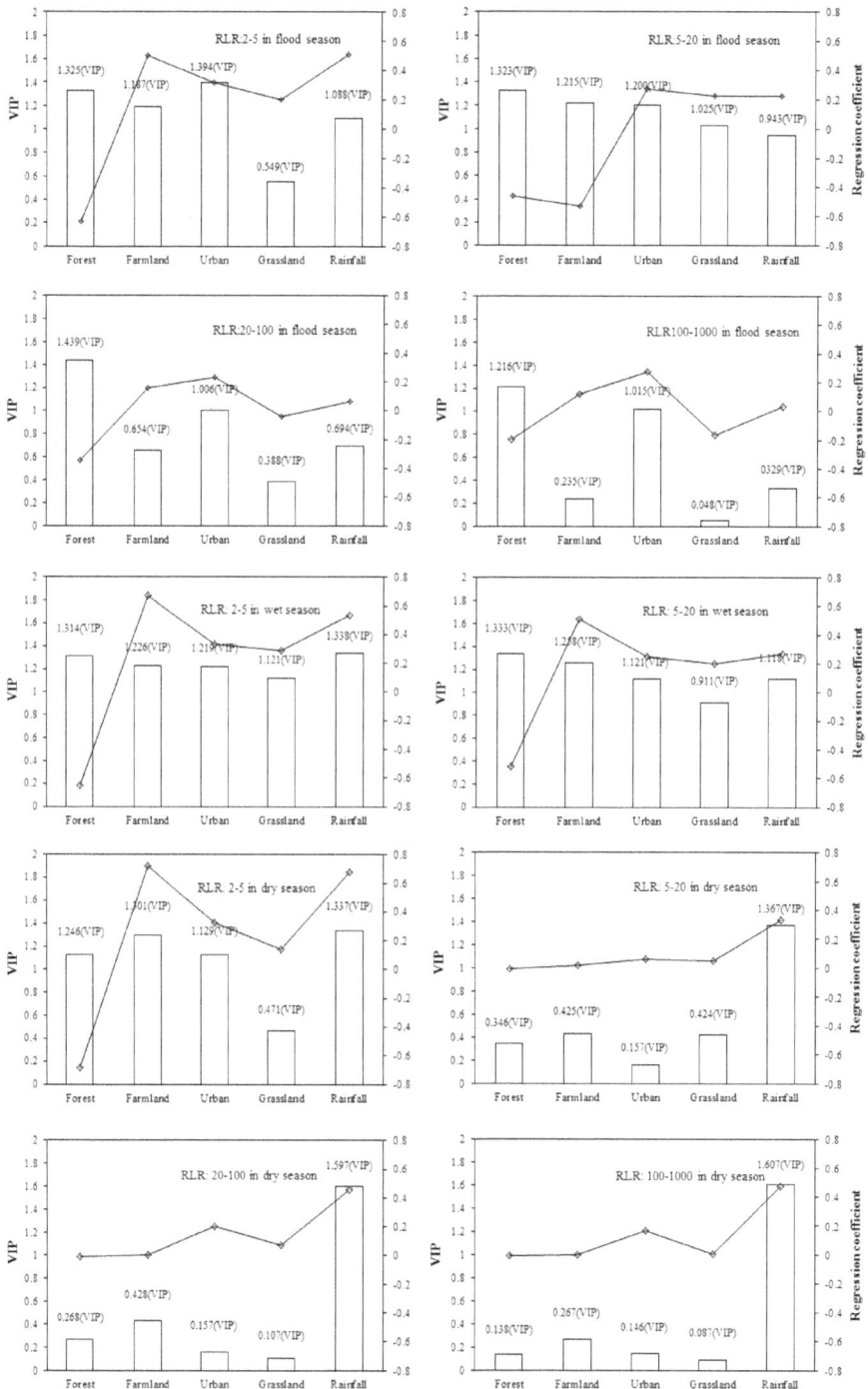

**Figure 7.** Regression coefficients (lines) and the variable influence on projection (VIP) (bars) of each factor.

## 4. Discussion

The ability of the SWAT model to simulate the daily streamflow data in the frequency analysis has been demonstrated by many papers [21,23,40]. A high goodness-of-fit value for the monthly and daily streamflow suggests good model performance in our study. Table 2 shows good modeling results for the monthly streamflow, with a maximum Nash–Sutcliffe efficiency $(E_{NS})$ of 0.80. In our study, daily streamflow $E_{NS}$ is less satisfactory than that of monthly streamflow, which is similar to other SWAT modeling studies [41,42]. For hydrologic evaluations performed at a monthly time interval, the model results are satisfactory when $E_{NS}$ values exceed 0.5. However, appropriate relaxing of the standard may be performed for daily time–step evaluations [32]. Thus, a daily $E_{NS}$ of 0.5 corresponds to a monthly $E_{NS}$ of approximately >0.8, as suggested by [43]. Therefore, the performance measures for the simulations range from satisfactory to good in all studied sub-watersheds according to [32]. This evaluation methodology is suggested by many studies [8,41,43,44]. Previously, we studied the Upper Du River watershed, which is a forest watershed similar to the Upper Honganjian watershed in China, using the same method and the results confirmed the satisfactory performance of SWAT [22,43,44], the results of this study are consistent with Liu [45], which was a study of a runoff simulation in the Upper Honganjian watershed.

We address the relative importance of precipitation and land use types for flood events in 33 sub-watersheds, the most convenient and comprehensive description of the relative importance of predictors can be derived from exploring their VIP values [26]. For different return periods in each season, relevant fluctuations in the total number of events in each decade are found. These fluctuations are consistent with the variations in flood events in the Upper Honganjian watershed, implying that a similar pattern of the flood events seasonal distribution exists in specific return periods. However, there are significantly different dynamics for return periods that exceed 100 years in the flood season, return periods that exceed 20 years in the wet season, and return periods that exceed five years in the dry season. These results may be related to the coupling of precipitation, land use types, and anthropogenic construction [5,46].

The results in Figure 7 show that precipitation changes will influence flood events, especially at a two- to five-year return period in the flood and wet seasons. Precipitation could trigger a higher risk of floods in a watershed, and many complex factors (temperature, reservoirs, and drainage system) may also influence the streamflow volume [47]. Similar conclusions were reported by Zhang et al. [5], who showed that precipitation is one of the pivotal factors triggering hydrological alterations of flood events, the flood dynamics at return periods > 5 years may be attributed to the hydrological regulations of water reservoirs. Anthropogenic construction and management, such as dams and reservoirs, river training, and human water use affect the seasonality of flood events in these periods [48,49]. Reservoirs, dams, irrigation flow diversions, and flood control structures have been developed and generate significant hydrogeomorphic alterations with impacts occurring in both streams and catchments of the watershed [50]. Reservoirs and dams result in increased evapotranspiration (ET) and lead to fewer flood events, while seasonal withdrawals affect the seasonality of flood events [48]. Small reservoirs may lose up to 50% of their stored volume due to evaporation in many regions due to the high ratio of surface/volume area. Evaporation constitutes a major component of the water balance in the reservoirs and may significantly decrease flood events [51]. Moreover, in order to sustain and maintain the ecological integrity of watershed, numerous watershed management measures, such as management of flood utilization, establishing and maintaining minimum flow releases, or permitting controlled "flushing" releases that establish the necessary high flows for sediment transport have been applied. These hydrogeomorphic and watershed management practice impacts have profoundly influenced flood evolution and frequency [50]. In the dry season, the correlations between flood events and the precipitation amount shift from low to high values as the return period increases. This anomaly can be interpreted as follows: for a longer return period, the flood frequency depends on the initial soil moisture conditions [21]; therefore, the precipitation amount in the dry season determines the initial soil moisture conditions and indirectly influences the flood events that occur in the wet and flood seasons.

For each season, the influence of land use on flood events decreases with an increasing return period. Water management facilities have led to variations in hydrological events for a longer return period [4]. Our results reveal a strong influence of land use types on flood events for specific return periods; the number of flood events is also highly influenced by land use type, namely, forest, urban, and farmland areas. Farmland and urban areas are found to increase the number of flood events, while forest land results in fewer flood events. Similar results were reported by [47,52]. Farmland reduces evapotranspiration, enhances infiltration, increases the initial moisture stored in the soil, eventually increases the number of flood events [21], the increased number of flood events associated with the expansion of urban areas can be explained as follows. With urbanization, infiltration is reduced by soil compaction and impervious surface additions, and water flushes more quickly through the watershed as a result of decreases in the hydraulic resistance of land surfaces and channels [53]. However, forest land increases evapotranspiration and tends to decrease the number of flood events [21]. For return periods of two to five years, the number of flood events is most closely related to the land use types of forest, urban, and farmland in the dry season, followed by the wet and flood seasons. This conclusion was also drawn by Liu et al. [54], who indicated that the impact of deforestation or reforestation on hydrological events is more significant in the dry season than in other seasons. In dry season, the effects of rainfall are greater than those of land use type. This phenomenon can be explained as follows, the dry season is not a growth period for most vegetation, and ET of forest and other land use types associated with flood events can be ignored [22]. In addition, compared with flood and wet seasons, interception of the forest canopy and undergrowth vegetation associated with throughfall in the dry season is generally lower, which mitigates the negative influences of forests on flood [53].

## 5. Conclusions

The SWAT model was used to estimate the daily streamflow for ungauged sub-watersheds. Exploratory data analysis and outlier detection were performed using box plots. Based on the maximum likelihood (ML) estimation method, Akaike information criterion (AIC), and Bayesian information criterion (BIC), the lognormal (LN2) distribution was preferable to the generalized extreme value (GEV) distribution to fit the partial time series streamflow data.

In low return periods, similar patterns existed for the flood event distribution. Significantly different patterns existed for return periods that exceeded 100 years in the flood season, return periods that exceeded 20 years in the wet season, and return periods that exceeded five years in the dry season. Rainfall and forest are pivotal factors triggering flood event alterations in lower return periods, and the flood events in higher return periods are attributed to the hydrological regulations of water management facilities. Farmland and urban areas were related to fewer flood events, while the presence of forest land was found to decrease the number of flood events. In the dry season, the influence of rainfall on flood events is much greater than that of land use.

The approach used in this study can help to easily select the optimal distributions for watersheds using ungauged sub-watersheds, the return periods and flood events can be simulated more precisely using optimal distributions. Moreover, flood-prone regions can be identified, which can provide a scientific foundation to determine flood-resistant measures by comparing the increased flood risk at different return levels.

**Author Contributions:** X.H. and P.H. conceived and designed the experiments; X.H. and L.W. performed the experiments; W.W. and P.H. analyzed the data; X.H. wrote the paper.

**Funding:** Financial support for this research was provided by the Project of Hydraulic Science and Technology of Jiang Xi province, China (KT201615); National Natural Science Foundation of China (No: 51509088); Henan province university scientific and technological innovation team (18IRTSTHN009); the Key Scientific Research Projects of Higher Education Institutions (18A170010); Henan Key Laboratory of Water Environment Simulation and Treatment (2017016).

**Acknowledgments:** We are truly grateful to editors and the anonymous reviewers for providing critical comments and thoughtful suggestions.

**Conflicts of Interest:** The authors declare no conflict of interest.

## Appendix

*Appendix A.1 Grubbs and Beck (1972) Statistical Test*

The observations are arranged in ascending order: $x_1 \leq x_2 \leq \dots \leq x_n$. Therefore, to test whether the largest observation, $x_n$, in normal samples is too large, we compute

$$T_n = (x_n - \bar{x})/S, \tag{A1}$$

and

$$S = \left[\sum (x_i - \bar{x})^2 / (n-1)\right]^{1/2} = \left\{\left[n\sum x_i{}^2 - \left(\sum x_i\right)^2\right] / n(n-1)\right\}^{1/2}, \tag{A2}$$

and refer the result to the table of Grubbs test [31], which provides various upper probability levels for $T_n$. A significance test of the smallest observation for normal samples is obtained by computing

$$T_1 = (\bar{x} - x_1)/S, \tag{A3}$$

To test the significance of the two largest observations, $x_{n-1}$ and $x_n$ we compute

$$S_{n-1,n}^2 / S_O^2, \tag{A4}$$

in which

$$S_O^2 = \sum_{i=1}^{n} (x_i - \bar{x})^2, \tag{A5}$$

$$S_{n-1,n}^2 = \sum_{i=1}^{n-2} (x_i - \bar{x}_{n-1,n})^2, \tag{A6}$$

in which

$$\bar{x}_{n-1,n} = \sum_{i=1}^{n-2} x_i / (n-2), \tag{A7}$$

*Appendix A.2 Wald–Wolfowitz Test*

This test is used to compare two unmatched, supposedly continuous distributions, and the null hypothesis is that the two samples are distributed identically [55]. A run is a set of sequential values that are either all above or below the mean. To simplify computations, the data are first centered around their mean. To carry out the test, the total number of runs is computed along with the number of positive and negative values. A positive run is a sequence of values greater than zero, and a negative run is a sequence of values less than zero. We can then test whether the numbers of positive and negative runs are distributed equally in time, the test statistic is asymptotically normally distributed, and therefore, this test computes $Z$, the large sample test statistic, as follows:

$$Z = \frac{R - \left(\frac{2n_1 n_2}{n_1 + n_2} + 1\right)}{\sqrt{\frac{2n_1 n_2 (2n_1 n_2 - n_1 - n_2)}{(n_1 + n_2)^2 (n_1 + n_2 - 1)}}}, \tag{A8}$$

in which $R$ is the number of runs.

The null and alternative hypotheses are as follows:

$H_0$: $X$ and $Y$ come from two identical populations [55].

When the large sample approach is in question, the test statistic calculated by the average of this formula will be compared with the values obtained from the standard normal table for the previously determined level of significance [55]. If the $Z$ value is lower than or equal to the table value, then the $H_0$ hypothesis must be rejected at the significance level of $\alpha$.

*Appendix A.3 Mann–Kendall Test*

The Mann–Kendall (MK) statistic, $S$, is defined as follows:

$$S = \sum_{i=1}^{n-1} \sum_{j=i+1}^{n} \text{sgn}(X_j - X_i), \tag{A9}$$

where $X_j$ represents the sequential data values, $n$ is the length of the dataset, and

$$\text{sgn}(\theta) = \begin{cases} 1 & \text{if } \theta > 0 \\ 0 & \text{if } \theta = 0 \\ -1 & \text{if } \theta = 0 \end{cases}, \tag{A10}$$

The statistic $S$ is approximately normally distributed when $n \geq 8$, with the mean and the variance as follows:

$$E[S] = 0, \tag{A11}$$

$$V(S) = \frac{n(n-1)(2n+5) - \sum_{i-1}^{n} t_i i(i-1)(2i+5)}{18}, \tag{A12}$$

where $t_i$ is the number of ties of extent $i$.

The standardized test statistic ($Z$) of the MK test and the corresponding $p$-value ($p$) for the one–tailed test are given by

$$Z = \begin{cases} \frac{S-1}{\sqrt{Var(S)}} & S > 0 \\ 0 & S = 0 \\ \frac{S+1}{\sqrt{Var(S)}} & S < 0 \end{cases}, \tag{A13}$$

$$p = 0.5 - \Phi(|Z|), \tag{A14}$$

$$\Phi(|Z|) = \frac{1}{\sqrt{2\pi}} \int_0^{|Z|} e^{-\frac{t^2}{2}} dt, \tag{A15}$$

If the $p$-value is small enough, then the trend is unlikely to be caused by random sampling. Positive and negative $Z$ values indicate upward and downward trends, respectively. At the significance level of 0.05, if $p \leq 0.05$, then the existing trend is considered statistically significant [56,57].

*Appendix A.4 GEV Distributions*

The distributions of extreme values (EV) were developed by Fisher and Tippett [58]. Families (Gumbel, Fréchet, and Weibull) of traditional EV were combined into the generalized extreme values (GEV) distribution with a cumulative distribution function by Jenkinson [59]:

$$F_{GEV}(x) = \begin{cases} \exp\left\{-(1 - \frac{k}{a}(x-u))^{1/k}\right\} & k \neq 0 \\ \exp[-\exp(-\frac{(x-u)}{a})] & k = 0 \end{cases}, \tag{A16}$$

where $u + a/k \leq x < +\infty$ when $k < 0$ (Fre'chet), $-\infty < x < +\infty$ when $k = 0$ (Gumbel) and $+\infty < x \leq u + a/k$ when $k > 0$ (Weibull). $u(\in \mathbb{R})$, $a$ (>0), and $k(\in \mathbb{R})$ are the location, the scale and the shape parameters, respectively.

*Appendix A.5 LN2*

In probability theory, a log-normal distribution (LN) is a continuous probability distribution of a random variable the logarithm of which is normally distributed, the probability density function of the two–parameter log–normal distribution (LN2) is:

$$f(k) = \frac{1}{\sqrt{2\pi}\sigma k} \exp\left[-\frac{1}{2}\left(\frac{\ln k - u}{\sigma}\right)^2\right] \quad k > 0 \tag{A17}$$

in which $-\infty < u < +\infty, 0 \leq \sigma < +\infty$.

## References

1. Eagleson, P.S. Dynamics of flood frequency. *Water Resour. Res.* **1972**, *8*, 878–898. [CrossRef]
2. Zaman, M.A.; Rahman, A.; Haddad, K. Regional flood frequency analysis in arid regions: A case study for Australia. *J. Hydrol.* **2012**, *475*, 74–83. [CrossRef]
3. O'Brien, N.L.; Burn, D.H. A nonstationary index-flood technique for estimating extreme quantiles for annual maximum streamflow. *J. Hydrol.* **2014**, *519*, 2040–2048. [CrossRef]
4. Haberlandt, U.; Radtke, I. Hydrological model calibration for derived flood frequency analysis using stochastic rainfall and probability distributions of peak flows. *Hydrol. Earth Syst. Sci.* **2014**, *18*, 353–365. [CrossRef]
5. Zhang, Q.; Gu, X.; Singh, V.P.; Xiao, M. Flood frequency analysis with consideration of hydrological alterations: Changing properties, causes and implications. *J. Hydrol.* **2014**, *519*, 803–813. [CrossRef]
6. Rao, A.R.; Hamed, K.H. *Flood Frequency Analysis*; CRC Press: New York, NY, USA, 2000; p. 355.
7. Wei, W.; Chen, L.D.; Fu, B.J.; Huang, Z.L.; Wu, D.P.; Gui, L.D, the effect of land uses and rainfall regimes on runoff and soil erosion in the semi-arid loess hilly area. China. *J. Hydrol.* **2007**, *335*, 247–258. [CrossRef]
8. Yan, B.; Fang, N.F.; Zhang, P.C.; Shi, Z.H. Impacts of land use change on watershed streamflow and sediment yield: An assessment using hydrologic modelling and partial least squares regression. *J. Hydrol.* **2013**, *484*, 26–37. [CrossRef]
9. Karim, F.; Hasan, M.; Marvanek, S. Evaluating annual maximum and partial duration series for estimating frequency of small magnitude floods. *Water* **2017**, *9*, 481. [CrossRef]
10. Benkhaled, A.; Higgins, H.; Chebana, F.; Necir, A. Frequency analysis of annual maximum suspended sediment concentrations in Abiod wadi, Biskra (Algeria). *Hydrol. Process.* **2014**, *28*, 3841–3854. [CrossRef]
11. Khaliq, M.N.; Ouarda, T.B.M.J.; Ondo, J.C.; Gachon, P.; Bobée, B. Frequency analysis of a sequence of dependent and/or non-stationary hydro-meteorological observations: A review. *J. Hydrol.* **2006**, *329*, 534–552. [CrossRef]
12. Schendel, T.; Thongwichian, R. Considering historical flood events in flood frequency analysis: Is it worth the effort? *Adv. Water Resour.* **2017**, *105*, 144–153. [CrossRef]
13. Obeysekera, J.; Salas, J.D. Quantifying the uncertainty of design floods under nonstationary conditions. *J. Hydrol. Eng.* **2013**, *19*, 1438–1446. [CrossRef]
14. She, D.X.; Xia, J.; Zhang, D.; Ye, A.Z.; Sood, A. Regional extreme-dry-spell frequency analysis using the L-moments method in the middle reaches of the Yellow River Basin, China. *Hydrol. Process.* **2014**, *28*, 4694–4707. [CrossRef]
15. Tramblay, Y.; St–Hilaire, A.; Ouarda, T.B.M.J. Frequency analysis of maximum annual suspended sediment concentrations in North America. *Hydrol. Sci. J.* **2008**, *53*, 236–252. [CrossRef]
16. Milly, P.C.D.; Julio, B.; Malin, F.; Robert, M.H.; Zbigniew, W.K.; Dennis, P.L.; Ronald, J.S. Stationarity is dead: Whither water management. *Science* **2008**, *319*, 573–574. [CrossRef] [PubMed]
17. Zhang, X.; Harvey, K.D.; Hogg, W.D.; Yuzuk, T.R. Trends in Canadian streamflow. *Water Resour. Res.* **2001**, *37*, 987–999. [CrossRef]
18. Leclerc, M.; Ouarda, T.B. Non-stationary regional flood frequency analysis at ungauged sites. *J. Hydrol.* **2007**, *343*, 254–265. [CrossRef]
19. Kovalets, I.V.; Kivva, S.L.; Udovenko, O.I. Usage of the WRF/DHSVM model chain for simulation of extreme floods in mountainous areas: A pilot study for the Uzh River Basin in the Ukrainian Carpathians. *Nat. Hazards* **2014**, *75*, 2049–2063. [CrossRef]

20. Behera, S.; Panda, R.K. Evaluation of management alternatives for an agricultural watershed in a sub-humid subtropical region using a physical process based model. *Agric. Ecosyst. Environ.* **2006**, *113*, 62–72. [CrossRef]
21. Ryu, J.H.; Lee, J.H.; Jeong, S.; Park, S.K.; Han, K, the impacts of climate change on local hydrology and low flow frequency in the Geum River Basin, Korea. *Hydrol. Process.* **2011**, *25*, 3437–3447. [CrossRef]
22. Huang, X.D.; Shi, Z.H.; Fang, N.F.; Li, X. Influences of land use change on baseflow in mountainous watersheds. *Forests* **2016**, *7*, 16. [CrossRef]
23. Dessu, S.B.; Melesse, A.M. Modelling the rainfall-runoff process of the Mara River basin using the Soil and Water Assessment Tool. *Hydrol. Process.* **2012**, *26*, 4038–4049. [CrossRef]
24. Artita, K.S.; Kaini, P.; Nicklow, J.W. Examining the possibilities: Generating alternative watershed-scale BMP designs with evolutionary algorithms. *Water Resour. Manag.* **2013**, *27*, 3849–3863. [CrossRef]
25. Zhou, F.; Xu, Y.; Chen, Y.; Xu, C.Y.; Gao, Y.; Du, J. Hydrological response to urbanization at different spatio-temporal scales simulated by coupling of CLUE-S and the SWAT model in the Yangtze River Delta region. *J. Hydrol.* **2013**, *485*, 113–125. [CrossRef]
26. Abdi, H.; Williams, L.J. Principal component analysis. *Wiley Interdiscip. Rev. Comput. Stat.* **2010**, *2*, 433–459. [CrossRef]
27. Buondonno, A.; Amenta, P.; Viscarra-Rossel, R.A.; Leone, A.P. Prediction of soil properties with plsr and vis-nir spectroscopy: Application to mediterranean soils from southern Italy. *Curr. Anal. Chem.* **2012**, *8*, 283–299.
28. Geladi, P.; Sethson, B.; Nyström, J.; Lillhonga, T.; Lestander, T.; Burger, J. Chemometrics in spectroscopy. *Spectrochim. Acta B At. Spectrosc.* **2004**, *59*, 1347–1357. [CrossRef]
29. Li, S.; Li, J.; Zhang, Q. Water quality assessment in the rivers along the water conveyance system of the Middle Route of the South to North Water Transfer Project (China) using multivariate statistical techniques and receptor modeling. *J. Hazard. Mater.* **2011**, *195*, 306–317. [CrossRef] [PubMed]
30. Soil Survey Staff. *Soil Taxonomy, a Basic System of Soil Classification for Making and Interpreting Soil Surveys*, 2nd ed.; Agriculture Handbook No. 436; USDA Natural Resources Conservation Service, U.S. Government Printing 23 Office: Washington, DC, USA, 1999; pp. 160–162, 494–495.
31. Duan, Q.D.; Gupta, V.K.; Sorooshian, S. Effective and efficient global optimization for conceptual rainfall-runoff models. *Water Resour. Res.* **1992**, *28*, 1015–1031. [CrossRef]
32. Moriasi, D.N.; Arnold, J.G.; Van Liew, M.W.; Bingner, R.L.; Harmel, R.D.; Veith, T.L. Model evaluation guidelines for systematic quantification of accuracy in watershed simulations. *Trans. ASABE* **2007**, *50*, 885–900. [CrossRef]
33. Grubbs, F.E.; Beck, G. Extension of sample sizes and percentage points for significance tests of outlying observations. *Technometrics* **1972**, *14*, 847–854. [CrossRef]
34. Cunderlik, J.M.; Ouarda, T.B. Regional flood–duration–frequency modeling in the changing environment. *J. Hydrol.* **2006**, *318*, 276–291. [CrossRef]
35. Akaike, H. A new look at the statistical model identification. *IEEE Trans. Automat. Cont.* **1974**, *19*, 716–723. [CrossRef]
36. Schwarz, G. Estimating the dimension of a model. *Ann. Stat.* **1978**, *6*, 461–464. [CrossRef]
37. Calenda, G.; Mancini, C.P.; Volpi, E. Selection of the probabilistic model of extreme floods: The case of the River Tiber in Rome. *J. Hydrol.* **2009**, *371*, 1–11. [CrossRef]
38. Laio, F.; Di Baldassarre, G.; Montanari, A. Model selection techniques for the frequency analysis of hydrological extremes. *Water Resour. Res.* **2009**, *45*, W07416. [CrossRef]
39. Sahu, N.; Behera, S.K.; Yamashiki, Y.; Takara, K.; Yamagata, T. IOD and ENSO impacts on the extreme stream-flows of Citarum river in Indonesia. *Clim. Dynam.* **2012**, *39*, 1673–1680. [CrossRef]
40. Wang, G.Q.; Yang, H.; Wang, L.; Xu, Z.; Xue, B. Using the SWAT model to assess impacts of land use changes on runoff generation in headwaters. *Hydrol. Process.* **2014**, *28*, 1032–1042. [CrossRef]
41. Seidou, O.; Ramsay, A.; Nistor, I. Climate change impacts on extreme floods I: Combining imperfect deterministic simulations and non-stationary frequency analysis. *Nat. Hazards* **2011**, *61*, 647–659. [CrossRef]
42. Fu, C.; James, A.L.; Yao, H. SWAT-CS: Revision and testing of SWAT for Canadian Shield catchments. *J. Hydrol.* **2014**, *511*, 719–735. [CrossRef]
43. Shi, Z.H.; Huang, X.D.; Ai, L.; Fang, N.F.; Wu, G.L. Quantitative analysis of factors controlling sediment yield in mountainous watersheds. *Geomorphology* **2014**, *226*, 193–201. [CrossRef]

44. Shi, Z.H.; Ai, L.; Li, X.; Huang, X.D.; Wu, G.L.; Liao, W. Partial least-squares regression for linking land-cover patterns to soil erosion and sediment yield in watersheds. *J. Hydrol.* **2013**, *498*, 165–176. [CrossRef]

45. Liu, H.M, the Runoff Simulation of SWAT Model Coupled with the ECMWF Dataset. Master Thesis, North China University of Water Resources and Electric Power, Zhengzhou, China, 2017.

46. Koutroulis, A.G.; Tsanis, I.K.; Daliakopoulos, I.N. Seasonality of floods and their hydrometeorologic characteristics in the island of Crete. *J. Hydrol.* **2010**, *394*, 90–100. [CrossRef]

47. Condon, L.E.; Gangopadhyay, S.; Pruitt, T. Climate change and non–stationary flood risk for the upper Truckee River basin. *Hydrol. Earth Syst. Sci.* **2015**, *19*, 159–175. [CrossRef]

48. Döll, P. Vulnerability to the impact of climate change on renewable groundwater resources: A global-scale assessment. *Environ. Res. Lett.* **2009**, *4*, 035006. [CrossRef]

49. Villarini, G.; Smith, J.A.; Serinaldi, F.; Ntelekos, A.A. Analyses of seasonal and annual maximum daily discharge records for central Europe. *J. Hydrol.* **2011**, *399*, 299–312. [CrossRef]

50. Magilligan, F.J.; Nislow, K.H. Long-term changes in regional hydrologic regime following impoundment in a humid-climate watershed. *J. Am. Water Resour. As.* **2001**, *37*, 1551–1569. [CrossRef]

51. Ashraf, M.; Kahlown, M.A.; Ashfaq, A. Impact of small dams on agriculture and groundwater development: A case study from Pakistan. *Agric. Water Manag.* **2007**, *92*, 90–98. [CrossRef]

52. Baratti, E.; Montanari, A.; Castellarin, A.; Salinas, J.L.; Viglione, A.; Bezzi, A. Estimating the flood frequency distribution at seasonal and annual time scales. *Hydrol. Earth Syst. Sci.* **2012**, *16*, 4651–4660. [CrossRef]

53. Price, K. Effects of watershed topography, soils, land use, and climate on baseflow hydrology in humid regions: A review. *Prog. Phys. Geogr.* **2011**, *35*, 465–492. [CrossRef]

54. Liu, Y.Y.; Zhang, X.N.; Xia, D.Z.; You, J.S.; Rong, Y.S.; Bakir, M. Impacts of land–use and climate changes on hydrologic processes in the Qingyi River watershed, China. *J. Hydrol. Eng.* **2013**, *18*, 1495–1512. [CrossRef]

55. Daniel, W.W. *Applied Nonparametric Statistics*, 2nd ed.; PWS-Kent: Boston, MA, USA, 1990.

56. Mann, H.B. Nonparametric tests against trend. *Econometrica* **1945**, *13*, 245–259. [CrossRef]

57. Kendall, M.G. *Rank Correlation Methods*; Oxford Univ. Press: New York, NY, USA, 1975.

58. Fisher, R.A.; Tippett, L.H.C. Limiting forms of the frequency distribution of the largest or the smallest member of a sample. *Math. Proc. Camb.* **1928**, *24*, 180–190. [CrossRef]

59. Jenkinson, A.F, the frequency distributions of the annual maximum (or minimum) values of meteorological elements. *Q. J. R. Meteorol. Soc.* **1955**, *81*, 145–158. [CrossRef]

*Article*

# Quantifying Impacts of Forest Recovery on Water Yield in Two Large Watersheds in the Cold Region of Northeast China

**Liangliang Duan** [ORCID] **and Tijiu Cai** *

Department of Forestry, School of Forestry, Northeast Forestry University, Harbin 150040, China;
liangliang.duan@nefu.edu.cn
* Correspondence: caitijiu1963@163.com; Tel.: +86-04-1882191821

Received: 12 June 2018; Accepted: 29 June 2018; Published: 3 July 2018

**Abstract:** In northern China, large-scale reforestations were implemented to restore the ecosystem functions (e.g., hydrology function). However, few studies have been conducted to quantify the relative contributions of forest recovery to water yield in boreal forest region across the globe. In this study, the impacts of forest recovery on the changes in mean annual water yield were assessed in two large forested watersheds in the boreal forest region of northeast China using three different approaches. As commonly considered, the results confirmed that forest recovery was the dominant driver of the reductions in annual water yield in the two watersheds in the past three decades (1987–2016), explaining 64.3% (15.4 mm) and 87.4% (40.7 mm) of variations in annual water yield for Upper Tahe watershed (UTH) and Xinancha watershed (XNC), respectively. By contrast, climate variability played minor role in annual water yield variation, explaining only 35.7% (8.5 mm) and 12.6% (7.2 mm) for UTH and XNC, respectively. The response differences between the two watersheds may mainly be attributed to differences in forest type, topography and climate regimes. This study provided important insight into sustainable forest and water resources management in the region.

**Keywords:** reforestation; annual water yield; forest hydrology; boreal forest; relative contribution

## 1. Introduction

The large-scale reforestation programs have greatly increased forest coverage in China since the 1980s, restoring ecosystem services and benefits of the communities [1]. Despite the positive effects of reforestation in ecosystems restoration, the massive afforestation was also considered a threat to the water resources or supply in some regions [2]. For example, Qiu [3] reported that reforestation may potentially contribute to the droughts in Southwest China. Although numerous existing studies were dedicated to assessing the impacts of reforestation on water yield globally with varying numbers of watersheds and watershed sizes, as reviewed by Zhang et al. [4] and Li et al. [5], there exists limited studies [6,7] examining the effects of afforestation and reforestation on water yield in boreal regions, and, particularly, a lack of studies in the boreal forest region in China [8]. This raises a critical need to study such effects in the boreal regions to enrich our knowledge of the relationship between forest recovery and water resources.

In addition to forest recovery, climate change is also a critical driver to the hydrology in large forested watersheds [9–11]. For instance, Li et al. [5] reported that forest and climate played a co-equal role in hydrological variations. Wei et al. [11] found that forest change can only explain 30% of hydrological variations in the global forested regions in the period of 2000–2011. This further highlighted the importance of considering both forest and climate change in watershed assessment.

To quantify the relative contribution of forest cover change to the changes in water yield, the effects of climate variability have to be removed or quantified firstly. Several recent studies have successfully separated the relative impacts of forest cover change and climate variability on water yield in large watersheds using different methods. For instance, Wei and Zhang [9] and Zhang and Wei [12] applied Modified Double-Mass Curve (MDMC) to quantify the relative contributions of the changes in forest cover to water yield in large forested watersheds in Canada and China, respectively. Zhao et al. [13] used Time Trend Analysis (TRA) and Sensitivity-Based Methods (SBM) to separate the effects of vegetation change and the effects of climate variability on streamflow in different catchments in New Zealand, South Africa and Australia, respectively. The Budyko Framework (BF) is also widely used to quantify the impacts of vegetation changes on streamflow [14–16]. Wei et al. [17] compared pros and cons of eight methods for quantifying the relative contribution of forest or land cover change and climatic variability to hydrology in large watersheds and concluded that each method must be supplemented by other methods to achieve a robust conclusion. In this study, we applied three methods including MDMC, TRA and SBM to get the reliable results.

The cold region of Northeast China can be typically represented by the boreal coniferous forest in the Da Hinggan Mountains [18] and the boreal/temperate transition mixed coniferous and broadleaved forest in the Xiao Hinggan Mountains [19], respectively. These forests experienced long-term timber harvesting from the 1960s to 1990s and a forest recovery period through natural forest protection projects in the late 1990s [20]. As a result, the forest coverage has greatly increased in the past three decades, which provides a suitable base for investigating the interactions between forest recovery and water yield in this region.

Two large forested watersheds including Upper Tahe (UTH) watershed and Xinancha (XNC) watershed in the cold region of Northeast China were selected for this study. Long-term (1987–2016) forest cover and hydrometeorological data were used: (1) to quantify the relative contributions of forest recovery to the long-term water yield; (2) to examine the sensitivity of annual water yield to forest recovery; and (3) to explore the implications of our research findings for watershed management in the cold region.

## 2. Materials and Methods

### 2.1. Study Watersheds

The UTH watershed and XNC watershed have drainage areas of 2359 and 2582 km$^2$, respectively, and are located in Heilongjiang Province in the high latitude cold region of Northeast China (Figure 1a). According to Chinese vegetation database administrated by Data Center for Resources and Environmental Sciences, Chinese Academy of Sciences (RESDC), the UTH watershed is zoned in boreal coniferous forests, while XNC watershed is zoned in the boreal/temperate transition mixed coniferous and broadleaved forest (Figure 1b). The two watersheds are both dominated by gentle hills with the average slope of 12.5° and 13.4°, respectively, and the elevations range from 96 to 1276 m above sea level (Figure 1c,d). Both watersheds are dominated by dark brown earths and brown coniferous forest soils. The main characteristics of two watersheds are summarized in Table 1. According to the Circum-Arctic Map of Permafrost and Ground Ice Conditions, Version 2 (http://nsidc.org/data/ggd318#) [21], the UTH watershed spans discontinuous and sporadic permafrost zones, while the XNC watershed spans seasonal frozen ground without permafrost. It should be noted that warming climate might further enhance permafrost thaw particularly in boreal regions [22,23]. The permafrost thaw can alter the regional hydrological processes [24,25]. Therefore, to minimize the effects of permafrost thawing on streamflow, the period from 1987 to 2016 with a relative stable mean annual air temperature, as found by Duan et al. [8] and Duan et al. [26], was selected to quantify the relative contribution of forest recovery on water yield in this study.

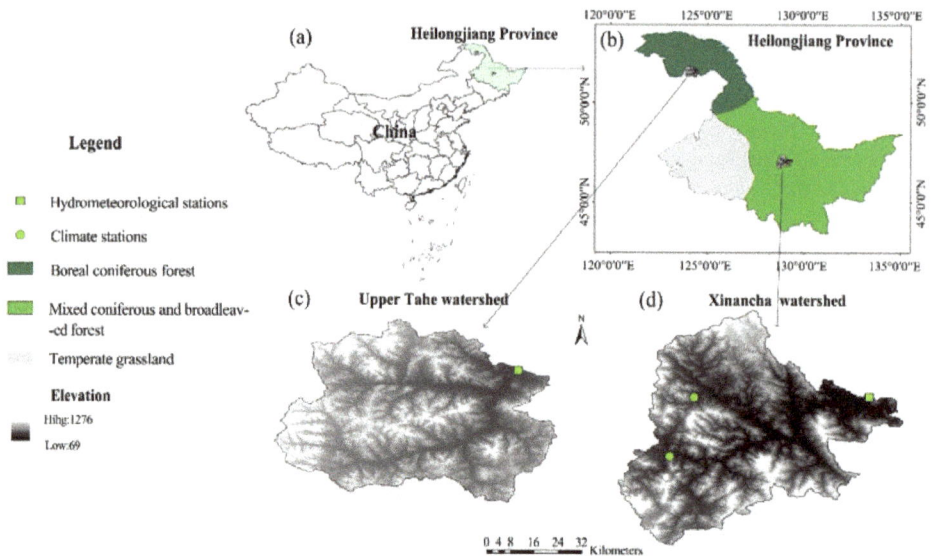

**Figure 1.** Locations (**a**); forest types (**b**); topography of the UTH watershed (**c**); and topography of the XNC watershed (**d**).

**Table 1.** Watershed characteristics for the UTH watershed and XNC watershed.

| Metrics | UTH Watershed | XNC Watershed |
|---|---|---|
| Drainage area (km$^2$) | 2359 | 2582 |
| Mainstream length (km) | 83 | 82 |
| Average elevation (m) | 783 | 501 |
| Elevation range (m) | 432–1276 | 69–1226 |
| Average slope (°) | 12.5 | 13.4 |
| Soil type | Dark brown earths and brown coniferous forest soil | Dark brown earths and brown coniferous forest soil |
| Annual mean precipitation (mm) | 534.8 | 706.0 |
| Annual mean PET (mm) | 520.5 | 711.8 |
| Annual mean air temperature (°C) | −2.1 | 2.4 |
| Annual mean flow (mm) | 297.3 | 298.7 |
| Average forest cover (%) | 75.6 | 87.7 |
| Forest type | Boreal coniferous forest | Mixed coniferous and broadleaved forest |
| Hydrometric station | Xinlin | Nancha |
| Climate stations | Xinlin | Nancha, Xiaobai, Nanlie |

The two study watersheds are characterized by a typical continental monsoon climate. Based on the climate data from 1987 to 2016, the average annual precipitation of two watersheds are 534.8 and 706.0 mm for the UTH and XNC, respectively, of which approximately 85% occurs as rain from May to September (the wet season) (Figure 2). The average annual air temperature for the UTH and XNC watersheds is −2.1 and 2.4 °C, respectively, with the average highest and lowest occurring in July and January, respectively (Figure 2). The UTH watershed is located in the boreal coniferous forest zone, where the native vegetation consists of forest communities dominated by larch (*Larix gmelinii*), along with broadleaf species, such as birch (*Betula platyphylla*). The XNC watershed is in the boreal/temperate transition mixed coniferous and broadleaved forest zone, where the flora is more diverse [18]. In this region, the coniferous forest is dominated by Korean pine (*Pinus koraiensis*), and the broadleaved species mainly include *Tilia amurensis, Fraxinus mandschurica,* and Mongolian oak (*Quercus mongolica*). The forests were mainly regenerated from the natural forest succession with the major secondary forests of birch in UTH watershed and Mongolian oak forest in XNC watershed.

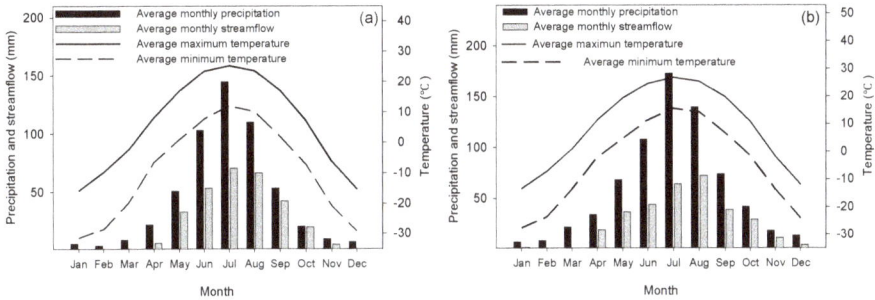

**Figure 2.** Average monthly streamflow, precipitation, and air temperature in the UTH (**a**) and XNC (**b**) watersheds from 1987 to 2016.

## 2.2. Data

### 2.2.1. Forest Cover Data

Annual forest cover data from 1987 to 2016 were obtained from the forest resource inventory database administrated by the Xinlin Forestry and Yichun Forestry Bureaus [8,27], respectively. Following the protocol of "Observation Methodology for Long-term Forest Ecosystem Research" of the National Standards of the People's Republic of China (GB/T 33027-2016), the forest coverage was calculated as percent ratio of all forest area with canopy coverage greater than 30% over the total area of the study watershed. As shown in Figure 3, the forest cover of two study watersheds showed a significantly positive trend ($p < 0.001$, Mann-Kendall test) during the entire study period. The Pettitt's Tests [28] indicated that the forest coverage of UTH and XNC watersheds had the statistically significant change points in 2002 and 2000 ($p < 0.05$), respectively. The forest coverage of UTH watershed did not show very much change from 1987 to 2002 with the average of 70.2%, but increased from 72.2% to 87.5% from 2003 to 2016 (Figure 3). The forest coverage of XNC watershed increased from 75.8% to 86.6% from 1987 to 2000, and showed a slightly increasing trend from 2001 to 2016 with the average of 88.4%.

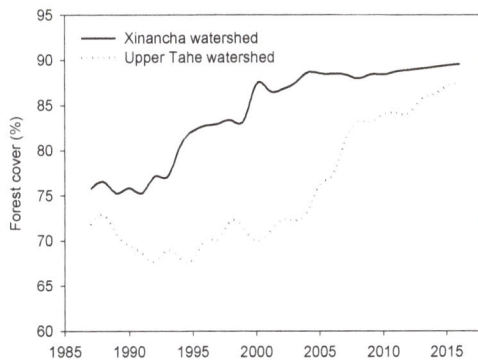

**Figure 3.** Forest cover (%) from 1987 to 2016 in UTH watershed and XNC watershed.

2.2.2. Hydrometeorological Data

The annual and seasonal streamflow data from 1987 to 2016 were both produced by the aggregation of daily streamflow data, which were calculated based on the relationship between streamflow and discharge stage height collected at the Xinlin and Nancha hydrometric stations (Figure 1), respectively. The details on annual streamflow, Annual peak flows, the monthly precipitation, monthly mean, maximum and minimum air temperatures, and monthly sunshine hours from 1987 to 2016 are shown in Table 1. The averaged watershed-based precipitation estimates were derived by the Thiessen polygon method for XNC watershed. To understand the relative contribution of seasonal streamflow variations to annual streamflow changes, streamflow data from the entire year was summarized into four seasons: spring (April to May), summer (June to August), autumn (September to October), and winter (November to March).

*2.3. Data Analysis Methods*

2.3.1. Trend Analysis on Hydrological and Climatic Series

Trend analysis is useful for understanding dynamics of hydrological and climatic variables over a long-term period [29,30]. A nonparametric trend test, namely the Mann-Kendall test [31,32], was adopted to detect whether significant trends exist in the long-term hydrological and climatic data [29]. The magnitude of the trend, β, or the slope (change per unit time) was used to describe the change rate of long-term hydrological and climatic variables, which was estimated using the Sen's slope method [33,34] as follows:

$$\beta = Median\left[\frac{X_j - X_i}{j - i}\right] \text{ for all } i < j \tag{1}$$

where $1 < i < j < n$ and is the median of all possible combinations of pairs for the entire data set.

2.3.2. Time Series Correlation Analysis

Annual time series of hydrological and forest cover data were used to conduct cross-correlation analyses for the entire study period and detect statistical relationships between these data sets. To eliminate the autocorrelations in the data series, all data series were pre-whitened using the best fitting autoregressive integrated moving average models (ARIMA) [35,36]. Then, residual series were used in the cross-correlation analysis. These analyses were conducted using the software STATISTICA 7 (StatSoft®, Palo Alto, CA, United States).

2.3.3. Separation of the Impacts of Climatic Variability and Forest Recovery on Annual Streamflow

Modified Double Mass Curve Method

Modified Double Mass Curve (MDMC) developed by Wei and Zhang [9] was firstly used to separate the relative contributions of climate variability and forest recovery on annual streamflow and detect the break point in annual streamflow caused by forest recovery. The modified double mass curve assumes that there is a linear relation between streamflow and effective precipitation, and plots accumulated annual streamflow versus accumulated annual effective precipitation. The annual effective precipitation refers to the difference between annual precipitation and annual evapotranspiration [9]. In the period when there are no significant impacts of forest cover change on annual streamflow, a straight line should be expected to describe the relation between annual streamflow and annual effective precipitation. However, with a management or other disturbances in the forest, a curve line with a turning point would be expected. The statistical significance of the break points was confirmed by the interrupted ARIMA model.

Once the statistical significance of the break point was confirmed, the whole study period was subsequently divided into reference (before the break point) and forest recovery (after the break point) periods. The difference between the average observed annual streamflow ($Q_2$) and the average annual streamflow predicted by the baseline in the forest recovery period can be attributed to the effects of forest recovery on annual streamflow ($\Delta Q_F$). The variation of annual streamflow caused by climate change ($\Delta Q_C$) can be determined as:

$$\Delta Q_C = \Delta Q - \Delta Q_\Gamma \tag{2}$$

where $\Delta Q$ is the deviation of average annual streamflow between forest recovery period and reference period:

$$\Delta Q = Q_2 - Q_1 \tag{3}$$

where $Q_1$ and $Q_2$ are the average annual streamflow in the reference and forest recovery periods, respectively.

The relative contributions of forest recovery and climate variability to the changes in mean annual streamflow can be estimated as:

$$R_F = \frac{|\Delta Q_F|}{|\Delta Q_F| + |\Delta Q_C|} \times 100\% \tag{4}$$

$$R_C = \frac{|\Delta Q_C|}{|\Delta Q_F| + |\Delta Q_C|} \times 100\% \tag{5}$$

In this study, potential evapotranspiration is estimated using the temperature-based Hamon method [37], which has been shown to provide reasonable potential evapotranspiration for forested regions [8,38,39]:

$$PET = 0.1651 \times D \times V_d \times K \times N \tag{6}$$

where $D$ is the monthly average time from sunrise to sunset (from the climate station) in multiples of 12 h, $V_d$ is the saturated vapor density (g m$^{-3}$) at the monthly mean air temperature (T, °C) as shown in Equation (7), and $N$ is the number of days in each month.

$$V_d = 216.7 \times \frac{V_s}{T + 273.3} \tag{7}$$

$V_S$ is the saturated vapor pressure (millibars), expressed as:

$$V_S = 6.108 \times \exp\left[17.26939 \times \frac{T}{T + 273.3}\right] \tag{8}$$

where K is a correction coefficient to adjust PET from the Hamon's method to reflect realistic values for PET. The reported K values range from 1.0 to 1.4 [37,39]. Due to the expected low potential evapotranspiration in the cold study watersheds, the K value was set to be 1.1 as suggested by Duan et al. [8] and Sun et al. [40].

In this study, annual evapotranspiration (ET) was firstly estimated using both the Budyko (Equation 8) and Zhang equations (Equation 9) and, then, compared with the difference between the long-term mean annual precipitation and mean annual streamflow, which is considered as the "actual evapotranspiration" because that it is reasonable to assume that changes in soil water storage are zero and the changes in the recharge to groundwater are small over a long period of time (i.e., 5–10 years) in the catchment water balance framework [41]. We found that the annual ET calculated by Budyko equation was closer to the actual evapotranspiration. Thus, the results from Budyko equation was used in this study.

$$ET = \{P[1 - \exp(-PET/P)] \times PET \times \tanh(P/PET)\}^{0.5} \tag{9}$$

$$ET = P[1 + w(PET/P)]/[1 + w(PET/P) + P/PET] \tag{10}$$

where w is plant available water coefficient, which is set to be 2 given the large proportion of forest cover in study watersheds [41]. P and PET represent annual precipitation and annual potential evapotranspiration, respectively.

Sensitivity-Based Method

The variations in mean annual streamflow attributed to climate variability ($\Delta Q_C$) is calculated from the changes of annual precipitation (P) and potential evapotranspiration (PET) between the different phases in the sensitivity-based method as follow [42,43]:

$$\Delta Q_C = \psi_P \Delta P + \psi_{PET} \Delta PET \tag{11}$$

where $\Delta P$ and $\Delta PET$ are the changes in precipitation P and PET between the forest recovery period and the reference period. $\psi_P$ and $\psi_{PET}$ are streamflow sensitivity coefficients to P and PET as expressed below [44]:

$$\psi_P = \frac{1 + 2x + 3wx}{(1 + x + wx^2)^2} \tag{12}$$

$$\psi_{PET} = -\frac{1 + 2wx}{(1 + x + wx^2)^2} \tag{13}$$

where x is the mean annual index of dryness and is equal to PET/P and w is plant available water coefficient the same as Equation (10).

Once $\Delta Q_C$ was estimated, the $\Delta Q_F$ can be determined by the equation as follow:

$$\Delta Q_F = \Delta Q - \Delta Q_C \tag{14}$$

Time Trend Analysis Method

In the time trend analysis method, the Kendall–Theil robust line method [45] was firstly used to create a linear equation between annual streamflow and annual effective precipitation in the reference period as Equation (15), and the equation was then applied in the forest recovery period to predict annual streamflow. Thus, the difference between mean annual streamflow ($Q_2$) and mean annual predicted streamflow ($Q_{2P}$) in the forest recovery period was considered the effects of forest recovery on annual streamflow ($\Delta Q_F$), as shown in Equation (16). Once the $\Delta Q_F$ was estimated, the $\Delta Q_C$ can be determined by Equation (2).

$$Q_t = aP_{et} + b \tag{15}$$

$$\Delta Q_F = Q_2 - Q_{2P} \tag{16}$$

where $Q_t$ and $P_{et}$ are annual streamflow and annual effective precipitation at the tth year in the reference period, respectively, and a and b are the constants of linear equation.

## 3. Results

### 3.1. Trends of Annual and Seasonal Hydrometeorological Variables

For annual hydrometeorological data series from 1987 to 2016 (Table 2), very few significant ($p < 0.05$) trends were found, mainly on annual PET, spring flow rate, and spring and winter precipitations in XNC watershed in comparison with none significant ($p > 0.05$) trend in UTH watershed. Although there was no significant trend in annual streamflow in UTH watershed, the annual flow rate decreased by 2.0 mm/year vs. 1.1 mm/year in XNC watersheds. However, the summer streamflow in UTH watershed was characterized by an apparent decreasing pattern (1.88 mm/year) comparing to the decreasing precipitation (1.90 mm/year).

**Table 2.** Results of Mann-Kendall trend tests on hydrometeorological variables in the UTH watershed and XNC watershed from 1987 to 2016.

| Period | Watershed | Q Slope [1] (mm/year) | p | P Slope (mm/year) | p | T Slope (°C/year) | p | PET Slope (mm/year) | p |
|---|---|---|---|---|---|---|---|---|---|
| Annual | UTH | −2.0 | 0.26 | −1.2 | 0.48 | 0.00 | 0.93 | 0.58 | 0.15 |
| | XNC | −1.1 | 0.67 | 0.25 | 0.92 | 0.00 | 0.75 | 0.59 * | 0.01 |
| Spring | UTH | 0.35 | 0.55 | 1.03 | 0.11 | 0.02 | 0.31 | 0.30 | 0.11 |
| | XNC | 1.09 * | 0.02 | 2.36 * | 0.01 | 0.02 | 0.33 | 0.25 | 0.17 |
| Summer | UTH | −1.88 | 0.32 | −1.90 | 0.32 | 0.02 | 0.16 | 0.58 | 0.08 |
| | XNC | −0.04 | 1.00 | 0.31 | 0.97 | 0.03 | 0.07 | 0.47 | 0.11 |
| Autumn | UTH | −0.29 | 0.62 | −0.16 | 0.86 | −0.02 | 0.41 | 0.00 | 0.94 |
| | XNC | −0.63 | 0.50 | −1.67 | 0.18 | 0.01 | 0.72 | 0.10 | 0.27 |
| Winter | UTH | −0.04 | 0.44 | −0.2 | 0.36 | −0.03 | 0.25 | −0.04 | 0.83 |
| | XNC | −0.03 | 0.86 | 1.18 * | 0.02 | −0.04 | 0.38 | −0.11 | 0.55 |

* values indicate the statistical significance at the level of 0.05. [1] The Slope was estimated using the nonparametric median-based slope method [33,34]. Q, streamflow; P, precipitation; T, mean annual air temperature; PET, potential evapotranspiration ET, evapotranspiration; Spring, April to May; Summer, June to August; Autumn, September to October; Winter, November to March.

## 3.2. Cross-Correlations between Forest Cover and Hydrological Variables

Cross-correlation analysis suggests that annual streamflow in UTH and XNC watersheds was significantly negatively correlated with annual forest cover from 1987 to 2016 (Table 3) except for winter season regardless of watersheds. However, there is a significantly positive correction between the streamflow and forest cover during spring season for both watersheds. In addition, the lags between hydrological variables and forest cover varied with watershed, 4-10 years for UTH and 0–5 years for XNC watersheds, respectively.

**Table 3.** Cross-correlations between forest cover and hydrological variables.

| Variables | Forest Cover of UTH (1,1,1) ARIMA [1] Model | Cross-Correlation Coefficient | p | Lag | Forest Cover of XNC (1,1,1) ARIMA Model | Cross-Correlation Coefficient | p | Lag |
|---|---|---|---|---|---|---|---|---|
| Annual | (1,0,0) | −0.51 ** | 0.005 | 9 | (1,0,0) | −0.54 ** | 0.001 | 5 |
| Spring | (1,0,0) | 0.46 * | 0.019 | 4 | (0,0,1) | 0.41 * | 0.026 | 0 |
| Summer | (0,0,1) | −0.40 * | 0.038 | 10 | (1,0,0) | −0.46 ** | 0.007 | 5 |
| Autumn | (1,0,1) | −0.41 * | 0.017 | 9 | (0,0,1) | −0.40 * | 0.025 | 5 |
| Winter | (1,0,0) | −0.28 | 0.170 | 5 | (1,0,0) | 0.16 | 0.404 | 0 |

[1] ARIMA is autoregressive integrated moving average. ** values indicate the statistical significance at the level of 0.01. * the statistical significance at the level of 0.05.

## 3.3. Separating the Relative Contributions of Forest Recovery and Climate Variability to the Changes in Annual Streamflow

Break points were detected in the modified double mass curves (MDMCs) in 2003 and 2001 for UTH and XNC watersheds (Figure 4), respectively. The fitted interrupted ARIMA model of the MDMCs slopes further confirmed that the break points were statistically significant ($p < 0.05$, Table 4). Thus, the entire study period (1987–2016) was then identified as two periods: the reference period (1987 to 2002 for UTH and 1987 to 2000 for XNC) and the forest recovery period (2003 to 2016 for UTH and 2001 to 2016 for XNC).

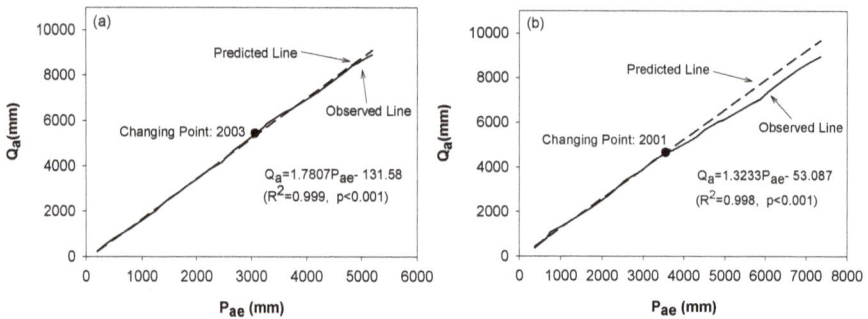

**Figure 4.** Double mass curve of accumulated annual streamflow ($Q_a$) and accumulated annual effective precipitation ($P_{ae}$) for: UTH (**a**); and XNC (**b**).

**Table 4.** Interrupted ARIMA models for slopes of MDMC in Figure 4.

| Model Input | Model Structure | Parameter Estimation q(1) [1] | $\Omega$ [2] |
|---|---|---|---|
| Slope of MDMC [3] of UTH in Figure 4a | Interrupted ARIMA: (0,0,1), intervention at year 2003 | $-0.73$ ($p < 0.05$) | $0.52$ ($p < 0.05$) |
| Slope of MDMC of XNC in Figure 4b | Interrupted ARIMA: (0,0,1), intervention at year 2001 | $-0.76$ ($p < 0.05$) | $0.73$ ($p < 0.05$) |

[1] q is moving average parameter, [2] $\Omega$ is intervention parameter for abrupt permanent intervention type, [3] is modified double mass curve.

Forest recovery decreased the mean annual streamflow by 16.9 and 43.4 mm from the reference period in UTH and XNC watersheds, respectively, the corresponding relative contributions to the changes in mean annual streamflow being 70.6% and 93.3%, respectively. By contrast, climate variability induced reductions in mean annual streamflow were 7.0 and 3.1 mm, respectively, with corresponding relative contributions being 29.4% and 6.7%, respectively.

The results of the sensitivity-based method (Table 5) indicate that climate variability decreased mean annual streamflow by 8.8 and 7.5 mm from the reference period to the forest recovery period in UTH and XNC watersheds, respectively, in comparison with the reduction of precipitation ($-11.3$ mm for UTH and $-7.7$ mm for XNC) and the increase of PET (7.0 mm for UTH and 7.9 mm for XNC). Thus, the contributions of forest recovery to the changes in mean annual streamflow were determined as $-15.1$ and $-39.0$ mm in UTH and XNC watersheds, respectively (Table 5).

**Table 5.** The relative contributions of climate variability and forest recovery calculated by the sensitivity-based method.

| Watershed | Sub-Periods | P (mm) | PET (mm) | $\Delta P$ (mm) | $\Delta PET$ (mm) | $\beta$ | $\gamma$ | $\Delta Q$ (mm) | $\Delta Q_C$ (mm) | $\Delta Q_F$ (mm) |
|---|---|---|---|---|---|---|---|---|---|---|
| UTH | 1987–2002 | 540.7 | 517.3 | $-11.3$ | 7.0 | 0.58 | $-0.33$ | $-23.9$ | $-8.8$ | $-15.1$ |
| | 2003–2016 | 529.5 | 524.3 | | | | | | | |
| XNC | 1987–2000 | 715.9 | 652.8 | $-7.7$ | 7.9 | 0.60 | $-0.36$ | $-46.5$ | $-7.5$ | $-39.0$ |
| | 2001–2016 | 708.2 | 660.7 | | | | | | | |

The results of time trend analysis method shown in Figure 5 indicated that the mean annual observed streamflow was 14.1 and 39.8 mm lower than the predicted values in forest recovery period in UTH and XNC watersheds, respectively. Thus, the relative contributions of climate variability were calculated as the difference between the total changes in mean annual streamflow and the changes attributed to forest recovery (Equation 1), and these values were −9.8 and −6.9 mm for UTH and XNC watersheds, respectively.

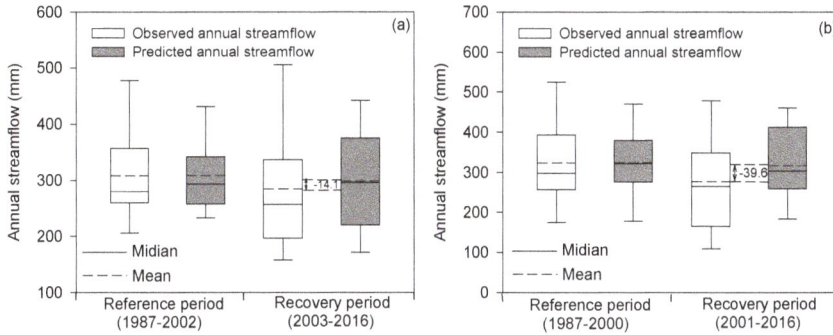

**Figure 5.** The box plot of observed and predicted annual streamflow calculated by time trend analysis method in: XNC (**a**); and UTH (**b**).

The total changes in mean annual streamflow ($\Delta Q$) and the associated components ($\Delta Q_C$ and $\Delta Q_F$) calculated by three independent methods in two watersheds are summarized in Table 6. The results from three methods produced similar results. The relative contributions of forest recovery to the reductions of mean annual streamflow were −15.4 mm (64.3%) and −40.7 mm (87.4%) in UTH and XNC watersheds, respectively, while the relative contributions of climate variability were −8.5 mm (35.7%) and −7.2 mm (12.6%), respectively. Overall, the impacts of forest recovery on long-term annual streamflow variations were much higher than those from climate variability, which indicated that the variations of water yield in the UTH and XNC watersheds were mainly controlled by forest recovery in the past three decades.

**Table 6.** Long-term annual streamflow changes and their components in UTH and XNC watersheds.

| Method | UTH (from 1987–2002 to 2003–2016) | | | | | XNC (from 1987–2000 to 2001–2016) | | | | |
|---|---|---|---|---|---|---|---|---|---|---|
| | Changes in Streamflow (mm) | | | Relative Contributions (%) | | Changes in Streamflow (mm) | | | Relative Contributions (%) | |
| | $\Delta Q$ | $\Delta Q_C$ | $\Delta Q_F$ | Climate | Forest | $\Delta Q$ | $\Delta Q_C$ | $\Delta Q_F$ | Climate | Forest |
| MDMC [1] | | −7.0 | −16.9 | 29.4 | 70.6 | | −3.1 | −43.4 | 6.7 | 93.3 |
| TRA [2] | | −9.8 | −14.1 | 40.9 | 59.1 | | −6.9 | −39.6 | 14.9 | 85.1 |
| SBM [3] | −23.9 | −8.8 | −15.1 | 36.8 | 63.2 | −46.5 | −7.5 | −39.0 | 16.1 | 83.9 |
| Average | | −8.5 | −15.4 | 35.7 | 64.3 | | −7.2 | −40.7 | 12.6 | 87.4 |

[1] MDMC is Modified Double Mass Curve method, [2] TRA is Time Trend Analysis method and [3] SBM is Sensitivity-based Method.

## 4. Discussion

### 4.1. The Effects of Forest Recovery on Water Yield

Our results indicated that the forest recovery was the dominant driver of the reduction in mean annual water yield, −15.4 and −40.7 mm for UTH and XNC watersheds, respectively. The negative effects of reforestation on water yield are consistent with many study findings in other regions. For instance, Liu et al. [46] found that the reforestation in the Meijiang watershed (6983 km²) covered by subtropical evergreen broad-leaved forest caused an average annual streamflow reduction of 51 mm in the reforestation period (1985–2006). Tuteja et al. [47] found that annual runoff reductions from pine plantations ranged from 22 to 52 mm/year in the different subcatchments of a large catchment in southeastern Australia. The negative effects of reforestation on water yield are possibly due to the increase in evapotranspiration resulting from vegetation development (e.g., increasing leaf areas and root systems) [46]. Forest recovery can also improve soil conditions and enhance the water storage in aquifers [48]. Such changes in hydrological processes consequently result in the reduction of water yield observed from stream. Such reductions in annual water yield caused by reforestation were also reviewed on 73 watershed studies across the globe by Li et al. [5].

Although the forest recovery in this study showed similar impacts on water yield to the previous studies, the sensitivity of annual water yield to forest cover change in these study watersheds is much greater than that in other studies. Our results indicated that 1% forest cover increase resulted in 0.7% of reduction in annual water yield in the boreal coniferous forest watershed (UTH), while 1.8% of reduction in the mixed coniferous and broadleaved forest watershed (XNC). After examining 61 large (>1000 km$^2$) watersheds across the world, Zhang et al. [4] found that in large mixed forest dominated watersheds, 1% forest cover change can result in 0.80% change in annual runoff vs. 0.24% for large coniferous forest dominated watersheds. The difference in the sensitivity of annual water yield response to forest cover change between our study and the study of Zhang et al. [4] may be due to the differences in climatic regimes. The watersheds in this study are located in the high latitude cold region of Northeast China, where the climate is characterized by a typical continental monsoon climate. Approximately 85% of precipitation occurring as rain during the growing season may have been evapotranspired and consequently enhance the negative effects of forest recovery on water yield, as reported by Zhang, Dawes and Walker [41]. This conclusion can be also reinforced by the fact that the total decreases in streamflow during growing season (from June to October) accounted for 84.4% and 77.9% of the total reductions of annual water yield in UTH and XNC watersheds, respectively. Thus, it is safe to say that the impacts of forest recovery on long-term water yield are closely related to its impacts during wet seasons (i.e., summer and autumn in study watersheds), a good consistence with other summer dominant rainfall regions as well [49–51]. In addition, the different contribution rates of forest recovery to streamflow change between two watersheds in our study may also be partially related to the water condition difference (Table 1).

### 4.2. The Effects of Forest Type and Topography on the Response Intensity of Water Yield to Forest Recovery

Our results indicated that 1% forest cover increase can result in 1.8% reduction in annual streamflow in the mixed coniferous and broadleaved forest watershed (XNC) and 0.7% in the boreal coniferous forest watershed (UTH). The distinct response between two watersheds may be because of the difference in forest types, topography and climate regimes. Firstly, forest type can significantly affect the hydrological response to forest recovery [52]. For instance, in the Puget Sound basin covered by mixed coniferous and deciduous forest in northwestern Washington, USA, the annual streamflow response to forest recovery was estimated as 6.8% [53]. However, in the large boreal coniferous forest watersheds in Northeastern Ontario, Canada, there was no definitive changes in annual water yield with forest cover ranging from 8.6% to 25.2% [54]. Meanwhile, Zhang et al. [4] found that the response intensity of annual streamflow to forest cover change in large mixed forest watersheds was approximately three time of that in large coniferous forest watersheds based on the data from 61 large watersheds across the world. These previous studies and ours all suggested that large boreal coniferous forest watersheds may have the relative higher hydrological elasticity in response to forest cover change than the large mixed forest watersheds. In this study, relatively lower transpiration of boreal coniferous forest than the broadleaved forest in the cold regions [55,56] may lead to smaller changes of evapotranspiration in response to forest recovery. In addition, the different mean annual rainfall between two sites in this study can have a strong influence on the change in annual water yield with forest cover change [41,57]. Bosch and Hewlett [58] and Farley, Jobbágy and Jackson [52] found that vegetation change has the largest absolute impacts on water yield in high-rainfall areas. This is confirmed by our findings that the XNC watershed has a higher magnitude of rainfall than

UTH watershed, therefore, a greater response intensity of annual streamflow to forest recovery as well. Additionally, the boreal forest watershed (UTH) has a shorter and cooler growing season than the mixed forest watershed (XNC) because of the higher latitude, which consequently results in a lower evapotranspiration in the watershed. All these different characteristics of forest types and interactions with climate in boreal forest watershed attenuate the strength of the impacts of forest cover change on long-term water yield. Nevertheless, more case and modeling studies can definitely help to examine how forest types affect water yield in large forested watersheds.

Topography also plays a critical role in determining hydrological responses to reforestation [59]. As shown in Table 7, the UTH watershed is characterized by a more gently topography than XNC watershed. XNC watershed has 68% of the watershed area covered by the slopes greater than 15° vs. 44% in UTH watershed, which may have made the UTH a greater hydrological elasticity in response to forest recovery. Although few studies have been dedicated to detect the impacts of different topography on annual water yield, the studies on the effects of topography on specific flow variable [60] can help understand how different topography affects annual water yield. Liu et al. [61] found that hydrological recovery is limited and slower with reforestation in the steeper watershed. Li et al. [62] also found that the topography indices including perimeter, slope length factor, surface area, openness, and topographic characteristic index can be responsive to the streamflow change, in particular, to the low flow variables in snow-dominated regions in the Southern Interior of British Columbia, Canada. This may be because a watershed with gentler topography would likely have a higher water retention ability due to longer flow paths and residence time and consequently enhance the hydrological elasticity [63,64]. Thus, the different response intensity of annual water yield to forest recovery between two study watersheds was partly explained by the hydrological elasticity differentiated by topography in the two watersheds.

**Table 7.** Averaged slopes in two studied watersheds.

| Watershed | Percentage of Watershed Area (%) | | | | | |
|---|---|---|---|---|---|---|
| | Slope >40° | 20–40° | 15–20° | 10–15° | 5–10° | <5° |
| UTH | 0.7 | 22.0 | 19.6 | 30.4 | 3.8 | 23.4 |
| XNC | 0.5 | 33.4 | 34.0 | 4.6 | 4.6 | 23.0 |

Our results of cross-correlations between forest cover and annual flow indicated that there were nine- and five-year lags between forest cover and annual water yield in the UTH and XNC watersheds, respectively (Table 3). Such lagging effects were mainly due to the delayed hydrological responses to forest recovery, because forest recovery may take years or decades to reach a new hydrological equilibrium [65], particularly in the boreal forest that takes much longer time to recover after disturbance than other forest ecosystem in warmer regions [66]. On the other hand, the longer lags in the boreal coniferous forest watershed (UTH) could also partly be due to the stronger hydrological elasticity discussed above.

*4.3. The Relative Contributions of Forest Recovery and Climate Variability to Water Yield Variations*

Our results from three independent methods indicated that forest recovery was the dominate driver with the relative contributions to the changes in water yield being 64.3% and 87.4% in the UTH and XNC watersheds, respectively. Consistent results were also found in other regions. For instance, Liu et al. [67] studied in the middle and lower reaches of the Yellow River, China, vegetation changes were the main cause for triggering annual runoff changes, which account for approximately 80% from baseline period to changeable period. Similar findings were also reported for the headwaters of the Yellow River basin by Zheng et al. [68]. By contrast, many more studies conclude climate variability has a similar or greater strength of impacts on water yield compared to forest cover change. For instance, the equal hydrological impacts of forest cover change and climate variability were found in large forested watersheds in the central interior of British Columbia, Canada [9], in the the Upper Minjiang River of Yangtze River basin, China [69] and in the upper reach of the Poyang Lake basin, China [46]. Li et al. [10] found that the relative contribution of forest disturbance was only 27% in the Upper Similkameen River watershed situated between Canada and the USA. Similarly, Shi et al. [70] found that the streamflow was more sensitive to climate variability than land cover change in the Upstream of Huai River, China. The relative contributions of forest cover change and climate variability are largely dependent on their magnitude and the characteristics of watersheds. More case studies would help explore how forest cover change and climate variability interactively affect hydrology in large watersheds.

### 4.4. Implications and Uncertainty

Boreal coniferous forest and boreal/temperate transition mixed coniferous and broadleaved forest are both located in the remote cold region of Northeast China, where the natural forest experienced prolonged timber harvesting during the 1960s to late 1990s before Natural Forest Protection Project was carried out, and experienced a recovery period in the past three decades. Since 2016, timber harvesting has been completely banned in the natural forest. Although such forest management policies have made forest ecosystems well restored, the potential changes in water resources shrink caused by the increasing forest cover had been ignored. Our results indicated that the mean annual water yield decreased by 8.0% and 15.6% due to the increases of 11.6% and 8.7% in forest cover in UTH and XNC watersheds, respectively, in the past three decades. Such great sensitivity of annual water yield to the increases in forest cover should be a great concern, especially in the region covered by mixed coniferous and broadleaved forest, as the downstream of this region is one of the most important crop-producing areas in China. The reduction of annual water yield in the streams may cause a shortage of water for irrigation in downstream. Thus, the policies of forest management should meet the water level that can maintain aquatic functions and ecosystem integrity as well as ensure the water supply for irrigation in downstream area, which need more future quantitative studies to provide more information. For the boreal forest watershed, the reduction of water yield may not be a serious issue due to the relatively low demands of water supply. Nevertheless, the reduction in water flow caused by forest cover change require further investigation, because they are critical for maintaining the dynamics of in-channel and floodplain habitats that play a critical role in sustaining native biodiversity and ecosystem integrity in rivers [71,72], which can be more important than water supply in the boreal forest region in China.

There are several uncertainties in this study. Firstly, although three independent methods were used to quantify the relative contribution of forest recovery to annual water yield and achieved relatively consistent results, the variations of hydrological processes over study period were rarely understood. For example, annual effective precipitation was used to minimize the impacts of annual precipitation on streamflow in the methods of MDMC and time trend analysis method. However, the changes in intra-annual [73] or seasonal climate patterns [74] can significantly affect inter-annual water yield. In addition, reforestation can potentially more or less increase regional precipitation and water availability, and consequently compensate water loss by increased forest evapotranspiration [1,75]. However, these meteorological and hydrological variations were not considered in our quantitative methods. Second, the PET values were estimated by the Hamon method, which is a widely used temperature-based method. However, relative humidity, wind speed, and solar radiation could be affected by reforestation and, in turn, affect the PET [76–78]. These factors were not included in the PET estimations in this study. Thirdly, permafrost thaw and seasonal frost changes caused by climate warming can happen even in a period with the relative stable temperature [79,80], and consequently affect the long-term regional water yield [8,25,81]. Although the period 1987 to 2016 had a relative stable temperature, many studies demonstrated that there was a significant warming trend in Northeast China in the past half century [82,83], thus climate change consideration would be necessary for a longer study. In particular, when permafrost warming or frozen ground degradation has already been observed in this region [84,85], which may completely change the response pattern of streamflow to forest recover. However, the impacts of permafrost thaw and frozen ground degradation on water yield were not considered in this study, which need future more process-based studies to investigate.

## 5. Conclusions

Based on data from two monitored large watersheds, this study proved that forest recovery was the dominant driver to the reduction of mean annual water yield, while the impacts of climate variability were relatively low in the two large forested watersheds in cold region of Northeast China during the past three decades. The relative contributions of forest recovery to the reductions in mean annual water yield were 64.3% (15.4 mm) and 87.4% (40.7 mm) in UTH and XNC watersheds, respectively, while the rest of the reductions in mean annual water yield were attributed to climate variability. We also found that the response intensity of annual water yield response to increasing forest cover in mixed coniferous and broadleaved forest watershed (XNC) was much greater than that in boreal coniferous forest watershed (UTH). It is well known that forest can conserve water and soil resources, therefore, reduce streamflow. However, the reduction of streamflow responding to the increasing vegetation recover may pose an additional issue to the downstream water supply for irrigating agricultural land. A proper trade-off between forest resource protection and proper downstream irrigation water supply must be sought in the future for an effective ecosystem management. These findings are of great importance for both water resource and forest management in large forested watersheds in Northeast China and similar watersheds in other cold regions.

**Author Contributions:** L.D and T.C. conceived and designed the research themes; L.D. wrote the paper. T.C. contributed to data preparation.

**Funding:** This research was funded by the National Science Foundation of China (Grant No. 31770488), and the CFERN & BEIJING TECHNO SOLUTIONS Award Funds on excellent academic achievements.

**Acknowledgments:** We acknowledge the financial support by the National Science Foundation of China (Grant No. 31770488). This work is also supported by CFERN & BEIJING TECHNO SOLUTIONS Award Funds on excellent academic achievements. We thank Zisheng Xing for improving the English of this article.

**Conflicts of Interest:** The authors declare no conflict of interest.

## References

1. Li, Y.; Piao, S.; Li, L.Z.X.; Chen, A.; Wang, X.; Ciais, P.; Huang, L.; Lian, X.; Peng, S.; Zeng, Z. Divergent hydrological response to large-scale afforestation and vegetation greening in China. *Sci. Adv.* **2018**, *4*, eaar4182. [CrossRef] [PubMed]
2. Zhang, S.; Yang, D.; Yang, Y.; Piao, S.; Yang, H.; Lei, H.; Fu, B. Excessive afforestation and soil drying on China's loess plateau. *J. Geophys. Res. Biogeosci.* **2018**, *123*, 923–935. [CrossRef]
3. Qiu, J. China drought highlights future climate threats. *Nature* **2010**, *465*, 142–143. [CrossRef] [PubMed]
4. Zhang, M.; Liu, N.; Harper, R.; Li, Q.; Liu, K.; Wei, X.; Ning, D.; Hou, Y.; Liu, S. A global review on hydrological responses to forest change across multiple spatial scales: Importance of scale, climate, forest type and hydrological regime. *J. Hydrol.* **2017**, *546*, 44–59. [CrossRef]
5. Li, Q.; Wei, X.; Zhang, M.; Liu, W.; Fan, H.; Zhou, G.; Giles-Hansen, K.; Liu, S.; Wang, Y. Forest cover change and water yield in large forested watersheds: A global synthetic assessment. *Ecohydrology* **2017**, *10*, e1838. [CrossRef]
6. Kiely, G.; Nadezhdina, N.; Zappa, M. The response of the water fluxes of the boreal forest region at the volga's source area to climatic and land-use changes. *Phys. Chem. Earth* **2002**, *27*, 675–690.
7. Wattenbach, M.; Zebisch, M.; Hattermann, F.; Gottschalk, P.; Goemann, H.; Kreins, P.; Badeck, F.; Lasch, P.; Suckow, F.; Wechsung, F. Hydrological impact assessment of afforestation and change in tree-species composition—A regional case study for the federal state of Brandenburg (Germany). *J. Hydrol.* **2007**, *346*, 1–17. [CrossRef]
8. Duan, L.L.; Man, X.L.; Kurylyk, B.L.; Cai, T.J.; Li, Q. Distinguishing streamflow trends caused by changes in climate, forest cover, and permafrost in a large watershed in northeastern China. *Hydrol. Process.* **2017**, *31*, 1938–1951. [CrossRef]

9.  Wei, X.H.; Zhang, M.F. Quantifying streamflow change caused by forest disturbance at a large spatial scale: A single watershed study. *Water Resour. Res.* **2010**, *46*. [CrossRef]
10. Li, Q.; Wei, X.; Zhang, M.; Liu, W.; Giles-Hansen, K.; Wang, Y. The cumulative effects of forest disturbance and climate variability on streamflow components in a large forest-dominated watershed. *J. Hydrol.* **2017**, *557*, 448–459. [CrossRef]
11. Wei, X.; Li, Q.; Zhang, M.; Giles-Hansen, K.; Liu, W.; Fan, H.; Wang, Y.; Zhou, G.; Piao, S.; Liu, S. Vegetation cover—Another dominant factor in determining global water resources in forested regions. *Glob. Chang. Biol.* **2017**, *24*, 786–795. [CrossRef] [PubMed]
12. Zhang, M.; Wei, X. The effects of cumulative forest disturbance on streamflow in a large watershed in the central interior of British Columbia, Canada. Hydrol. *Earth Syst. Sci.* **2012**, *16*, 2021–2034. [CrossRef]
13. Zhao, F.F.; Zhang, L.; Xu, Z.X.; Scott, D.F. Evaluation of methods for estimating the effects of vegetation change and climate variability on streamflow. *Water Resour. Res.* **2010**, *46*. [CrossRef]
14. Zhang, S.; Yang, Y.; Mcvicar, T.R.; Yang, D. An analytical solution for the impact of vegetation changes on hydrological partitioning within the budyko framework. *Water Resour. Res.* **2018**, *54*, 519–537. [CrossRef]
15. Peña-Arancibia, J.L.; Dijk, A.I.J.M.V.; Guerschman, J.P.; Mulligan, M.; Bruijnzeel, L.A.; Mcvicar, T.R. Detecting changes in streamflow after partial woodland clearing in two large catchments in the seasonal tropics. *J. Hydrol.* **2012**, *416*, 60–71. [CrossRef]
16. Beck, H.E.; Bruijnzeel, L.A.; Dijk, A.I.J.M.V.; Mcvicar, T.R. The impact of forest regeneration on streamflow in 12 meso-scale humid tropical catchments. *Hydrol. Earth Syst. Sci.* **2013**, *17*, 2613–2635. [CrossRef]
17. Wei, X.H.; Liu, W.F.; Zhou, P.C. Quantifying the relative contributions of forest change and climatic variability to hydrology in large watersheds: A critical review of research methods. *Water* **2013**, *5*, 728–746. [CrossRef]
18. Liu, Z.; Yang, J.; Chang, Y.; Weisberg, P.J.; He, H.S. Spatial patterns and drivers of fire occurrence and its future trend under climate change in a boreal forest of northeast china. *Glob. Chang. Biol.* **2012**, *18*, 2041–2056. [CrossRef]
19. Chen, X. Modeling the effects of global climatic change at the ecotone of boreal larch forest and temperate forest in northeast china. *Clim. Change* **2002**, *55*, 77–97. [CrossRef]
20. Liu, J.; Zhang, Z.; Xu, X.; Kuang, W.; Zhou, W.; Zhang, S.; Li, R.; Yan, C.; Yu, D.; Wu, S. Spatial patterns and driving forces of land use change in china during the early 21st century. *J. Geog. Sci.* **2010**, *20*, 483–494. [CrossRef]
21. Brown, J.; Ferrians Jr, O.; Heginbottom, J.; Melnikov, E. *Circum-Arctic Map of Permafrost and Ground-Ice Conditions, Version 2*; NSIDC: National Snow and Ice Data Center: Boulder, CO, USA, 2002.
22. Serreze, M.; Walsh, J.; Chapin Iii, F.; Osterkamp, T.; Dyurgerov, M.; Romanovsky, V.; Oechel, W.; Morison, J.; Zhang, T.; Barry, R. Observational evidence of recent change in the northern high-latitude environment. *Clim. Change* **2000**, *46*, 159–207. [CrossRef]
23. Group, M.I.E.W.; Pepin, N.; Bradley, R.S.; Diaz, H.F.; Baraer, M.; Caceres, E.B.; Forsythe, N.; Fowler, H.; Greenwood, G.; Hashmi, M.Z. Elevation-dependent warming in mountain regions of the world. *Nat. Clim. Chang.* **2015**, *5*, 424–430.
24. Woo, M.K.; Kane, D.L.; Carey, S.K.; Yang, D.Q. Progress in permafrost hydrology in the new millennium. *Permafr. Periglac. Process.* **2008**, *19*, 237–254. [CrossRef]
25. Walvoord, M.A.; Kurylyk, B.L. Hydrologic impacts of thawing permafrost-a review. *Vadose Zone J.* **2016**, *15*. [CrossRef]
26. Duan, L.; Man, X.; Kurylyk, B.L.; Cai, T. Increasing winter baseflow in response to permafrost thaw and precipitation regime shifts in northeastern China. *Water* **2017**, *9*, 25. [CrossRef]
27. Yao, Y.; Cai, T.; Ju, C.; He, C. Effect of reforestation on annual water yield in a large watershed in northeast China. *J. For. Res.* **2015**, *26*, 697–702. [CrossRef]
28. Pettitt, A. A non-parametric approach to the change-point problem. *Applied Statistics* **1979**, *28*, 126–135. [CrossRef]
29. Yue, S.; Pilon, P.; Cavadias, G. Power of the Mann-Kendall and spearman's rho tests for detecting monotonic trends in hydrological series. *J. Hydrol.* **2002**, *259*, 254–271. [CrossRef]
30. Shadmani, M.; Marofi, S.; Roknian, M. Trend analysis in reference evapotranspiration using mann-kendall and spearman's rho tests in arid regions of Iran. *Water Resour. Manag.* **2012**, *26*, 211–224. [CrossRef]
31. Mann, H.B. Nonparametric tests against trend. *Econometrica J. Econom. Soc.* **1945**, *13*, 245–259. [CrossRef]
32. Kendall, M.G. *Rank Correlation Measures*; Charles Griffin: London, UK, 1975; p. 202.

33. Sen, P.K. Estimates of the regression coefficient based on Kendall's tau. *J. Am. Stat. Assoc.* **1968**, *63*, 1379–1389. [CrossRef]

34. Hirsch, R.M.; Slack, J.R.; Smith, R.A. Techniques of trend analysis for monthly water-quality data. *Water Resour. Res.* **1982**, *18*, 107–121. [CrossRef]

35. Box, G.E.; Jenkins, G.M. Time series analysis: Forecasting and control. In *Holden-Day Series in Time Series Analysis*; Holden-Day: San Francisco, CA, USA, 1976.

36. Jassby, A.D.; Powell, T.M. Detecting changes in ecological time-series. *Ecology* **1990**, *71*, 2044–2052. [CrossRef]

37. Zhou, G.Y.; Wei, X.H.; Chen, X.Z.; Zhou, P.; Liu, X.D.; Xiao, Y.; Sun, G.; Scott, D.F.; Zhou, S.Y.D.; Han, L.S.; et al. Global pattern for the effect of climate and land cover on water yield. *Nat. Commun.* **2015**, *6*, 5918. [CrossRef] [PubMed]

38. Vörösmarty, C.J.; Federer, C.A.; Schloss, A.L. Potential evaporation functions compared on us watersheds: Possible implications for global-scale water balance and terrestrial ecosystem modeling. *J. Hydrol.* **1998**, *207*, 147–169. [CrossRef]

39. Pyzoha, J.E.; Callahan, T.J.; Sun, G.; Trettin, C.C.; Miwa, M. A conceptual hydrologic model for a forested Carolina bay depressional wetland on the coastal plain of south Carolina, USA. *Hydrol. Process.* **2008**, *22*, 2689–2698. [CrossRef]

40. Sun, G.; Zuo, C.; Liu, S.; Liu, M.; Mcnulty, S.G.; Vose, J.M. Watershed evapotranspiration increased due to changes in vegetation composition and structure under a subtropical climate1. *J. Am. Water Resour. Assoc.* **2008**, *44*, 1164–1175. [CrossRef]

41. Zhang, L.; Dawes, W.R.; Walker, G.R. Response of mean annual evapotranspiration to vegetation changes at catchment scale. *Water Resour. Res.* **2001**, *37*, 701–708. [CrossRef]

42. Koster, R.D.; Suarez, M.J. A simple framework for examining the interannual variability of land surface moisture fluxes. *J. Clim.* **1999**, *12*, 1911–1917. [CrossRef]

43. Milly, P.C.D.; Dunne, K.A. Macroscale water fluxes—2. Water and energy supply control of their interannual variability. *Water Resour. Res.* **2002**, *38*, 1–9. [CrossRef]

44. Li, L.J.; Zhang, L.; Wang, H.; Wang, J.; Yang, J.W.; Jiang, D.J.; Li, J.Y.; Qin, D.Y. Assessing the impact of climate variability and human activities on streamflow from the wuding river basin in China. *Hydrol. Process.* **2007**, *21*, 3485–3491. [CrossRef]

45. Helsel, D.R.; Hirsch, R.M. *Statistical methods in water resources*; US Geological Survey: Reston, VA, USA, 2002; Volume 323.

46. Liu, W.; Wei, X.; Liu, S.; Liu, Y.; Fan, H.; Zhang, M.; Yin, J.; Zhan, M. How do climate and forest changes affect long-term streamflow dynamics? A case study in the upper reach of Poyang river basin. *Ecohydrology* **2015**, *8*, 46–57. [CrossRef]

47. Tuteja, N.K.; Vaze, J.; Teng, J.; Mutendeudzi, M. Partitioning the effects of pine plantations and climate variability on runoff from a large catchment in southeastern Australia. *Water Res. Res.* **2007**, *43*, 199–212. [CrossRef]

48. Zhou, G.Y.; Wei, X.H.; Luo, Y.; Zhang, M.F.; Li, Y.L.; Qiao, Y.N.; Liu, H.G.; Wang, C.L. Forest recovery and river discharge at the regional scale of Guangdong province, China. *Water Resour. Res.* **2010**, *46*. [CrossRef]

49. Gafur, A.; Jensen, J.R.; Borggaard, O.K.; Petersen, L. Runoff and losses of soil and nutrients from small watersheds under shifting cultivation (Jhum) in the Chittagong hill tracts of Bangladesh. *J. Hydrol.* **2003**, *274*, 30–46. [CrossRef]

50. Scott, D.; Smith, R. Preliminary empirical models to predict reductions in total and low flows resulting from afforestation. *Water SA.* **1997**, *23*, 135–140.

51. Bruijnzeel, L.A. Hydrology of moist tropical forests and effects of conversion: A state of knowledge review. *J. Hydrol.* **1990**, *129*, 397–399.

52. Farley, K.A.; Jobbágy, E.G.; Jackson, R.B. Effects of afforestation on water yield: A global synthesis with implications for policy. *Glob. Chang. Biol.* **2005**, *11*, 1565–1576. [CrossRef]

53. Lan, C.; Lettenmaier, D.P.; Alberti, M.; Richey, J.E. Effects of a century of land cover and climate change on the hydrology of the puget sound basin. *Hydrolog. Process.* **2009**, *23*, 907–933.

54. Buttle, J.M.; Metcalfe, R.A. Boreal forest disturbance and streamflow response, northeastern Ontario. *Can. J. Fish. Aquat. Sci.* **2000**, *57*, 5–18. [CrossRef]

55. Mu, T. The estimation of transpiration of main tree species and water consumption of larch in Da Hinggan Moutains. *Inn. Mong. For. Sci. Technol.* **1980**, *2*, 3.

56. Komatsu, H.; Tanaka, N.; Kume, T. Do coniferous forests evaporate more water than broad-leaved forests in Japan? *J. Hydrol.* **2007**, *336*, 361–375. [CrossRef]

57. Vertessy, R.; Zhang, L.; Dawes, W.R. Plantations, river flows and river salinity. *Aust. For.* **2003**, *66*, 55–61. [CrossRef]

58. Bosch, J.M.; Hewlett, J. A review of catchment experiments to determine the effect of vegetation changes on water yield and evapotranspiration. *J. Hydrol.* **1982**, *55*, 3–23. [CrossRef]

59. Moore, I.D.; Grayson, R.B.; Ladson, A.R. Digital terrain modelling: A review of hydrological, geomorphological, and biological applications. *Hydrol. Process.* **1991**, *5*, 3–30. [CrossRef]

60. Karlsen, R.H.; Grabs, T.; Bishop, K.; Buffam, I.; Laudon, H.; Seibert, J. Landscape controls on spatiotemporal discharge variability in a boreal catchment. *Water Resour. Res.* **2016**, *52*. [CrossRef]

61. Liu, W.; Wei, X.; Li, Q.; Fan, H.; Duan, H.; Wu, J.; Giles-Hansen, K.; Zhang, H. Hydrological recovery in two large forested watersheds of southeastern China: Importance of watershed property in determining hydrological responses to reforestation. *Hydrol Earth Syst. Sci.* **2016**, *20*, 4747–4756. [CrossRef]

62. Li, Q.; Wei, X.; Yang, X.; Giles-Hansen, K.; Zhang, M.; Liu, W. Topography significantly influencing low flows in snow-dominated watersheds. *Hydrol. Earth Syst. Sci.* **2018**, *22*, 1947. [CrossRef]

63. Price, K. Effects of watershed topography, soils, land use, and climate on baseflow hydrology in humid regions: A review. *Prog. Phys. Geog.* **2011**, *35*, 465–492. [CrossRef]

64. McGuire, K.J.; McDonnell, J.J.; Weiler, M.; Kendall, C.; McGlynn, B.L.; Welker, J.M.; Seibert, J. The role of topography on catchment-scale water residence time. *Water Resour. Res.* **2005**, *41*, 302–317. [CrossRef]

65. Brown, A.E.; Zhang, L.; McMahon, T.A.; Western, A.W.; Vertessy, R.A. A review of paired catchment studies for determining changes in water yield resulting from alterations in vegetation. *J. hydrol.* **2005**, *310*, 28–61. [CrossRef]

66. Gauthier, S.; Vaillancourt, M.A. *Ecosystem Management in the Boreal Forest*; Presses de l'Université du Québec: Quebec, QC, Canada, 2009.

67. Liu, Q.; Yang, Z.; Cui, B.; Sun, T. Temporal trends of hydro-climatic variables and runoff response to climatic variability and vegetation changes in the yiluo river basin, China. *Hydrol. Process.* **2010**, *23*, 3030–3039. [CrossRef]

68. Zheng, H.X.; Zhang, L.; Zhu, R.R.; Liu, C.M.; Sato, Y.; Fukushima, Y. Responses of streamflow to climate and land surface change in the headwaters of the yellow river basin. *Water Resour. Res.* **2009**, *45*. [CrossRef]

69. Zhang, M.F.; Wei, X.H.; Sun, P.S.; Liu, S.R. The effect of forest harvesting and climatic variability on runoff in a large watershed: The case study in the upper Minjiang river of Yangtze river basin. *J. Hydrol.* **2012**, *464*, 1–11. [CrossRef]

70. Shi, P.; Ma, X.; Hou, Y.; Li, Q.; Zhang, Z.; Qu, S.; Chen, C.; Cai, T.; Fang, X. Effects of land-use and climate change on hydrological processes in the upstream of Huai river, China. *Water Resour. Manag.* **2013**, *27*, 1263–1278. [CrossRef]

71. Poff, N.L.; Allan, J.D.; Bain, M.B.; Karr, J.R.; Prestegaard, K.L.; Richter, B.D.; Sparks, R.E.; Stromberg, J.C. The natural flow regime. *Bioscience* **1997**, *47*, 769–784. [CrossRef]

72. Poff, N.L.; Zimmerman, J.K.H. Ecological responses to altered flow regimes: A literature review to inform the science and management of environmental flows. *Freshwater Biol.* **2010**, *55*, 194–205. [CrossRef]

73. Zanardo, S.; Harman, C.J.; Troch, P.A.; Rao, P.S.C.; Sivapalan, M. Intra-annual rainfall variability control on interannual variability of catchment water balance: A stochastic analysis. *Water Resour. Res.* **2012**, *48*. [CrossRef]

74. Bruijnzeel, L.A. Hydrological functions of tropical forests: Not seeing the soil for the trees? *Agric. Ecosyst. Environ.* **2004**, *104*, 185–228. [CrossRef]

75. Ellison, D.; Futter, M.N.; Bishop, K. On the forest cover–water yield debate: From demand- to supply-side thinking. *Glob. Chang. Biol.* **2012**, *18*, 806–820. [CrossRef]

76. Valipour, M. Importance of solar radiation, temperature, relative humidity, and wind speed for calculation of reference evapotranspiration. *Arch. Agron. Soil Sci.* **2015**, *61*, 239–255. [CrossRef]

77. Liu, X.; Zhang, D. Trend analysis of reference evapotranspiration in northwest china: The roles of changing wind speed and surface air temperature. *Hydrol. Process.* **2013**, *27*, 3941–3948. [CrossRef]

78. Valipour, M. Study of different climatic conditions to assess the role of solar radiation in reference crop evapotranspiration equations. *Arch Agron. Soil Sci.* **2015**, *61*, 679–694. [CrossRef]

79. Kurylyk, B.L. Discussion of 'a simple thaw-freeze algorithm for a multi-layered soil using the Stefan equation' by Xie and Gough (2013). *Permafrost Periglac.* **2015**, *26*, 200–206. [CrossRef]

80. Kurylyk, B.L.; Hayashi, M.; Quinton, W.L.; McKenzie, J.M.; Voss, C.I. Influence of vertical and lateral heat transfer on permafrost thaw, peatland landscape transition, and groundwater flow. *Water Resour. Res.* **2016**, *52*, 1286–1305. [CrossRef]

81. Wang, T.; Yang, H.; Yang, D.; Qin, Y.; Wang, Y. Quantifying the streamflow response to frozen ground degradation in the source region of the yellow river within the budyko framework. *J. Hydrol.* **2018**, *558*, 301–313. [CrossRef]

82. Ding, Y.H.; Ren, G.Y.; Zhao, Z.C.; Xu, Y.; Luo, Y.; Li, Q.P.; Zhang, J. Detection, causes and projection of climate change over china: An overview of recent progress. *Adv. Atmos. Sci.* **2007**, *24*, 954–971. [CrossRef]

83. Liu, Y.; Zhuoxin, G.U.; Wang, X. Impact of simulated climate warming on the radial growth of larix gmelinii in northeast china. *Acta Ecologica Sinica* **2017**, *37*, 2684–2693.

84. Jin, H.J.; Yu, Q.H.; Lii, L.Z.; Guo, D.X.; He, R.X.; Yu, S.P.; Sun, G.Y.; Li, Y.W. Degradation of permafrost in the Xing'anling mountains, northeastern china. *Permafrost Periglac.* **2007**, *18*, 245–258. [CrossRef]

85. Chang, X.; Jin, H.; He, R.; Yang, S.; Yu, S.; Lv, L.; Guo, D.; Wang, S.; Kang, X. Advances in permafrost and cold regions environments studies in the da Xing'anling (Da Hinggan) mountains, northeastern China. *J. Glaciol Geocryol* **2008**, *30*, 176–182.

*forests*

**MDPI**

*Article*

# Evaluation of the Water-Storage Capacity of Bryophytes along an Altitudinal Gradient from Temperate Forests to the Alpine Zone

Yoshitaka Oishi [iD]

Center for Arts and Sciences, Fukui Prefectural University, 4-1-1 Kenjojima, Matsuoka, Eiheiji-cho, Yoshida-gun, Fukui 910-1195, Japan; oishiy@fpu.ac.jp; Tel.: +81-76-61-6000

Received: 10 May 2018; Accepted: 16 July 2018; Published: 18 July 2018

**Abstract:** Forests play crucial roles in regulating the amount and timing of streamflow through the water storage function. Bryophytes contribute to this increase in water storage owing to their high water-holding capacity; however, they might be severely damaged by climate warming. This study examined the water storage capacity (WSC) of bryophytes in forests in the mountainous areas of Japan. Sampling plots (100 m$^2$) were established along two mountainous trails at 200-m altitude intervals. Bryophytes were sampled in these plots using 100-cm$^2$ quadrats, and their WSC was evaluated according to the maximum amount of water retained in them (WSC-quadrat). The total amount of water in bryophytes within each plot (WSC-plot) was then calculated. The WSC-quadrat was affected by the forms of bryophyte communities (life forms) and their interactions, further influencing soil moisture. The WSC-quadrat did not show any significant trend with altitude, whereas, the highest WSC-plot values were obtained in subalpine forests. These changes to WSC-plot were explained by large differences in bryophyte cover with altitude. As the WSC controlled by the life forms might be vulnerable to climate warming, it can provide an early indicator of how bryophyte WCS and associated biological activities are influenced.

**Keywords:** climate warming; East Asia; forest floor; forest hydrology; subalpine forest

---

## 1. Introduction

Forests play crucial roles in regulating the amount and timing of streamflow by mitigating the effects of precipitation via their water storage function [1,2]. The influence of afforestation on streamflow has been shown by many studies; for example, afforestation of grasslands and shrublands resulted in a loss of one-third to three-quarters of streamflow on average [3]. These roles contribute to the water cycle in forest ecosystem that has diverse ecological functions, such as a source of water supply, flood and erosion control, conservation of biodiversity, and climate stabilization [4–6].

The storage of water in forests is achieved by the interaction of plant groups with vertical stratification [7]. Forest canopies intercept rainfall and reduce the impact of rain drops on the ground [8]. Under these canopies, epiphytes on trees store rainfall and fog temporarily [9], while the roots of trees and grasses improve the infiltration of soil, reducing runoff and soil erosion [10,11]. In areas with less developed forest canopies, grass and biological soil crusts (assemblages of bryophytes, lichens, algae, cyanobacteria, and fungi) strongly affect the process of water infiltration into the soil, and partly control the amount of run off [12,13].

In recent years, there is serious concern about the changes of water storage in forests because of the influence of climate warming on vegetation [14–16]. The rapid shift in species distribution by the global rise in temperature is evidenced by species both expanding to newly favorable locations and declining in unfavorable areas. The estimated shift of species to higher elevations is at a median rate of 11.0 m per decade, while the shift to higher latitudes is estimated at 16.9 km per decade [17]. Under current climate change scenarios, one-tenth to one–half of global land might be highly or very highly vulnerable [18]. Temperate mixed forests, boreal conifer forests, tundra, and alpine biomes are considered the most vulnerable biomes to these changes [18]. The changes of forest ecosystems in response to climate warming might alter water yield, impacting water supply for human consumption [16].

The response of vegetation to climate warming differs according to plant groups. Among plant groups, serious damage to bryophytes may arise due to their sensitivity to these changes because of their poikilohydric properties [19]. The water content of bryophytes is highly dependent on their external environment, decreasing rapidly when temperature rises and humidity drops [19,20]. The decrease in the water content leads to a shorter period of metabolic activity and tissue damage caused by drought stress [20]. As a result, bryophytes, especially those currently growing in environments with low drought stress, are sensitive to climatic warming that causes both thermal and drought stress [20,21].

Although bryophytes are vulnerable to climate warming, they play important roles in increasing the water storage capacity of forest ecosystems [9,22–27]. Because of their poikilohydric properties, bryophytes can retain relatively high amount of water within the community, ranging between approximately 200% and 3000% of their dry mass [25,28,29]. Their water storage capacity is severely reduced once they are water saturated [9]; however, they contribute towards buffering the influence of rainfall on forest ecosystems, especially at the beginning of rainfall events [9]. This buffering function might be more important as the frequency of heavy rainfall in short term are expected to increase due to climatic change [30]. Regarding the water storage function of bryophytes, this is well documented in epiphytes occupying montane cloud forests [9,22–27]. For example, in tropical cloud forests, epiphytic bryophytes are estimated to store ca. 3–3.5 mm of rainfall [23,25] whereas epiphytes (bryophytes and lichens) store 1.2–1.4 mm in an old Douglas-fir forest [22]. The contribution of bryophytes to total rainfall interception is estimated to be 6% in tropical montane forests [24].

Unlike epiphytic bryophytes, the water storage function of forest floor bryophytes (bryophytes on the ground or on logs) is poorly known. Nevertheless, these bryophytes exhibit higher maximum values of water storage when compared to epiphytic bryophytes [22], contributing to the forest water cycle. In addition, the cover of forest floor bryophytes reduces soil temperature, while it improves the retention of soil moisture by decreasing evapotranspiration [31]. Lower soil temperature subsequently limits the decomposition of fresh litter, causing organic carbon to accumulate in the soil [32]. When considering these ecological functions of forest floor bryophytes, it is important to evaluate their water storage capacity to reveal the influence of climate warming on the hydrological processes of forest ecosystems.

In this study, the water storage capacity of bryophytes, including forest floor species, was evaluated in the montane forests of Japan, where high bryophyte diversity is harbored [33,34]. For this purpose, this study focused on the altitudinal patterns of bryophytes because altitude is one of the major factors that determine the behavior of bryophytes in montane forest ecosystems [35,36]. The results of this study will advance our understanding of how bryophytes contribute to the hydrological processes in forest ecosystems, which may implicate changes in their water storage function in response to climate warming.

## 2. Materials and Methods

### 2.1. Study Site

The study site is located on the Yatsugatake Mountains, in central Japan (Figure 1). These mountains stretch ca. 30 km from north to south and 15 km from east to west. The highest peak is Mt. Akadake (2899 m). The vegetation is roughly grouped into four types; temperate broadleaved forests (below ca. 1800 m), subalpine conifer forests (ca. 1800–2600 m), stone pine forests (ca. 2600–2800 m), and alpine meadows (ca. 2800–2900 m). Annual mean temperature and precipitation from 1981 to 2010, measured at the closest weather station (Nobeyama; 1350 m alt.), were 6.9 °C and 1439 mm, respectively [37]. The highest temperatures are recorded in August (19.2 °C) and the lowest are in January (−5.3 °C) [38]. Precipitation also changes seasonally; being highest in September (210.5 mm/month) and lowest in December (38.4 mm/month) [38].

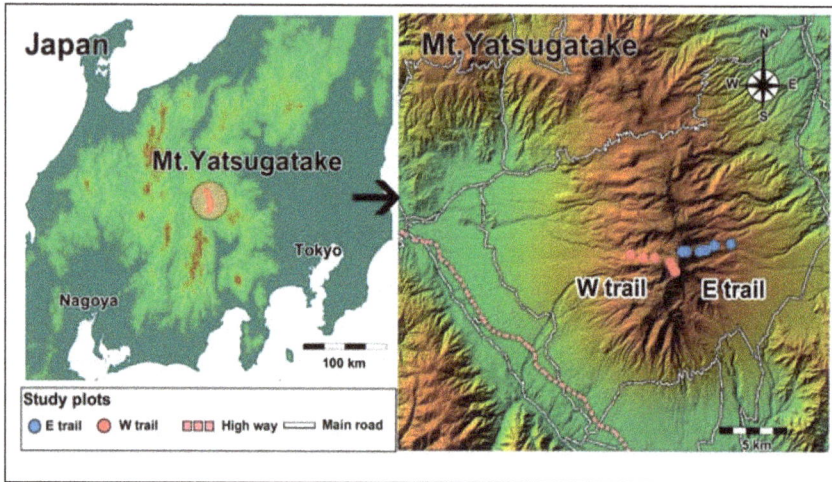

**Figure 1.** Study site, Mt. Yatsugatake, central Japan. The location of Mt. Yatsugatake is shown by the red circles in the slide on the left. Study plots were established along the E and W trails at 200-m altitude intervals, from 1800 to 2800 m. The left-hand panel was adapted from Figure 1 of Oishi [39]. The right-hand panel was created using global information system data provided by the Ministry of Land and Geospatial Information Authority of Japan.

*2.2. Sampling and Water Storage Capacity*

Twelve 10 m × 10 m study plots were selected at 200-m altitude intervals from 1800 to 2800 m along two trails (eastern trail; E trail, western trail; W trail) extending from the east to the west side of Mt. Yatsugatake (Figure 1). The plot of the E trail at 1800 m belongs to temperate-subalpine mixed forests, while the E trail extending from ca. 2000 m to 2600 m and the W trail extending from ca. 1800–2400 m altitude are classified as subalpine forests. Other plots at higher altitudes belong to stone pine forests or alpine meadows. In these study plots, three to four samples of dominant bryophyte communities were collected from each substrate [soil (including humus), rock, logs, and tree trunks] using sampling quadrats (10 cm × 10 cm). When dominant bryophyte communities consisted of more than one species, the ratios of each species in the collected samples were adjusted to those observed in the field. When the largest bryophyte cover was smaller than 100 cm × 100 cm on each substrate, this substrate was not included in the sampling. The life forms of sampled species were also recorded according to the life form classification of Bates [40]. To estimate maximum water storage capacity (WSC) of bryophytes under field conditions, sampling was completed during August 2015, when no rain occurred for five consecutive days during the summer. Substrate and the bryophyte cover were measured in these plots by 10% increments; however, when the cover was less than 10%, they were recorded by 5% increments. Percentage values of these covers were then transformed to m$^2$ values.

Collected samples were placed in sealed plastic bags to keep their community structure as intact as possible and were transported to the laboratory. The soils, litter, and other small mixed species were cleaned from the samples, and these samples were weighed (fresh weight; Fw). After weighing, the samples were dipped in a water container to represent the state of bryophytes when fully saturated by heavy rainfall. Samples were taken and then placed for 10 min to remove the extra external water, which was not tightly connected with the shoots. Then, these samples were weighed again (saturated weight; Sw). After these procedures, bryophyte samples were oven-dried at 80 °C for 48 h, and then weighed again (dry weight; Dw). Using Fw and Dw values, two types of water storage capacity (WSC) of bryophytes were calculated: WSC of fresh samples ($WSC_f$) and WSC of oven-dried samples ($WSC_d$). $WSC_f$ represents the possible maximum amount of water absorbed by bryophytes under field conditions. However, dried samples are often used to estimate bryophyte WSC [9,22,25,27,28]; therefore, $WSC_d$ was also used for comparison with other studies.

To examine the influence of bryophytes on soil moisture, three soil samples were collected using a soil core sampler (100 cm$^3$) in each plot. The collected soil samples were preserved in sealed plastic bags and were weighed in the laboratory. After oven-drying at 80 °C for 48 h, these oven dried samples were weighed again. The differences in weight before and after the soil samples were dried were used as the soil moisture (g/100 cm$^3$).

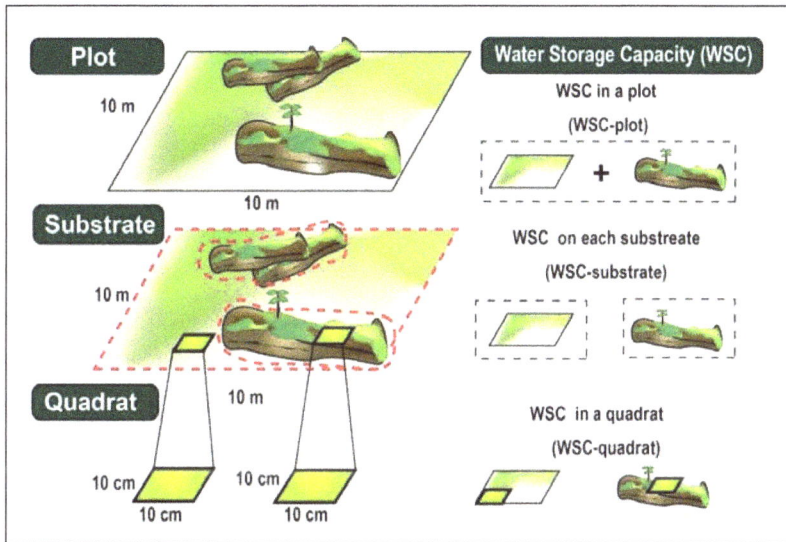

**Figure 2.** Schematic showing the hieratical evaluation of the water storage capacity of bryophytes.

*2.3. Water Storage Capacity at Quadrat, Substrate, and Plot Scales*

Bryophyte WSC was assessed hierarchically at three scales: quadrat, substrate, and plot scales (Figure 2). The values of $WSC_f/WSC_d$ at sampling quadrats ($WSC_f$-quadrat/$WSC_d$-quadrat; g/100 cm$^2$) were calculated by subtracting Fw/Dw from Sw as follows:

$$WSC_f\text{-quadrat (g/100 cm}^2) = Sw - Fw \tag{1}$$

$$WSC_d\text{-quadrat (g/100 cm}^2) = Sw - Dw. \tag{2}$$

Using these values, the total $WSC_f/WSC_d$ of bryophytes on each substrate within a plot ($WSC_f$-substrate/$WSC_d$-substrate) was estimated according to the following equations:

$$WSC_f\text{-substrate (K) (L/100 m}^2) = WSC_f\text{-quadrat (K)} \times Cov\ (K) \times 10^{-1} \tag{3}$$

$$WSC_d\text{-substrate (K) (L/100 m}^2) = WSC_d\text{-quadrat (K)} \times Cov\ (K) \times 10^{-1} \tag{4}$$

where, K = substrate types (soil, rocks, logs, and tree trunk) and Cov (K) = total bryophyte cover on substrate (K).

Then, total $WSC_f/WSC_d$ within each plot ($WSC_f$-plot/$WSC_d$-plot; $L/100\ m^2$) was evaluated using the values of $WSC_f$-substrate/$WSC_d$-substrate as follows:

$$WSC_{f\text{-plot}}\ (L/100\ m^2) = \sum_{k=1}^{4} [WSC_{f\text{-substrate}}\ (K)] \tag{5}$$

$$WSC_{d\text{-plot}}\ (L/100\ m^2) = \sum_{k=1}^{4} [WSC_{d\text{-substrate}}\ (K)] \tag{6}$$

where, k means substrate type (k = 1; soil, k = 2; rocks, k = 3; logs, k = 4; tree trunk).

*2.4. Modeling*

The difference in bryophyte WSC-quadrat between substrate types and between life form types was examined by *t*-test or Tukey's multiple comparison test. The influence of forest floor bryophytes on soil moisture was then examined by Pearson product-moment correlation between the WSC-quadrat values of bryophytes on soil and soil moisture. As the texture of soils largely differed between alpine and below alpine areas, the alpine data were not included in the calculation.

At substrate scales, the values of bryophyte WSC can be influenced by environmental (e.g., substrate type and cover) and ecological factors (e.g., types of bryophyte community). To reveal these influences, linear models were used that correlated the values of $WSC_f$-substrate/$WSC_d$-substrate with these variables. This modeling was performed for fresh and dried bryophyte samples, respectively. The environmental and ecological variables used in the modeling were substrate type, cover of each substrate within a plot ($m^2$), Fw/Dw, $WSC_f$-quadrat/$WSC_d$-quadrat, total bryophyte cover on each substrate ($m^2$), and water uptake strategies of the dominant bryophyte communities (ectohydric, endohydric, and mixed types). The types of substrate and water uptake strategies were adopted as categorical variables. The best-fit models were identified using the step Akaike information criterion (AIC) function. For these models, only variables that were significantly correlated with $WSC_f$-substrate/$WSC_d$-substrate were used. In addition to these models, linear mixed models were constructed with the E/W trails as nested variables to reflect the influence of each trail on bryophyte WSC. The best-fit models were selected using the same procedures as for the linear models. All calculations were performed with R software [41].

## 3. Results

*3.1. Water Storage Capacity of Bryophyte Communities at the Quadrat Scale*

### 3.1.1. Comparison of Water Storage Capacity of Bryophyte Communities

In the 12 study plots, bryophyte cover larger than 100 cm × 100 cm was only found on the soil and on logs; hence, the bryophyte communities on these substrates were sampled. In total, 62 samples were collected from these plots and 16 bryophyte communities consisting of 11 bryophyte species were recorded (Table 1). These species included *Codriophorus fasicularis* (Hedw.) Bednarek-Ochyra & Ochyra (*Cod fas*), *Dicranum majus* Turner (*Dic maj*), *Dicranum nipponense* Besch. (*Dic nip*), *Heterophyllium affine* (Hook.) M. Fleisch. (*Het aff*), *Hylocomium splendens* (Hedw.) Schimp. (*Hyl spl*), *Hypnum plicatulum* (Lindb.) A. Jaeger (*Hyp pli*), *Nowellia curvifolia* (Dicks.) Mitt. (*Now cur*), *Pleurozium schreberi* (Willd. ex Brid.) Mitt. (*Ple shr*), *Pogonatum contortum* (Menzies ex Brid.) Lesq. (*Pog con*), *Pogonatum japonicum* Sull. & Lesq. (*Pog jap*), and *Rigodiadelphus robustus* (Lindb.) Nog. (*Rig rob*). Among the 16 bryophyte communities, several small liverworts (e.g., *Cephaloziea* sp.) were often found; however, their influence on the WSC of these bryophyte communities was not considered because they represented low biomass and a small number of shoots. The life form types recorded in the samples were tall turfs (T), large cushions (Cu), smooth mats (Sm), rough mats (Rm), thread-like forms (Tl), and wefts (W).

The bryophyte communities were grouped into three types by the number of species. Bryophyte communities consisted of one species (e.g., *Cor fas*), two species (e.g., *Dic maj–Hyl spl*), or three species (e.g., *Dic maj–Het aff–Ple shr*). Among the collected species, *Hyl spl* and *Ple shr* had the highest occurrence in these dominant communities (*Hyl spl;* 11 times, *Ple shr;* 11 times), ranging from 1800 to 2600 m altitude. The average ± standard deviation (*SD*) of Fw, Dw, and Sw were 13.78 ± 7.91, 4.98 ± 2.86, and 35.16 ± 15.46 g/100 cm$^2$, respectively. Using these values, WSC$_f$-quadrat/WSC$_d$-quadrat was calculated. The values of WSC$_f$-quadrat were 19.72 ± 9.10 g/100 cm$^2$ for the soil and 23.69 ± 11.94 g/100 cm$^2$ for logs. The higher WSC$_f$-quadrat values were measured in *Pog jap*, *Cor fac*, *Hyl spl–Ple shr*, and *Rig rob–Ple shr* communities (average values; >30 g/100 cm$^2$), while *Hyl spl–Ple shr* communities also had larger variation in these values, ranging from 12.21 to 36.00 g/100 cm$^2$ on average, across substrates (Table 1). The values of the WSC$_d$-quadrat demonstrated similar trends to those of the WSC$_f$-quadrat and were significantly correlated with those of the WSC$_f$-quadrat ($r = 0.915$, $n = 62$, $p <$ 0.01). The values of the WSC$_d$-quadrat were 29.80 ± 13.11 g/100 cm$^2$ for the soil and were 30.72 ± 13.11 g/100 cm$^2$ for logs.

**Table 1.** Water storage capacity of dominant bryophyte communities at the quadrat scale (100 cm$^2$), and their total cover on each substrate within the study plots.

| Altitude (m) | Trail [1] | Bryophyte Community (Life form Type) | $n$ | FW (g/100 cm$^2$) | DW (g/100 cm$^2$) | WSC$_f$-q (g/100 cm$^2$) | WSC$_d$-q (g/100 cm$^2$) | Cover (m$^2$/plot) |
|---|---|---|---|---|---|---|---|---|
| **Soil** | | | | | | | | |
| 1800 | E | *Dic nipn–Pog con* (T–T: T) | 3 | 8.05 ± 1.37 | 2.80 ± 0.39 | 13.79 ± 4.04 | 19.04 ± 4.17 | 5 |
| 2000 | E | *Hyl spl–Ple shr* (W–W: W) | 3 | 9.28 ± 2.97 | 3.54 ± 0.74 | 20.27 ± 5.86 | 26.03 ± 8.08 | 7 |
| 2200 | E | *Pog con–Hyl spl* (T–W) | 3 | 17.12 ± 1.07 | 5.28 ± 0.96 | 14.11 ± 4.08 | 25.95 ± 3.93 | 24 |
| 2400 | E | *Pog con* (T) | 3 | 9.71 ± 1.47 | 3.38 ± 0.47 | 10.18 ± 1.96 | 16.51 ± 2.47 | 32 |
| 2600 | E | *Pog jap* (T) | 3 | 38.96 ± 9.52 | 14.81 ± 3.65 | 32.89 ± 3.94 | 57.03 ± 7.08 | 24 |
| 2800 | E | *Dic maj–Het aff–Ple shr* (T–Sm–W) | 3 | 7.21 ± 4.47 | 2.53 ± 1.47 | 12.11 ± 7.95 | 16.79 ± 10.32 | 10 |
| 1800 | W | *Hyl spl–Ple shr* (W–W: W) | 3 | 15.90 ± 5.37 | 5.23 ± 0.19 | 24.78 ± 5.40 | 35.47 ± 0.20 | 63 |
| 2000 | W | *Hyl spl* (W) | 3 | 10.03 ± 0.96 | 4.57 ± 0.52 | 24.39 ± 6.44 | 30.18 ± 6.90 | 51 |
| 2200 | W | *Pog jap–Hyl spl–Ple shr* (T–W–W: T–W) | 3 | 18.83 ± 6.51 | 5.56 ± 1.75 | 16.45 ± 9.42 | 29.73 ± 8.59 | 40 |
| 2400 | W | *Pog jap–Dic maj–Ple shr* (T–T–W: T–W) | 3 | 11.69 ± 6.43 | 3.90 ± 1.83 | 16.86 ± 10.54 | 24.65 ± 14.56 | 48 |
| 2600 | W | *Dic maj–Hyl spl* (T–W) | 3 | 15.12 ± 3.90 | 4.53 ± 1.16 | 19.79 ± 3.75 | 30.38 ± 6.21 | 50 |
| 2800 | W | *Cor fas* (Cu) | 3 | 22.24 ± 3.00 | 7.34 ± 1.85 | 31.02 ± 11.71 | 45.92 ± 7.03 | 5 |
| **Logs** | | | | | | | | |
| 2000 | E | *Hyl spl–Ple shr* (W–W: W) | 3 | 6.20 ± 0.73 | 2.91 ± 0.51 | 12.21 ± 5.60 | 15.49 ± 5.59 | 3 |
| 2200 | E | *Het aff–Ple shr* (Sm–W) | 3 | 14.85 ± 3.33 | 4.87 ± 0.47 | 25.51 ± 2.24 | 35.49 ± 1.78 | 32 |
| 2400 | E | *Now cur–Ple shr* (Tl–W) | 3 | 5.45 ± 2.16 | 2.00 ± 1.37 | 10.71 ± 7.20 | 14.16 ± 8.00 | 8 |
| 2600 | E | *Rig rob–Ple shr* (Rm–W) | 3 | 13.35 ± 2.62 | 6.52 ± 0.08 | 35.47 ± 9.30 | 42.30 ± 8.77 | 4 |
| 1800 | W | *Hyp pli–Hyl spl–Ple shr* (Sm–W–W: Sm–W) | 4 | 12.87 ± 3.75 | 5.29 ± 1.05 | 29.20 ± 10.44 | 36.78 ± 12.15 | 8 |
| 2000 | W | *Het aff–Hyl spl* (W–W: W) | 4 | 9.05 ± 1.87 | 3.24 ± 0.69 | 18.59 ± 8.12 | 24.40 ± 5.64 | 24 |
| 2200 | W | *Hyl spl–Ple shr* (W–W: W) | 3 | 17.40 ± 2.88 | 5.38 ± 1.67 | 21.66 ± 11.15 | 33.68 ± 7.42 | 40 |
| 2400 | W | *Hyl spl–Ple shr* (W–W: W) | 3 | 13.84 ± 3.90 | 6.32 ± 1.20 | 36.00 ± 16.55 | 43.52 ± 17.07 | 16 |

[1] Trails are represented in Figure 1. Abbreviations are as follows: *n*; number of quadrats, FW; fresh weight, DW; dry weight, WSC$_f$-q; water storage capacity of fresh samples within a quadrat, WSC$_d$-q; water storage capacity of dried samples within a quadrat, Cover; total cover on each substrate within a plot, E; E trail, W; W trail, *Cod fas; Codriophorus fasicularis, Dic maj; Dicranum majus, Dic nip; Dicranum nipponense, Het aff; Heterophyllium affine, Hyl spl; Hylocomium splendens, Hyp pli; Hypnum plicatulum, Now cur; Nowellia curvifolia, Ple shr; Pleurozium schreberi, Pog con; Pogonatum contortum, Pog jap; Pogonatum japonicum, Rig rob; Rigodiadelphus robustus,* T; tall turfs, Cu; large cushions, Rm; rough mats, Sm; smooth mats, Tl; thread-like forms, and W; wefts.

To examine the influence of life forms on bryophyte WSC, the WSC of each life form type was calculated (Table 2). In fresh samples, higher WSC was measured in Rm-W, followed by Cu, Sm-W, and W forms. In contrast, Tl-W, T-Sm-W, and T-W forms had lower values. The dry samples showed similar results. Using multiple comparison, significant differences were found between Rm-W and Tl-W in fresh samples and between Cu and Tl-W in dry samples ($p < 0.05$).

**Table 2.** Water storage capacity of bryophyte communities at the quadrat scale (100 cm$^2$) among life form types.

| Life Form | T | Cu | W | Rm–W | Sm–W | T–W | Tl–W | T–Sm–W |
|-----------|-----|-----|-----|------|------|-----|------|--------|
| *n* | 9 | 3 | 21 | 3 | 8 | 12 | 3 | 3 |
| | | | WSC$_f$-q (g/100 cm$^2$) | | | | | |
| Average | 18.95 | 31.02 | 23.55 | 35.47 | 23.89 | 16.80 | 10.71 | 12.11 |
| *SD* | 10.99 | 11.71 | 9.98 | 9.30 | 9.45 | 6.8 | 7.20 | 7.95 |
| Significance | ab | ab | ab | b | ab | ab | a | ab |
| | | | WSC$_d$-q (g/100 cm$^2$) | | | | | |
| Average | 30.86 | 45.92 | 31.40 | 42.30 | 30.59 | 27.68 | 14.16 | 16.79 |
| *SD* | 20.12 | 7.03 | 10.94 | 8.77 | 10.34 | 8.26 | 8.01 | 10.32 |
| Significance | ab | b | ab | ab | ab | ab | a | ab |

Abbreviations are as follows: T; tall turfs, Cu; large cushions, W; Wefts, Rm; rough mats, Sm; smooth mats, Tl; thread-like-forms, *n*; number of quadrats, WSC$_f$-q; water storage capacity of fresh samples within a quadrat, WSC$_d$-q; water storage capacity of dried samples within a quadrat, *SD*; standard deviation, Significance; different letters show significant differences by multiple comparisons (*p* < 0.05).

### 3.1.2. Water Storage Capacity of Bryophytes and Soil Moisture

The influence of bryophyte WSC on soil moisture was examined using the data except for alpine areas. The values of soil moisture were 13.82 ± 5.27 g/100 cm$^3$ (Average ± *SD*). These soil moistures were significantly and positively correlated with WSC$_d$-quadrat (*n* = 9, *r* = 0.778, *p* = 0.014) and were strongly, but not significantly, correlated with WSC$_f$-quadrat (*n* = 9, *r* = 0.613, *p* = 0.079).

### 3.1.3. Altitudinal Patterns of Water Storage Capacity of Bryophyte Communities at the Quadrat Scale

The values of WSC$_f$-quadrat/WSC$_d$-quadrat were compared in relation to substrate type and altitude. The differences in WSC$_f$-quadrat between the soil and logs were not statistically significant (WSC$_f$-quadrat; *t*-values = 1.486, *df* = 60, *p* = 0.143/WSC$_d$-quadrat; *t*-values = 0.270, *df* = 60, *p* = 0.788). Furthermore, the changes in the WSC$_f$-quadrat/WSC$_d$-quadrat with altitude were also not significant for the soil or logs (Figure 3). The Pearson correlation coefficient between WSC$_f$-quadrat/WSC$_d$-quadrat and altitude on soil was 0.118 (*n* = 62, *p* = 0.43)/0.227 (*n* = 62, *p* = 0.18), while that on logs was 0.209 (*n* = 62, *p* = 0.31)/0.193 (*n* = 62, *p* = 0.35).

### 3.2. Water Storage Capacity of Bryophyte Communities on Each Substrate

Bryophyte cover was recorded on each substrate to calculate WSC$_f$-substrate/WSC$_d$-substrate. Bryophyte cover differed greatly among study plots. The average and *SD* of total bryophyte cover within each plot was 24.70 ± 18.86 m$^2$/100 m$^2$. The cover showed the highest values in subalpine forests for both E and W trails and on the soil and logs (Table 1). The E trail had the highest values at 2400 m on soil and at 2200 m on logs, whereas the W trail had the highest value at 1800 m on soil and at 2200 m on logs. Then, the values of WSC$_f$-substrate/WSC$_d$-substrate were calculated according to Equation (3) and (4) for each substrate. The resulting values of WSC$_f$-substrate on soil were 60.02 ± 46.86 (L/100 m$^2$), while those on logs were 26.63 ± 31.34 (L/100 m$^2$). Using these values, linear models for the WSC$_f$-substrate/WSC$_d$-substrate were constructed based on the environmental and ecological variables. Among these variables, only bryophyte cover was significantly correlated with the values of the WSC$_f$-plot/WSC$_d$-plot. Hence, this variable was adopted as an explanatory variable for the linear models. The constructed models for both WSC$_f$-substrate (Y = 2.101X − 1.395, $R^2$ = 0.856) and WSC$_d$-substrate (Y = 3.088X − 2.167, $R^2$ = 0.866) fitted well. In comparison, no significant linier mixed models were constructed for each variable.

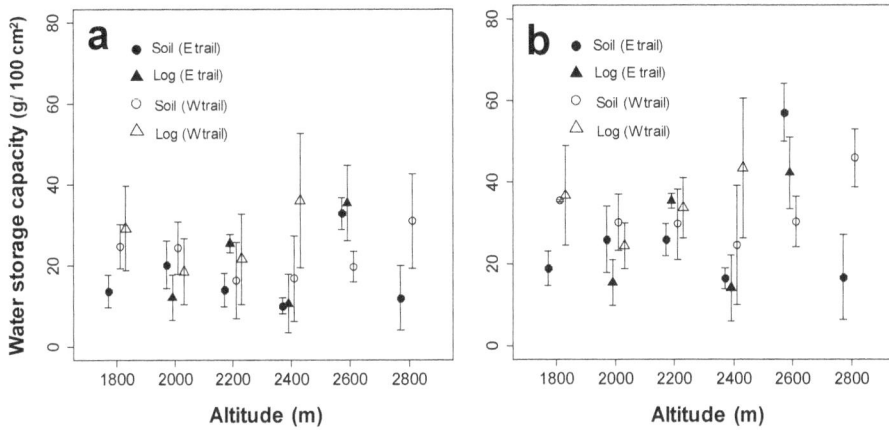

**Figure 3.** Changes to the water storage capacity of bryophyte communities at the quadrat scale with respect to altitude: (**a**) fresh samples and (**b**) dry samples; the error bar indicates standard deviation.

## 3.3. Water Storage Capacity of Bryophyte Communities per Plot

The values of the WSC$_f$-plot/WSC$_d$-plot were calculated based on Equations (5) and (6) (Table 3, Figure 4). The values of the WSC$_f$-plot (Average ± SD) were 86.71 ± 35.90 L/100 m$^2$ (equivalent to the increase of 0.8671 ± 0.3590 mm of rainfall interception), while those of the WSC$_d$-plot were 123.51 ± 52.59 L/100 m$^2$ (=1.2351 ± 0.5259 mm). Both WSC$_f$-plot/WSC$_d$-plot of E trail and W trail had the highest values in subalpine forests; however, the altitude of the plots with the highest WSC values differed for fresh and dry samples. The WSC$_f$-plot along the E trail had the highest values at 2200 m (115.51 L/100 m$^2$), while the highest value of the WSC$_d$-plot was observed at 2600 m (153.81 L/100 m$^2$). The values of the WSC$_f$-plot along the W trail were highest at 1800 m (179.45 L/100 m$^2$), whereas WSC$_d$-plot values were highest at 2200 m (253.60 L/100 m$^2$).

Regarding the changes in the WSC$_f$-plot/WSC$_d$-plot along the altitudinal gradient, the WSC$_f$-plot data on the E trail fitted a negative quadratic curve ($Y = -2.87 \times 10^{-4}X^2 + 1.34X - 1494.51$, $R^2 = 0.510$), while that on the W trail was fitted by liner regression ($Y = -1.49 \times 10^{-1}X + 468.59$, $R^2 = 0.841$). Similarly, the WSC$_d$-plot data on trails E and W were fitted by the negative quadratic curve ($Y = -3.73 \times 10^{-4}X^2 + 1.77X - 1979.54$, $R^2 = 0.515$) and by the linear regression ($Y = -2.00 \times 10^{-1}X + 639.25$, $R^2 = 0.747$), respectively.

**Table 3.** Bryophyte water storage capacity per plot (100 m$^2$).

| Altitude (m) | E Trail | | W Trail | |
| --- | --- | --- | --- | --- |
| | WSC$_f$-Plot (L/100 m$^2$) | WSC$_d$-Plot (L/100 m$^2$) | WSC$_f$-Plot (L/100 m$^2$) | WSC$_d$-Plot (L/100 m$^2$) |
| 1800 | 6.895 | 9.522 | 179.453 | 252.866 |
| 2000 | 17.853 | 22.850 | 168.982 | 212.466 |
| 2200 | 115.507 | 133.275 | 152.469 | 253.600 |
| 2400 | 40.459 | 64.939 | 138.523 | 187.947 |
| 2600 | 93.133 | 153.808 | 98.967 | 151.917 |
| 2800 | 12.107 | 16.790 | 15.508 | 22.958 |
| Average ± SD | 47.77 ± 41.89 | 66.73 ± 57.31 | 125.65 ± 55.53 | 180.29 ± 78.84 |

WSC$_f$-plot; water storage capacity of fresh samples in a plot, WSC$_d$-plot; water storage capacity of dried samples in a plot, *SD*; standard deviation.

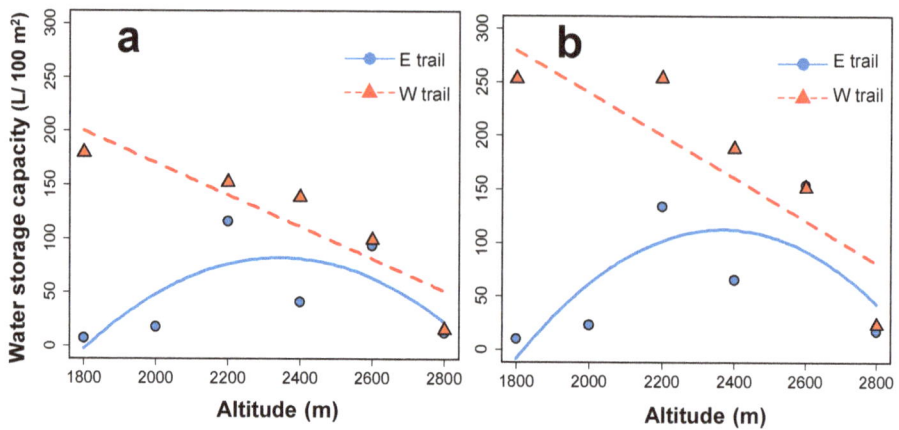

**Figure 4.** Change to water storage capacity of bryophyte communities within each plot along an altitudinal gradient. (**a**) Fresh samples, (**b**) Dry samples.

## 4. Discussion

The WSC quadrat of bryophytes was influenced by the life form types and their interactions, further affecting soil moisture; however, it did not vary with altitude or exhibit significant differences between the substrate types. At the plot scale, the WSC plot significantly correlated with bryophyte cover, with the highest in subalpine forests.

### 4.1. Water Storage Capacity at the Quadrat Scale

Comparison of the $WSC_f$-quadrat/$WSC_d$-quadrat showed higher values in *Hyl spl–Ple shr*, *Ple shr–Rig rob*, *Pog jap*, and *Cor fac* communities (Table 1). These results are explained by the community structure of bryophytes. Bryophytes forming compact mats had higher WSC because the spaces between individual shoots retain additional external water [29]. In the plots of the present study, these compact mats were formed by weft-forming mosses (W form; *Hyl spl* and *Ple shr*) and large cushion moss (Cu form; *Cor fas*), which contributed to the higher values of the $WSC_f$-quadrat/$WSC_d$-quadrat. In comparison, the community of *Pog jap* forms tall turfs (T form) that physically increases the amount of water held in these communities. Regarding the *Hyl spl–Ple shr* community, this moss community had large differences in the $WSC_f$-quadrat/$WSC_d$-quadrat, ranging from 12.21–36.00/15.49–43.52 g/100 cm$^2$. These results are attributed to difference in compactness or shoot density, which are reflected by the wider range of their FW/DW values (6.20–17.40/2.91–6.32 g/100 cm$^2$).

Some combinations of life forms (T–W and T–Sm–W forms) had lower WSC values on average than those of less mixed or single forms (T or W and T or Sm–W forms), despite there being no differences among the species in these communities (Table 2). The decreased WSC of these mixed life forms could be explained by their poor ability to form tight communities with neighboring species, due to differences in the characteristics of the life forms. The upright T form is largely different from that of the creeping Sm and W forms in their morphology. Moreover, the T form species include endohydric bryophytes (*Polytrichaceae*; *P. controtum* and *P. japonium*), which develop internal water conducting tissues and mainly absorb water from substrates [42]. In contrast, the Sm and W forms are ectohydric species without such conducting tissues, and exclusively rely on external capillary water [42]. These results are supported by an experiment that showed a reduced WSC in the mixture of bryophytes with different life forms and water uptake system [43].

In comparison, the results of this study demonstrated no statistical differences in WSC$_f$-quadrat/WSC$_d$-quadrat values between substrates, nor any change in these values with altitude (Figure 3). This is because logs largely covered by bryophytes were almost decayed and the surface material was similar to that of humus soil. These similarities in substrate surface could result in the development of similar bryophyte communities between logs and the soil. Furthermore, the differences of dominant bryophytes along the altitudinal gradient were less clear at the study sites, as several species occurred at a wider altitudinal range (e.g., *Hyl spl*; 1800–2600 m, and *Ple shr*; 1800–2600 m), which could reduce the magnitude of change to the WSC-quadrat with altitude.

### 4.2. Influence of Forest Floor Bryophytes on Below Ground Processes

The values of WSC$_f$-quadrat/WSC$_d$-quadrat were positively correlated with soil moisture. An increase in soil moisture by bryophyte cover has been reported because the evapotranspiration rate of bryophytes is lower than that of grasses [31] and they are able to retain a large amount of water during wet periods [44,45]. In addition, this study suggests that bryophytes with higher WSC have a larger influence on the increase in soil moisture. This influence could be related to the transport of larger amounts of moisture from bryophytes with higher WSC to the soil surface during evapotranspiration processes. Besides, bryophytes with higher WSC might further reduce water evaporation from the soil surface, as these bryophytes often retained water for longer periods [29].

### 4.3. Water Storage Capacity at the Substrate/Plot Scale

The WSC values of bryophytes were affected by biomass (plant tissues mass), species type, and growth form [29,43]; however, the constructed liner models revealed that the values of the WSC$_f$-substrate/WSC$_d$-substrate were largely dependent on total bryophyte cover on each substrate, regardless of the type of species and their substrates. These results are explained by the large differences in bryophyte cover among the study plots (24.70 $\pm$ 18.86 m$^2$/100 m$^2$), which decrease the relative influence of other factors (e.g., bryophyte community type) on biomass and makes cover a useful substitute for biomass and the associated WSC-plot.

Due to this strong significant correlation in bryophyte cover with biomass, the altitudinal patterns of the WSC$_f$-plot/WSC$_d$-plot closely fitted a negative quadratic curve (E trail) or a linear regression (W trail), with the highest values in subalpine forests where the highest bryophyte cover was recorded (Figure 4). In general, bryophyte cover on the forest floor changes with altitude. Subalpine conifer forests had higher cover due to the favorable environment for bryophyte growth, such as low temperature and high occurrence of fog, and less influence from fallen leaves [46]. In contrast, bryophyte cover tends to decline in temperate broadleaved forests because fallen leaves shade bryophytes on the forest floor and inhibit their photosynthesis [35]. A decline in bryophyte cover has also been reported in alpine zones due to the lack of forest canopies to provide suitable habitats for bryophytes [36].

Interestingly, the plots with the highest WSC-plot values differed between fresh (WSC$_f$-plot) and dried samples (WSC$_d$-plot) on both E and W trails (Table 3). These differences were attributed to larger differences between WSC$_f$-quadrat and WSC$_d$-quadrat in endohydric species (*Polytrichaceae* sp.). Despite the bryophyte samples being collected during the dry period (no rainfall), endohydric species still had higher water retention status because they absorb water from the soil; whereas the water content of ectohydric species was severely reduced. This retained water was completely lost during the oven-drying process, which increased the amount of water absorbed by the dry samples of endohydric species compared to ectohydric bryophytes. These differences in WSC between fresh and dry samples should be carefully considered when one estimates the WSC of bryophytes under field conditions, as the estimated WSC of endohydric bryophytes might be relatively higher than that of ectohydric species if dried samples are used for the calculation.

Due to the higher cover by bryophytes in subalpine forests, the estimated WSC$_d$-plot had a maximum of 2.5 mm extra rainfall interception (Table 3), which was almost equivalent to the values reported for the WSC of epiphytic bryophytes on trees in montane cloud forests (3.0–3.5 mm) [23,25]. These results underline the importance of forest floor bryophytes for the overall hydrological processes of subalpine forests. Furthermore, considering the influence of snowmelt and cloud water deposition on these bryophytes, their contribution to forest hydrology could be more significant than expected from their interception of additional rainfall alone. Snow pack is a key factor that determines the water dynamics in subalpine forests [47]. After thawing of snow, forest floor bryophytes absorb snowmelt and affect forest soil hydrology by increasing soil infiltration [48]. Like snow packs, cloud water deposition is an important water supply for forests at higher altitude, due to the frequent occurrence of fog [49]. Therefore, forest floor bryophytes, especially ectohydric species that largely rely on atmospheric water, might contribute to the water cycle in this ecosystem through the interception of fog and dew.

*4.4. Changes to Water Storage Capacity by Climatic Change*

Regarding the forest water storage for which bryophytes are responsible, the influence of climate warming might be more serious in subalpine forests where bryophyte WSC showed highest values (Figure 4). Given that global environmental changes seem to affect ecosystems more strongly at the community level than at the individual species level [50], the structure of bryophyte communities is more strongly influenced by climate warming than the species level [43]. For example, severe drought stress caused by climate warming [51] might facilitate the dominance of endohydric T form species (e.g., *Pog jap*) over ectohydric W form species (e.g., *Hyl spl*); because these T form species can be less affected by drought stress owing to their capacity to absorb water from soil. However, as this study revealed, these changes in the structure of bryophyte communities (i.e., dominant life form types and their interactions) influence the WSC, which also affects the soil moisture that determines soil carbon and nitrogen cycling [52,53]. Hence, climate warming might strongly affect bryophyte WSC controlled by life forms, further causing changes to the biological activity and nutrient cycling of soil in forest ecosystems.

## 5. Conclusions

The bryophyte WSC-plot changed with altitude and was highest in subalpine forests. This altitudinal pattern was explained by bryophyte cover, which could be used as a substitute for bryophyte biomass and its associated WSC. At the quadrat scale, the WCS of bryophytes was related to life form type and their interactions. The WSC further had a positive impact on soil moisture important for soil biological activities. Of importance, bryophyte WSC controlled by the life forms might be strongly affected by climate warming. Thus, changes to the dominance of bryophyte life forms might serve as an early indicator of how bryophyte WCS and associated biological activities are influenced by climate warming.

**Funding:** This research was supported by the Japan Society for the Promotion of Science—Grant-in-Aid for Young Scientists (B) (grant number 24710029) and Grant-in-Aid for Scientific Research (C) (grant number 16K00566).

**Conflicts of Interest:** The author declares no conflict of interest.

## References

1.  Vose, J.M.; Sun, G.; Ford, C.R.; Bredemeier, M.; Otsuki, K.; Wei, X.; Zhang, Z.; Zhang, L. Forest ecohydrological research in the 21st century: What are the critical needs? *Ecohydrology* **2011**, *4*, 146–158. [CrossRef]
2.  Brown, A.E.; Zhang, L.; McMahon, T.A.; Western, A.W.; Vertessy, R.A. A review of paired catchment studies for determining changes in water yield resulting from alterations in vegetation. *J. Hydrol.* **2005**, *310*, 28–61. [CrossRef]
3.  Farley, K.A.; Jobbagy, E.G.; Jackson, R.B. Effects of afforestation on water yield: A global synthesis with implications for policy. *Glob. Chang. Biol.* **2005**, *11*, 1565–1576. [CrossRef]

4. Postel, S.L.; Thompson, B.H. Watershed protection: Capturing the benefits of nature's water supply services. *Nat. Resour. Forum* **2005**, *29*, 98–108. [CrossRef]

5. Guo, Z.; Xiao, X.; Gan, Y.; Zheng, Y. Ecosystem functions, services and their values—A case study in Xingshan County of China. *Ecol. Econ.* **2001**, *38*, 141–154. [CrossRef]

6. Mashayekhi, Z.; Panahi, M.; Karami, M.; Khalighi, S.; Malekian, A. Economic valuation of water storage function of forest ecosystems (case study: Zagros Forests, Iran). *J. For. Res.* **2010**, *21*, 293–300. [CrossRef]

7. Ataroff, M.; Rada, F. Deforestation impact on water dynamics in a Venezuelan Andean cloud forest. *AMBIO* **2000**, *29*, 440–444. [CrossRef]

8. Keim, R.F.; Skaugset, A.E.; Weiler, M. Storage of water on vegetation under simulated rainfall of varying intensity. *Adv. Water Resour.* **2006**, *29*, 974–986. [CrossRef]

9. Veneklaas, E.J.; Zagt, R.; Van Leerdam, A.; Van Ek, R.; Broekhoven, A.; Van Genderen, M. Hydrological properties of the epiphyte mass of a montane tropical rain forest, colombia. *Vegetatio* **1990**, *89*, 183–192. [CrossRef]

10. Zhang, L.; Wang, J.; Bai, Z.; Lv, C. Effects of vegetation on runoff and soil erosion on reclaimed land in an opencast coal-mine dump in a loess area. *Catena* **2015**, *128*, 44–53. [CrossRef]

11. Le Bissonnais, Y.; Lecomte, V.; Cerdan, O. Grass strip effects on runoff and soil loss. *Agronomie* **2004**, *24*, 129–136. [CrossRef]

12. Chamizo, S.; Cantón, Y.; Rodríguez-Caballero, E.; Domingo, F.; Escudero, A. Runoff at contrasting scales in a semiarid ecosystem: A complex balance between biological soil crust features and rainfall characteristics. *J. Hydrol.* **2012**, *452–453*, 130–138. [CrossRef]

13. Belnap, J. The potential roles of biological soil crusts in dryland hydrologic cycles. *Hydrol. Process.* **2006**, *20*, 3159–3178. [CrossRef]

14. Sala, O.E.; Chapin, F.S.; Armesto, J.J.; Berlow, E.; Bloomfield, J.; Dirzo, R.; Huber-Sanwald, E.; Huenneke, L.F.; Jackson, R.B.; Kinzig, A. Global biodiversity scenarios for the year 2100. *Science* **2000**, *287*, 1770–1774. [CrossRef] [PubMed]

15. Root, T.L.; Price, J.T.; Hall, K.R.; Schneider, S.H.; Rosenzweig, C.; Pounds, J.A. Fingerprints of global warming on wild animals and plants. *Nature* **2003**, *421*, 57–60. [CrossRef] [PubMed]

16. Creed, I.F.; Spargo, A.T.; Jones, J.A.; Buttle, J.M.; Adams, M.B.; Beall, F.D.; Booth, E.G.; Campbell, J.L.; Clow, D.; Elder, K.; et al. Changing forest water yields in response to climate warming: Results from long-term experimental watershed sites across North America. *Glob. Chang. Biol.* **2014**, *20*, 3191–3208. [CrossRef] [PubMed]

17. Chen, I.-C.; Hill, J.K.; Ohlemüller, R.; Roy, D.B.; Thomas, C.D. Rapid range shifts of species associated with high levels of climate warming. *Science* **2011**, *333*, 1024–1026. [CrossRef] [PubMed]

18. Gonzalez, P.; Neilson, R.P.; Lenihan, J.M.; Drapek, R.J. Global patterns in the vulnerability of ecosystems to vegetation shifts due to climate change. *Glob. Ecol. Biogeogr.* **2010**, *19*, 755–768. [CrossRef]

19. Proctor, M.C.; Tuba, Z. Poikilohydry and homoihydry: Antithesis or spectrum of possibilities? *New Phytol.* **2002**, *156*, 327–349. [CrossRef]

20. He, X.; He, K.S.; Hyvönen, J. Will bryophytes survive in a warming world? *Perspect. Plant. Ecol. Evol. Syst.* **2016**, *19*, 49–60. [CrossRef]

21. Oishi, Y. Urban heat island effects on moss gardens in Kyoto, Japan. *Lands. Ecol. Eng.* **2018**. [CrossRef]

22. Pypker, T.G.; Unsworth, M.H.; Bond, B.J. The role of epiphytes in rainfall interception by forests in the Pacific Northwest. I. Laboratory measurements of water storage. *Can. J. For. Res.* **2006**, *36*, 809–818. [CrossRef]

23. Pócs, T. The epiphytic biomass and its effect on the water balance of two rain forest types in the Uluguru Mountains (Tanzania, East Africa). *Acta Bot. Acad. Sci. Hung.* **1980**, *26*, 143–167.

24. Hölscher, D.; Köhler, L.; van Dijk, A.I.; Bruijnzeel, L.S. The importance of epiphytes to total rainfall interception by a tropical montane rain forest in Costa Rica. *J. Hydrol.* **2004**, *292*, 308–322. [CrossRef]

25. Ah-Peng, C.; Cardoso, A.W.; Flores, O.; West, A.; Wilding, N.; Strasberg, D.; Hedderson, T.A.J. The role of epiphytic bryophytes in interception, storage, and the regulated release of atmospheric moisture in a tropical montane cloud forest. *J. Hydrol.* **2017**, *548*, 665–673. [CrossRef]

26. Chang, S.-C.; Lai, I.-L.; Wu, J.-T. Estimation of fog deposition on epiphytic bryophytes in a subtropical montane forest ecosystem in Northeastern Taiwan. *Atmos. Res.* **2002**, *64*, 159–167. [CrossRef]

27. Köhler, L.; Tobón, C.; Frumau, K.A.; Bruijnzeel, L.S. Biomass and water storage dynamics of epiphytes in old-growth and secondary montane cloud forest stands in Costa Rica. *Plant. Ecol.* **2007**, *193*, 171–184. [CrossRef]
28. Michel, P.; Payton, I.J.; Lee, W.G.; During, H.J. Impact of disturbance on above-ground water storage capacity of bryophytes in New Zealand indigenous tussock grassland ecosystems. *N. Z. J. Ecol.* **2013**, 114–126.
29. Elumeeva, T.G.; Soudzilovskaia, N.A.; During, H.J.; Cornelissen, J.H. The importance of colony structure versus shoot morphology for the water balance of 22 subarctic bryophyte species. *J. Veg. Sci.* **2011**, *22*, 152–164. [CrossRef]
30. Easterling, D.R.; Meehl, G.A.; Parmesan, C.; Changnon, S.A.; Karl, T.R.; Mearns, L.O. Climate extremes: Observations, modeling, and impacts. *Science* **2000**, *289*, 2068–2074. [CrossRef] [PubMed]
31. Zimov, S.A.; Chuprynin, V.; Oreshko, A.; Chapin III, F.; Reynolds, J.; Chapin, M. Steppe-tundra transition: A herbivore-driven biome shift at the end of the Pleistocene. *Am. Nat.* **1995**, *146*, 765–794. [CrossRef]
32. Hobbie, S.E.; Schimel, J.P.; Trumbore, S.E.; Randerson, J.R. Controls over carbon storage and turnover in high-latitude soils. *Glob. Chang. Biol.* **2000**, *6*, 196–210. [CrossRef]
33. Tan, B.C.; Iwatsuki, Z. Hot spots of mosses in East Asia. *Anal. Inst. Biol. Ser. Bot.* **1996**, *67*, 159–167.
34. Geffert, J.L.; Frahm, J.-P.; Barthlott, W.; Mutke, J. Global moss diversity: Spatial and taxonomic patterns of species richness. *J. Bryol.* **2013**, *35*, 1–11. [CrossRef]
35. Sun, S.-Q.; Wu, Y.-H.; Wang, G.-X.; Zhou, J.; Yu, D.; Bing, H.-J.; Luo, J. Bryophyte species richness and composition along an altitudinal gradient in Gongga Mountain, China. *PLoS ONE* **2013**, *8*, e58131. [CrossRef] [PubMed]
36. Grau, O.; Grytnes, J.A.; Birks, H. A comparison of altitudinal species richness patterns of bryophytes with other plant groups in Nepal, Central Himalaya. *J. Biogeogr.* **2007**, *34*, 1907–1915. [CrossRef]
37. Japan Meteorological Agency. Past Meteorological Data (Nobeyama). Available online: http://www.data.jma.go.jp/obd/stats/etrn/view/nml_amd_ym.php?prec_no=48&block_no=0415&year=&month=&day=&view= (accessed on 12 April 2018).
38. Japan Meteorological Agency. Top Ten in Recorded History. Available online: http://www.data.jma.go.jp/obd/stats/etrn/view/rank_a.php?prec_no=48&block_no=0415&year=&month=&day=&view= (accessed on 12 April 2018).
39. Oishi, Y. Comparison of moss and pine needles as bioindicators of transboundary polycyclic aromatic hydrocarbon pollution in central japan. *Environ. Pollut.* **2018**, *234*, 330–338. [CrossRef] [PubMed]
40. Bates, J. Is life-form a useful concept in bryophyte ecology? *Oikos* **1998**, 223–237. [CrossRef]
41. R Core Team. *R: A Language and Environment for Statistical Computing*; R Foundation for Statistical Computing: Vienna, Austria, 2018; Available online: http://www.R-project.org/ (accessed on 12 April 2018).
42. Proctor, M.C. The bryophyte paradox: Tolerance of desiccation, evasion of drought. *Plant. Ecol.* **2000**, *151*, 41–49. [CrossRef]
43. Michel, P.; Lee, W.G.; During, H.J.; Cornelissen, J.H.C. Species traits and their non-additive interactions control the water economy of bryophyte cushions. *J. Ecol.* **2012**, *100*, 222–231. [CrossRef]
44. Maestre, F.T.; Huesca, M.; Zaady, E.; Bautista, S.; Cortina, J. Infiltration, penetration resistance and microphytic crust composition in contrasted microsites within a Mediterranean semi-arid steppe. *Soil Biol. Biochem.* **2002**, *34*, 895–898. [CrossRef]
45. Eldridge, D.; Rosentreter, R. Morphological groups: A framework for monitoring microphytic crusts in arid landscapes. *J. Arid Environ.* **1999**, *41*, 11–25. [CrossRef]
46. Nakatsubo, T. The role of bryophytes in terrestrial ecosystems with special reference to forests and volcanic deserts. *Jpn. J. Ecol.* **1997**, *47*, 43–54. [CrossRef]
47. LaMalfa, E.M.; Ryle, R. Differential snowpack accumulation and water dynamics in aspen and conifer communities: Implications for water yield and ecosystem function. *Ecosystems* **2008**, *11*, 569–581. [CrossRef]
48. Beringer, J.; Lynch, A.H.; Chapin, F.S., III; Mack, M.; Bonan, G.B. The representation of arctic soils in the land surface model: The importance of mosses. *J. Clim.* **2001**, *14*, 3324–3335. [CrossRef]
49. Kalina, M.F.; Stopper, S.; Zambo, E.; Puxbaum, H. Altitude-dependent wet, dry and occult nitrogen deposition in an alpine region. *Environ. Sci. Pollut. Res.* **2002**, *9*, 16–22. [CrossRef]
50. Suding, K.N.; Lavorel, S.; Chapin, F.S.; Cornelissen, J.H.C.; DÍAz, S.; Garnier, E.; Goldberg, D.; Hooper, D.U.; Jackson, S.T.; Navas, M.-L. Scaling environmental change through the community-level: A trait-based response-and-effect framework for plants. *Glob. Chang. Biol.* **2008**, *14*, 1125–1140. [CrossRef]

51. Aiguo, D. Drought under global warming: A review. *WIREs Clim. Chang.* **2011**, *2*, 45–65. [CrossRef]
52. Robinson, C.; Wookey, P.; Parsons, A.; Potter, J.; Callaghan, T.; Lee, J.; Press, M.; Welker, J. Responses of plant litter decomposition and nitrogen mineralisation to simulated environmental change in a high arctic polar semi-desert and a subarctic dwarf shrub heath. *Oikos* **1995**, 503–512. [CrossRef]
53. Fisk, M.C.; Schmidt, S.K.; Seastedt, T.R. Topographic patterns of above-and belowground production and nitrogen cycling in alpine tundra. *Ecology* **1998**, *79*, 2253–2266. [CrossRef]

![forests logo] *forests*

MDPI

Article

# The Hydrological Impact of Extreme Weather-Induced Forest Disturbances in a Tropical Experimental Watershed in South China

Yiping Hou [1], Mingfang Zhang [1,2,*], Shirong Liu [3], Pengsen Sun [3], Lihe Yin [4], Taoli Yang [1], Yide Li [5], Qiang Li [6] and Xiaohua Wei [7]

[1]  School of Resources and Environment, University of Electronic Science and Technology of China, Chengdu 611731, China; yiping_hou@163.com (Y.H.); yangtl@uestc.edu.cn (T.Y.)
[2]  Center for Information Geoscience, University of Electronic Science and Technology of China, Chengdu 611731, China
[3]  Research Institute of Forest Ecology, Environment and Protection, Chinese Academy of Forestry, Beijing 100091, China; liusr@caf.ac.cn (S.L.); sunpsen@caf.ac.cn (P.S.)
[4]  Xi'an Center of Geological Survey, China Geological Survey, Xi'an 710054, China; ylihe@cgs.cn
[5]  Research Institute of Tropical Forestry, Chinese Academy of Forestry, Guangzhou 510520, China; liyide@126.com
[6]  Department of Civil Engineering, University of Victoria, Victoria BC V8W 2Y2, Canada; liqiang1205@gmail.com
[7]  Department of Earth and Environmental Sciences, University of British Columbia (Okanagan), 3333 University Way, Kelowna V1X1R3, Canada; adam.wei@ubc.ca
*  Correspondence: mingfangzhang@uestc.edu.cn; Tel.: +86-028-61830963

Received: 31 October 2018; Accepted: 22 November 2018; Published: 24 November 2018

**Abstract:** Tropical forests are frequently disturbed by extreme weather events including tropical cyclones and cold waves, which can not only yield direct impact on hydrological processes but also produce indirect effect on hydrology by disturbing growth and structures of tropical forests. However, the hydrological response to extreme weather-induced forest disturbances especially in tropical forested watersheds has been less evaluated. In this study, a tropical experimental watershed in Hainan Province, China, was selected to investigate the hydrological responses to extreme weather-induced forest disturbances by use of a single watershed approach and the paired-year approach. Key results are: (1) extreme weather-induced forest disturbances (e.g., typhoon and cold wave) generally had a positive effect on streamflow in the study watershed, while climate variability either yielded a negative effect or a positive effect in different periods; (2) the response of low flows to forest discussion was more pronounced; (3) the relative contribution of forest disturbances to annual streamflow (48.6%) was higher than that of climate variability (43.0%) from 1995 to 2005. Given the increasing extreme weather with climate change and their possible catastrophic effects on tropical forests and hydrology in recent decades, these findings are essential for future adaptive water resources and forest management in the tropical forested watersheds.

**Keywords:** forest disturbances; climate variability; extreme weather events; streamflow; low flows

## 1. Introduction

The past decades witnessed numerous studies on the hydrological impact of forest disturbances [1–8]. Most studies focus on anthropogenic disturbances (e.g., logging, road construction, dams, afforestation and reforestation) while the effect of natural disturbances (e.g., extreme weather events, wildfire, and insect infestation) on hydrology has been studied less [9–12]. In recent decades, natural disturbances including extreme weather events (e.g., cyclone, typhoon, hurricane, heat wave and cold wave), drought, flood, insect infestation and wildfire are intensified by climate change [13–18]. The large-scale outbreak of the mountain pine beetle around the year 2003 in the BC (British Columbia) interior of Canada is a good example of global warming-induced widespread insect infestation since warm winters are more favorable for the survival of beetles [19–22]. Therefore, in view of intensified natural disturbances due to climate change and their associated catastrophic effects on water, more studies on quantifying hydrological responses to natural disturbances are necessary for water resources and forest management to mitigate negative effects of climate change on both the ecosystem and human society.

Tropical forests are frequently disturbed by extreme weather including typhoons, hurricanes, droughts, and cold waves. These extreme weather events can not only yield impact on hydrological processes but also on forest growth, structure and species composition [23]. Changes of tropical forests due to extreme weather events can further affect hydrology mainly by altering evapotranspiration and canopy interception [24–27]. For example, short-term extreme weather events such as hurricanes and typhoons can lead to downed, snapped and dead trees and productivity loss in coastal forests, which consequently cause a reduction in evapotranspiration and canopy interception and an increase in streamflow [26,28–31]. However, rapid hydrological recovery may be observed several years later since the disturbed tropical forests can recover quickly due to the rapid regrowth of understory vegetation [5]. Although the effects of extreme weather events on either hydrological or ecological processes have often been studied [32,33], the indirect hydrological responses to extreme weather-induced forest disturbances especially the cold wave have been less examined [27,34]. This is mainly due to the great challenge in separating hydrological changes attributed to extreme weather induced-forest disturbances and climate variability. The traditional paired watershed experiment may fail to work since both control and treated watershed always experience disturbances such as hurricanes, droughts or cold waves simultaneously [27]. The hydrological modelling can also be used to quantify hydrological impact of extreme weather-induced forest disturbances [26,35]. However, the difficulties in collecting long-term detailed data on hydrology, climate, vegetation, soil and disturbance history as well as time-consuming model calibration impede the application of hydrological modelling [36,37]. This calls for the development of more efficient methods to quantify hydrological response to extreme weather-induced forest disturbances.

In this study, LAI (Leaf area index) was used as an integrated indicator of forest disturbance level. LAI as an important biophysical variable relating to photosynthesis, transpiration and energy balance can be a better indicator than disturbed area or forest coverage to express extreme weather-induced forest change [38,39]. Here, the No.1 experimental watershed in the Jianfengling National Forest Park, Hainan Province, China that perennially disturbed by extreme weather such as typhoon and cold wave, was used as an example. The major objectives of this study were: (1) to assess annual and seasonal streamflow responses to extreme weather-induced forest disturbances by adopting an improved single watershed approach combining modified double mass curve (MDMC) and multivariate Autoregressive Integrated Moving Average (ARIMAX), climate variability and other factors; (2) to quantify the effect of extreme weather-induced forest disturbances on low flows and high flows by the paired-year approach. Given the increasing extreme weather with climate change and their possible catastrophic effects on tropical forests and hydrology in recent decades, studies on hydrological response to extreme weather-induced forest disturbances are essential for future adaptive water resources and forest management.

## 2. Materials and Methods

### 2.1. Study Watersheds

This study was conducted in the No.1 experimental watershed located within the Jianfengling National Forest Park (452.67 km$^2$, latitude: 18°40′ N–18°57′ N, longitude: 108°41′ E–109°12′ E) in the Southwest of Hainan Province, China (Figure 1). The drainage area of the study watershed is 3.01 ha, which is fully covered with secondary tropical forests.

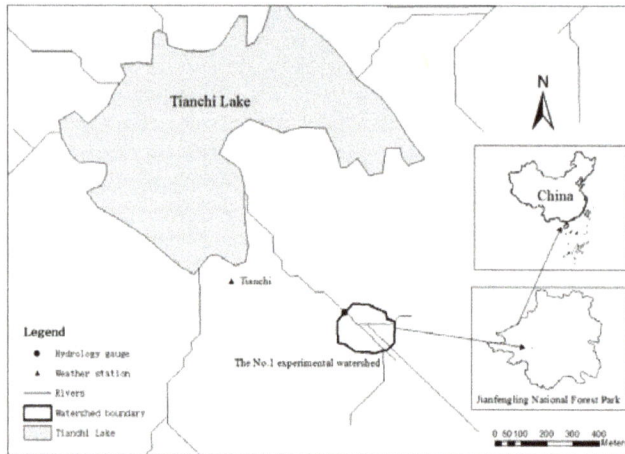

**Figure 1.** Location of the No.1 experimental watershed.

The No.1 experimental watershed lies in the tropical monsoon climate zone with distinct wet and dry seasons [40]. The wet season starts from May to October influenced by frequent cyclones or typhoons with high intensity rainfall [41]. The long-term mean annual precipitation is 2541 mm, of which 87% (2207 mm) falls in the wet season (May to October) and 13% (334 mm) in the dry season (November to April), respectively (Figure 2). The annual mean temperature in this watershed is 19.8 °C, and the maximum mean temperature reaches 23.3 °C (June), while the minimum mean temperature is 14.8 °C (January).

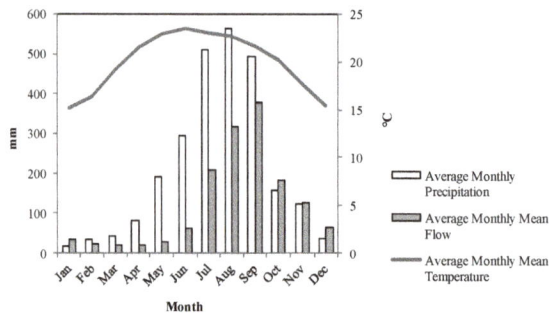

**Figure 2.** The averages of monthly mean precipitation, flow, and temperature from 1990 to 2005 in the No.1 experimental watershed.

The No.1 experimental watershed was originally covered by natural tropical montane rainforests which were gradually replaced by naturally regenerated secondary tropical rainforests after a clear-cut in 1965 [25]. The commercial harvesting in this experimental watershed stopped in 1993 when the Jianfengling National Forest Park was established. The No.1 experimental watershed as a part of the Jianfengling Park experienced human activities including infrastructure or road construction and recreation since then. The dominant vegetation types include *Clerodendrum canescens* Wall., *Litsea glutinosa* (Lour.) C.B.Rob., *Cyclobalanopsis kerrii* (Craib) Hu, *Eurya nitida* Korthals, *Mallotus paniculatus* (Lamb.) Muell. Arg., *Trema orientalis* (L.) Bl., *Microcos paniculata* L., *Sterculia lanceolata* Cav., *Litsea monopetala* (Roxb.) Pers., *Schima superba* Gardn. et Champ. and *Machilus bombycina* King ex Hook.f. [42].

As a tropical coastal watershed, forests in this area are frequently disturbed by typhoons, tropical cyclones and cold waves. Typhoons or tropical cyclones are associated with heavy rain, leading to more than 1000 mm rainfall (one third of annual precipitation) in a few days [43]. During the study period, the most severe typhoon was Lewis occurred in July 1993. It struck the South Hainan Island Coast with its eye passing through the Jianfengling National Forest Park at a wind speed of 41 m/s. The No.1 experimental watershed, located 20 km from the coast, was in close proximity to the path taken by the storm's eye and received severe damage. The tropical forests in the experimental watershed also suffered from a severe cold wave in December 1999, the strongest cold wave in recent 50 years. The extreme low temperature in the No.1 experimental watershed was 6.4 °C in December 1999 (Figure 3), 5.0 °C below the long-term average extreme low temperature of December (11.7 °C).

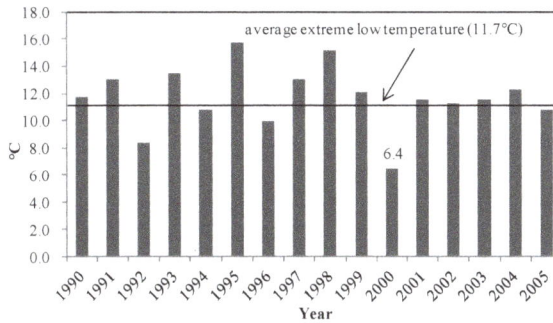

**Figure 3.** Extreme low temperature in December from 1990 to 2005 in the study watershed.

*2.2. Data*

One weir was built in 1989 at the outlet of the No.1 experimental watershed (latitude: 18°44′ N, longitude: 108°51′ E). Hydrological records including flow velocity and water table were continuously measured for a period from 1990 to 2005 in the No.1 experimental watershed [44]. Annual and seasonal (dry and wet season) streamflow, high flows and low flows, and precipitation were calculated based on daily flow records. The hydrological year (November–October) was divided into dry season (November–April) and wet season (May–October).

Climate data were obtained from the Tianchi weather station established near the Tianchi Lake (latitude: 18°43′ N, longitude: 108°52′ E, elevation: 880 m) in 1980, which is 1 km away from the study watershed (latitude: 18°44′ N, longitude: 108°51′ E). Daily temperature data from this station were used in this study. From 1980 to 1988, temperature was measured manually at 2:00 a.m., 8:00 a.m., 14:00 p.m. and 20:00 p.m. every day. The original weather station was then replaced by an automatic one in 1989, and climate data were recorded every 30 min since then. In this study, monthly temperature data from 1990 to 2005 were used.

Forest data used in this study mainly include LAI (Leaf area index, defined as one half of the total green leaf area per unit of horizontal ground surface area) data from the Global Land Surface Satellite Products (GLASS: http://glass-product.bnu.edu.cn/) between 1990 and 2005 [45]. The GLASS LAI product generated from Advanced Very High Resolution Radiometer (AVHRR) reflectance data is available with a temporal resolution of eight days and a spatial resolution of 0.05° from 1982 to 2015 [46–49]. By use of the GLASS LAI product, we generated two data series of LAI: dry season LAI (mean value of the 1st to 177th day in a year) and wet season LAI (mean value of the 185th to 361th day in a year) from 1990 to 2005 for the No.1 experimental watershed.

*2.3. Methods*

2.3.1. Quantification of Forest Disturbances

Forest disturbances such as logging can be simply described by logged area or forest coverage change since trees are normally removed out of forests. However, such an indicator is inappropriate for extreme weather-induced forest disturbances given that downed, snapped and dead trees remaining in the disturbed forests as well as a large number of trees with loss of branches and leaves. In this study, we selected LAI as an indicator for forest disturbances. LAI is considered to be an important biophysical variable influencing vegetation photosynthesis, transpiration, and land surface energy balance, and thus a good indicator of canopy structure and biomass to reflect vegetation change [38,39,50]. Figure 4 shows annual and seasonal LAI from 1990 to 2005 in the No.1 experimental watershed. The annual LAI varied between 3.965 m$^2$/m$^2$ (2000) and 4.928 m$^2$/m$^2$ (2003), with an average of 4.733 m$^2$/m$^2$.

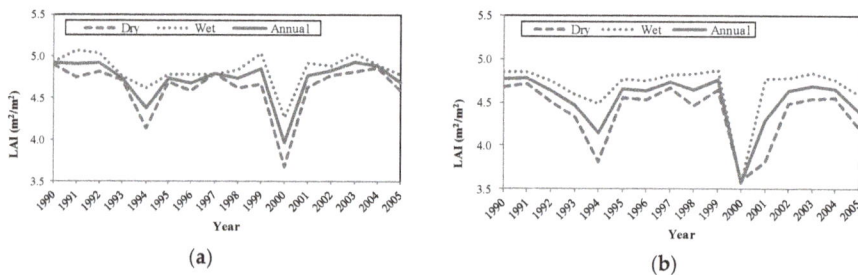

**Figure 4.** Annual and seasonal LAI in the (**a**) No.1 experimental watershed and (**b**) Jianfengling National Forest Park from 1990 to 2005.

2.3.2. Trend Analysis

Trend analysis was used in this study to detect whether the climate, hydrological and forest disturbance data have statistical significant upward or downward trends on multiple temporal scales (annual, dry season, wet season) [51,52]. Non-parametric tests, Kendall tau and Spearman rho tests were widely used for trend detection in climate and hydrology due to their fewer assumptions and the ability to eliminate the influences of outliers. For example, Kendall tau and Spearman rho tests are available for both normal and non-normal distribution data series [52].

2.3.3. Quantifying the Effects of Climate Variability, Forest Disturbances and Other Factors on Streamflow

The No.1 experimental watershed was perennially affected by extreme weather events and other factors such as human activities. An improved single watershed approach combining the modified double mass curve (MDMC) and time series ARIMAX model was applied to quantify the annual/seasonal streamflow responses to climate variability, forest disturbances and other factors [53].

Wei and Zhang (2010) developed the MDMC to exclude the influence of climate variability on streamflow, which has been successfully applied in many watersheds in Canada and China [22,54–57]. The MDMC was designed for a single watershed study with fewer data requirements in comparison with the traditional DMC used in the paired watershed studies. In a MDMC, accumulated annual/seasonal effective precipitation was plotted versus accumulated annual/seasonal streamflow, where annual/seasonal effective precipitation ($P_e$) was the difference between annual/seasonal precipitation ($P$) and annual/seasonal evapotranspiration ($E$). Since the effective precipitation indicates available water for streamflow generation, a consistent relationship between streamflow and effective precipitation can be observed in a watershed during a period with limited disturbances. Thus the MDMC is normally a straight line if the effect of non-climate factors (e.g., forest disturbances induced by extreme weather events and other factors) on streamflow is insignificant. In other words, streamflow variation is only determined by climate variability during undisturbed or less disturbed period. Once non-climatic factors such as forest disturbances and human activities produce noticeable impact on streamflow, a breakpoint in the modified double mass curve can be found [22]. Statistical tests, including ARIMA (Autoregressive Integrate Moving Average) Intervention and non-parametric tests (Wilcoxon test and Sign test) were applied to confirm the statistical significance of the breakpoint [58,59]. The period before the breakpoint is defined as the reference period (a period without significant hydrological alteration) when streamflow variation has a consistent linear relation with climate variability. And the period after the breakpoint is named as the disturbed period, a period with significant hydrological alteration. Then, the predicted accumulated seasonal streamflow were estimated by a linear regression model established by observed accumulated seasonal streamflow and accumulated seasonal effective precipitation during the reference period. In this way, the difference between the observed line and predicted line after the breakpoint (disturbed period) can be attributed to accumulated streamflow variation attributed to non-climate factors ($\Delta Q_{anc}$), and the effect of climate variability on accumulated streamflow ($\Delta Q_{ac}$) can be estimated accordingly (Equations (1) and (2)).

$$\Delta Q_{anc} = \Delta Q_a - \Delta Q_{a0} \tag{1}$$

$$\Delta Q_{an} = \Delta Q_a - \Delta Q_{anc} \tag{2}$$

where $Q_a$ and $Q_{a0}$ are observed accumulated seasonal streamflow, and predicted accumulated seasonal streamflow by the linear regression model after breakpoint, respectively; $\Delta Q_{anc}$ stands for accumulated seasonal streamflow variation attributed to non-climate factors; $\Delta Q_{ac}$ and $\Delta Q_a$ represent accumulated seasonal streamflow variation attributed to climate variability and streamflow variation, respectively.

We then applied the Multivariate ARIMA (ARIMAX) model to quantify streamflow responses to extreme weather-induced forest disturbances by establishing a quantitative relationship between accumulated seasonal streamflow variation attributed to non-climatic factors ($\Delta Q_{anc}$) and seasonal $\Delta LAI_a$ (accumulated seasonal LAI variation) during the disturbed period. The ARIMAX model is a typical ARIMA model with one or multiple external variables to improve the accuracy of simulation [60], which was widely used in analyzing auto-correlated data series [61]. This method was successfully applied to identify streamflow variation attributed to vegetation change and other factors from non-climate factor in a comparative study in China [53]. An ARIMAX model fitting $\Delta Q_{anc}$ with accumulated seasonal LAI variation ($\Delta LAI_a$) as regressor during the disturbed period was established using SAS version 9.4 (SAS Institute, Inc., Cary, NC, USA). If all parameters in the established ARIMAX model were significant, we can obtain the predicted accumulated streamflow variation attributed to seasonal non-climatic factors ($\Delta Q_{anc0}$) from the selected ARIMAX model. The differences between $\Delta Q_{anc}$ and $\Delta Q_{anc0}$ ($\Delta Q_{ad}$) can be viewed as accumulated seasonal streamflow variation attributed to other factors and statistical errors (Equation (3)). Here, the 95% confidence interval (CI) of $\Delta Q_d$ (seasonal differences of observed and predicted values in the selected ARIMAX model) was used to represent the margins of statistical errors ($\Delta Q_{se}$) (Equation (4)). If data points are located within 95% CI, $\Delta Q_d$ only indicates statistical errors and other factors yield an insignificant effect on seasonal streamflow. However, for those data points distributed beyond 95% CI, other factors produced significant impact on seasonal streamflow. In this way, seasonal streamflow variation attributed to other factors ($\Delta Q_o$) can be estimated and the response due to extreme weather-induced forest disturbances ($\Delta Q_f$) can be computed accordingly (Equation (5)).

$$\Delta Q_{ad} = \Delta Q_{anc} - \Delta Q_{anc0} \tag{3}$$

$$\Delta Q_o = \Delta Q_d - \Delta Q_{se} \tag{4}$$

$$\Delta Q_f = \Delta Q_{nc} - \Delta Q_o \tag{5}$$

where $\Delta Q_{anc}$, and $\Delta Q_{anc0}$ stand for observed and predicted accumulated seasonal streamflow variation attributed to non-climatic factors, $\Delta Q_{ad}$ is accumulated seasonal streamflow variation from others. $\Delta Q_d$, $\Delta Q_f$ and $\Delta Q_o$ represent seasonal streamflow variations attributed to others, forest disturbances and other factors; $\Delta Q_{se}$ is statistical errors from the ARIMAX model.

### 2.3.4. Quantifying the Effect of Forest Disturbances on High Flows and Low Flows

Flow duration curve (FDC), a widely used hydrograph, shows the percentage of time that streamflow equals or exceeds a given amount over a time interval, for example, annually or monthly [6]. Flows at a given percentile (denoted as $Q_p$%) can be derived from FDC. In this study, high flows ($Q_h$) refer to flows equal to or beyond $Q_5$%, and low flows ($Q_l$) are defined as flows equal to or below $Q_{95}$%, where $Q_5$% and $Q_{95}$% are flow exceeded at 5% and 95% of the time in a given year. According to definitions above, annual data series of high and low flows were generated.

The paired-year approach was then used to assess the changes of magnitude in high flows and low flows [62]. In the paired-year approach, a reference year (before the breakpoint of MDMC) was paired with a disturbed year (after the breakpoint of MDMC) according to their similarities in annual mean temperature and precipitation, where the effect of climate variability on streamflow can be eliminated [62]. To precisely assess the effect of extreme weather-induced forest disturbances on high flows and low flows, we also consider if extreme weather events happened around selected disturbed years. Based on the criteria above, we identified two pairs in this study (Table 1). Mann–Whitney U test was performed to detect the statistical significance of differences in the medians of high flows/low flows between the reference year and disturbed year for each pair. In this way, the effects of forest disturbances on high flows/low flows were eventually quantified.

**Table 1.** Selected pairs by paired-year approach. LAI: Leaf area index.

| Pair | Year | Type | T (°C) | P (mm) | LAI (m²/m²) | △LAI (%) | Disturbed Type |
|------|------|------|--------|--------|-------------|----------|----------------|
|      | 1992 | Reference | 19.6 | 2581.2 | 4.93 | | |
| #1   | 1995 | Disturbed | 19.9 | 2471.2 | 4.74 | 3.85 | Typhoon |
| #2   | 2000 | Disturbed | 19.8 | 2341.2 | 3.97 | 19.47 | Cold wave |

## 3. Results

### 3.1. Trend Analysis of Hydrological, Climatic and Forest Disturbance Variables

From 1990 to 2005, annual streamflow ranged from 477 mm to 2516 mm, with an average of 1465 mm. Mean annual precipitation reached 2524 mm (1252–3948 mm). Calculated by Thornthwaite method and Zhang's equation (a modification of Budyko's evaporation), annual evapotranspiration is much lower than annual streamflow, varied from 432 mm to 570 mm, with an average value of 503 mm, suggesting the No.1 experimental watershed a high water yield ecosystem. According to the trend analysis (Table 2), a significant downward trend ($\alpha = 0.05$) was detected in wet season evapotranspiration, whilst significant upward tendency in temperature (annual, dry season and wet season) was identified due to global warming [63,64]. In addition, no significant trend in other variables were found.

**Table 2.** Trend analysis of climate, hydrological and forest disturbance variable from 1990 to 2005.

| Variables | Kendall Tau | Spearman Rho |
|-----------|-------------|--------------|
| Annual precipitation | 0.17 | 0.22 |
| Dry season precipitation | −0.10 | −0.19 |
| Wet season precipitation | 0.10 | 0.14 |
| Annual temperature | **0.44 *** | **0.62 *** |
| Dry season temperature | **0.40 *** | **0.59 *** |
| Wet season temperature | **0.34 *** | **0.44 *** |

**Table 2.** *Cont.*

| Variables | Kendall Tau | Spearman Rho |
|---|---|---|
| Annual evapotranspiration | −0.25 | −0.36 |
| Dry season evapotranspiration | −0.50 | −0.09 |
| Wet season evapotranspiration | −0.45 * | −0.56 * |
| Annual streamflow | 0.23 | 0.31 |
| Dry season streamflow | 0.13 | 0.17 |
| Wet season streamflow | 0.07 | 0.09 |
| Annual LAI | −0.05 | −0.12 |
| Dry season LAI | −0.12 | −0.16 |
| wet season LAI | 0.01 | −0.06 |

* Significant at $\alpha = 0.05$.

## 3.2. Effects of Forest Disturbances on Annual and Seasonal Streamflow

### 3.2.1. Annual and Seasonal Streamflow Variations Attributed to Non-Climatic Factors

A breakpoint occurred in 1995 was found in modified double mass curve (Figure 5). The ARIMA intervention test of the MDMC slopes and non-parametric tests (Wilcoxon test and Sign test) both confirmed statistical significance of the breakpoint ($\alpha = 0.05$) (Tables 3 and 4). As estimated, accumulated seasonal streamflow variation attributed to non-climatic factors were from 145.7 mm to 3270.5 mm while accumulated seasonal streamflow variation attributed to climate variability varied from −210.3 mm to −2468.7 mm during 1995–2005.

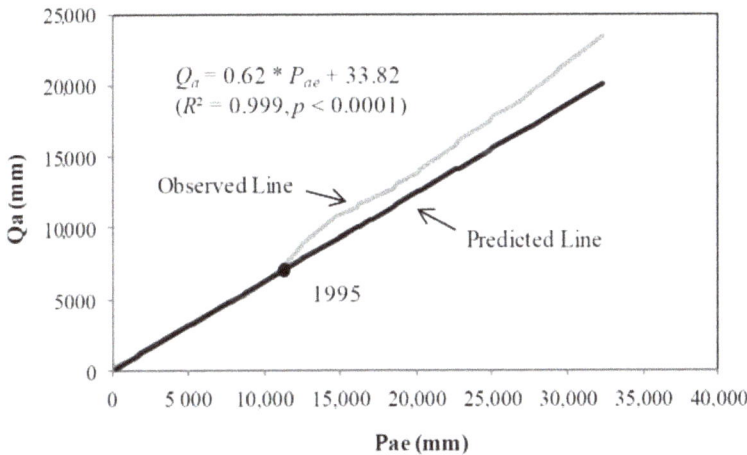

**Figure 5.** Modified double mass curve of accumulated seasonal streamflow ($Q_a$) and accumulated seasonal effective precipitation ($P_{ae}$).

**Table 3.** ARIMA Intervention for slope of MDMC (modified double mass curve).

| AR Part | Int Part | MA Part | Intervention Part | | | Model Structure | MS |
|---|---|---|---|---|---|---|---|
| | | | Change Type | CP (1995) | | | |
| $p(1)$ | $d(1)$ | $q(1)$ | | $\Omega(1)$ | $\triangle(1)$ | | |
| 0 | 1 | 0.78 ($p = 0.000$) | GP | 1.12 ($p = 0.011$) | −1.00 ($p = 0.000$) | Ln(x)(0,1,1) | 0.42 |

Note: AR, Int and MA part refer to autoregressive, integrated and moving average part, respectively; $p$, $d$, and $q$ are parameters for autoregression, differencing, and moving average; $\Omega$ and $\triangle$ are parameters for intervention; CP, GP and MS refer to the change point, gradual permanent change, and model residual, respectively.

**Table 4.** Wilcoxon test and Sign test for predicted and observed accumulated streamflow in reference and disturbed periods.

| Period | Wilcoxon Test | Sign Test |
|---|---|---|
| Reference period (1990–1994) | 0.46 ($p$ = 0.65) | −0.32 ($p$ = 0.75) |
| Disturbed period (1995–2005) | 4.11 * ($p$ = 0.00) | 4.48 * ($p$ = 0.00) |

\* Significant at $\alpha$ = 0.05.

### 3.2.2. Annual and Seasonal Streamflow Variations Attributed to Forest Disturbances

Table 5 showed the model structure and parameters of the best fitted ARIMAX model. The significant differences between observed accumulated seasonal streamflow variation attributed to non-climatic factors and its predicted values are associated with statistical errors and other factors (Figure 6). As shown in Figure 7, 17 data points falling outside the 95% CI (statistical errors) of predicted seasonal streamflow variation to non-climatic factors were identified as seasonal streamflow. This indicated that seasonal flows in these seasons were significantly affected by other factors and forest disturbances. On the contrary, the remaining five data points (1995 dry season and wet season, 1999 dry season, 2004 wet season and 2005 dry season) fell within the 95% CI, suggesting that streamflow variation attributed to other factors in these seasons is minor.

**Table 5.** ARIMAX (Autoregressive Integrated Moving Average) model structure and parameters.

| Model Input | Parameter Estimation | | | |
|---|---|---|---|---|
| | $c$ | $q$(1) | $Q$(1) | LAI (lag(2)) |
| ln $\Delta Q_{anc}$: | 7.208 | −0.652 | −0.601 | −0.273 |
| ARIMA (0,0,1) (0,0,1) + $\Delta LAI_a$ (lag(2)) | ($p$ < 0.0001) | ($p$ = 0.0073) | ($p$ = 0.0347) | ($p$ = 0.0401) |

Note: $\Delta LAI_a$ represents accumulated LAI variation; $c$, $q$ and $Q$ are constant, moving average parameter and seasonal moving average parameter.

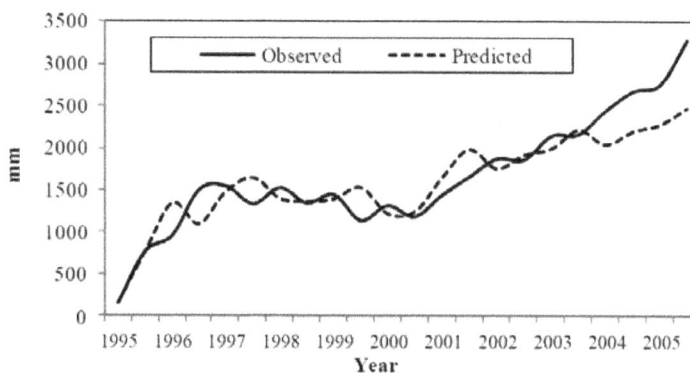

**Figure 6.** Comparison on observed and predicted accumulated seasonal streamflow variation from non-climate factors.

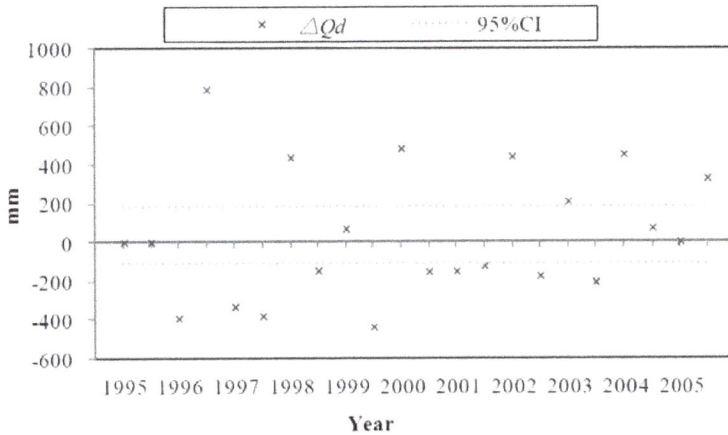

**Figure 7.** The distribution of seasonal streamflow variation attributed to others ($\Delta Q_d$) and its 95% CI.

According to the final analysis, forest disturbances mainly increased annual/dry/wet season streamflow while climate variability decreased annual/dry/wet season streamflow during 1995–2005 (Figure 8). Average annual/dry/wet season streamflow variations attributed to forest disturbances over the disturbed period were 126.7 mm, 127.0 mm, and 126.3 mm, respectively, and the relative contributions of forest disturbances to annual/dry/wet season streamflow were 48.6%, 44.7%, and 50.0%. Average annual/dry/wet season streamflow variations attributed to climate variability from 1995 to 2005 were −112.2 mm, −105.5 mm, and −118.9 mm, respectively, and the relative contributions of climate variability to annual/dry/wet season streamflow were 43.0%, 37.2%, and 47.1%. Other factors were less influential on streamflow in the study watershed. Average annual/dry/wet season streamflow variations attributed to other factors from 1995–2005 were only 22.0 mm, 51.4 mm, and −7.4 mm, respectively and the relative contributions of other factors to annual/dry/wet season streamflow were 8.4%, 18.1%, and 2.9% (Table 6).

From 1995–1999, forest disturbances led to a greater increment of streamflow in dry season while during 2000–2005 wet season streamflow response to forest disturbances was higher. Average dry season streamflow variation attributed to forest disturbances was 184.6 mm and wet season streamflow response was 95.9 mm from 1995–1999 (Table 6). On the contrary, average dry season streamflow variation attributed to forest disturbances was 79.0 mm and wet season streamflow response was up to 151.7 mm from 2000 to 2005. Unlike forest disturbances, climate variability can yield different effects on streamflow in different periods. From 1995–1999, climate variability produced negative effects on streamflow and average annual/dry/wet season streamflow variations attributed to climate variability were −259.8 mm, −159.1 mm, and −360.4 mm, respectively. However, climate variability had positive influence on annual and wet season streamflow during 2000–2005. Average annual and wet season streamflow variations attributed to climate variability were 10.8 mm and 82.4 mm then (Table 6).

(a)

(b)

(c)

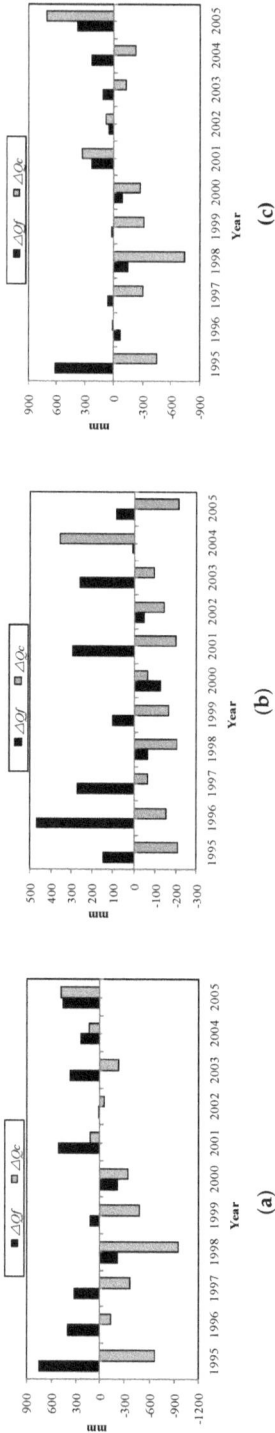

**Figure 8.** Streamflow variation attributed to forest disturbances ($\triangle Q_f$) and climate variability ($\triangle Q_c$) in (a) annual; (b) dry season and (c) wet season.

**Table 6.** Annual and seasonal streamflow variations to climate variability, forest disturbances and other factors in different phases.

| Phase | $\Delta Q$ (mm) | $\Delta Q_c$ (mm) | $\Delta Q_f$ (mm) | $\Delta Q_o$ (mm) | $\Delta Q_c$ (%) | $\Delta Q_f$ (%) | $\Delta Q_o$ (%) | $\Delta Q$ (%) | $R_c$ (%) | $R_f$ (%) | $R_o$ (%) | LAI (m²/m²) | P (mm) | DI | T (°C) |
|---|---|---|---|---|---|---|---|---|---|---|---|---|---|---|---|
| Dry season 1995–1999 | −24.2 ± 19.3 | −159.1 ± 26.8 | 184.6 ± 89.5 | −49.6 ± 95.1 | −54.9 ± 9.2 | 63.6 ± 30.9 | −17.1 ± 32.8 | −8.3 ± 6.7 | 40.5 ± 10.5 | 46.9 ± 7.0 | 12.6 ± 10.2 | 4.7 | 231.6 | 1.00 | 17.8 |
| Dry season 2000–2005 | 153.7 ± 105.0 | −60.9 ± 86.3 | 79.0 ± 68.1 | 135.6 ± 63.5 | −21.0 ± 29.8 | 27.2 ± 23.5 | 46.7 ± 21.9 | 53.0 ± 36.2 | 22.1 ± 8.8 | 28.7 ± 10.5 | 49.2 ± 11.4 | 4.6 | 409.2 | 0.73 | 17.6 |
| Dry season 1995–2005 | 72.8 ± 62.1 | −105.5 ± 49.0 | 127.0 ± 54.8 | 51.4 ± 60.0 | −36.4 ± 16.9 | 43.8 ± 18.9 | 17.7 ± 20.7 | 25.1 ± 21.4 | 37.2 ± 6.4 | 44.7 ± 6.4 | 18.1 ± 7.4 | 4.6 | 328.4 | 0.83 | 17.7 |
| Wet season 1995–1999 | −269.12 ± 272.0 | −360.4 ± 154.0 | 95.9 ± 133.4 | −4.6 ± 165.7 | −30.7 ± 10.4 | 8.2 ± 11.4 | −0.4 ± 14.1 | −22.9 ± 23.1 | 78.2 ± 12.4 | 20.8 ± 9.8 | 1.0 ± 16.3 | 4.8 | 1880.3 | 0.22 | 22.4 |
| Wet season 2000–2005 | 224.3 ± 237.4 | 82.4 ± 122.7 | 151.7 ± 66.7 | −9.7 ± 34.1 | 7.0 ± 13.1 | 12.9 ± 5.7 | −0.8 ± 2.9 | 19.1 ± 20.2 | 33.8 ± 4.5 | 62.2 ± 4.2 | 4.0 ± 5.3 | 4.8 | 2404.9 | 0.16 | 22.3 |
| Wet season 1995–2005 | 0.0 ± 186.7 | −118.9 ± 118.6 | 126.3 ± 67.3 | −7.4 ± 72.9 | −10.1 ± 10.1 | 10.8 ± 5.7 | −0.6 ± 6.2 | 0.0 ± 15.9 | 47.1 ± 5.9 | 50.0 ± 5.3 | 2.9 ± 8.3 | 4.8 | 2166.5 | 0.19 | 22.4 |
| Annual 1995–1999 | −146.7 ± 134.9 | −259.8 ± 68.1 | 140.3 ± 77.1 | −27.1 ± 90.4 | −35.4 ± 9.4 | 19.2 ± 10.9 | −3.8 ± 10.7 | −20.0 ± 19.1 | 60.8 ± 9.1 | 32.8 ± 8.1 | 6.4 ± 7.8 | 4.8 | 2111.9 | 0.29 | 20.1 |
| Annual 2000–2005 | 189.0 ± 124.2 | 110.8 ± 86.9 | 115.3 ± 46.8 | 62.9 ± 40.7 | 1.4 ± 8.2 | 15.8 ± 7.9 | 8.6 ± 4.2 | 25.8 ± 13.8 | 5.7 ± 4.7 | 61.0 ± 0.1 | 33.3 ± 10.6 | 4.7 | 2814.1 | 0.20 | 20.0 |
| Annual 1995–2005 | 36.4 ± 96.3 | −112.2 ± 62.6 | 126.7 ± 42.3 | 22.0 ± 46.5 | −15.4 ± 8.3 | 17.2 ± 6.3 | 3.0 ± 5.4 | 5.0 ± 13.1 | 43.0 ± 5.0 | 48.6 ± 6.5 | 8.4 ± 6.5 | 4.7 | 2494.9 | 0.24 | 20.1 |

Note: $\Delta Q\%$, $\Delta Q_f\%$, $\Delta Q_c\%$ and $\Delta Q_o\%$ are relative annual/seasonal streamflow variation, relative annual/seasonal streamflow variation attributed to forest disturbances, climate variability and other factors, respectively ($\Delta Q\% = \Delta Q/Q \times 100\%$, $\Delta Q_c\% = \Delta Q_c/Q \times 100\%$, $\Delta Q_f\% = \Delta Q_f/Q \times 100\%$, $\Delta Q_o\% = \Delta Q_o/Q \times 100\%$, Q is average annual/seasonal streamflow from 1990 to 2005). $R_f = |\Delta Q_f|/(|\Delta Q_f| + |\Delta Q_c| + |\Delta Q_o|) \times 100\%$; $R_c = 100 \times |\Delta Q_c|/(|\Delta Q_f| + |\Delta Q_c| + |\Delta Q_o|) \times 100\%$; $R_o = |\Delta Q_o|/(|\Delta Q_f| + |\Delta Q_c| + |\Delta Q_o|) \times 100\%$.

## 3.3. Effects of Forest Disturbances on High Flows and Low Flows

Figure 9 shows flow duration curves (FDCs) for the two paired years. As suggested by Box-plot and Mann–Whitney *U* test (Figure 10 and Table 7), for the # 2 pair the median of low flows in 2000 (the disturbed year) was significantly higher than that in 1992 (the reference year) at $\alpha$ = 0.05 while insignificant differences in the median of high flows were detected between them. The median of low flows in the reference year (1992) was 9.8 m$^3$/s, ranging from 0.7 m$^3$/s to 11.9 m$^3$/s, while in the year 2000, the median of low flows reached 13.1 m$^3$/s (from 0.0 m$^3$/s to 15.6 m$^3$/s). The differences in the medians of high flows or low flows between the disturbed year 1995 and the reference year 1992 were statistically insignificant at $\alpha$ = 0.05.

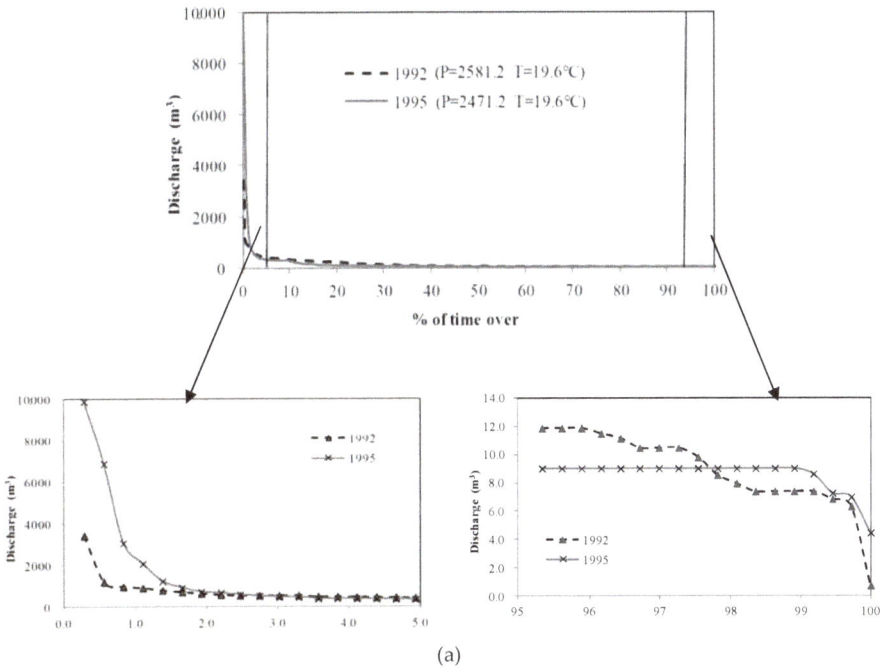

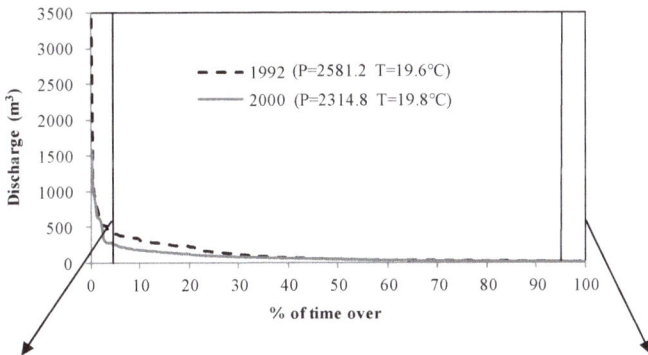

**Figure 9.** *Cont.*

(b)

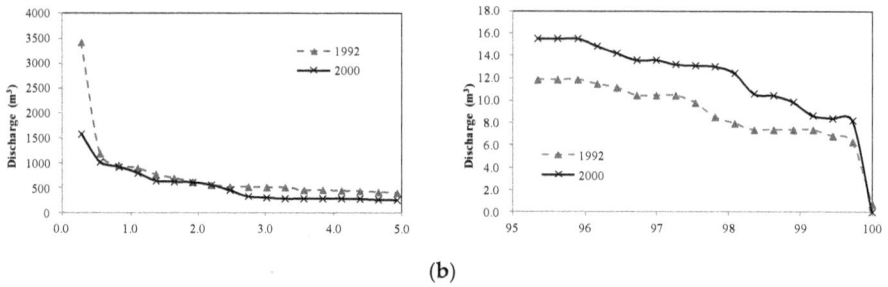

**Figure 9.** Flow duration curves (FDCs) for the selected pairs: (**a**) 1992 vs. 1995; and (**b**) 1992 vs. 2000.

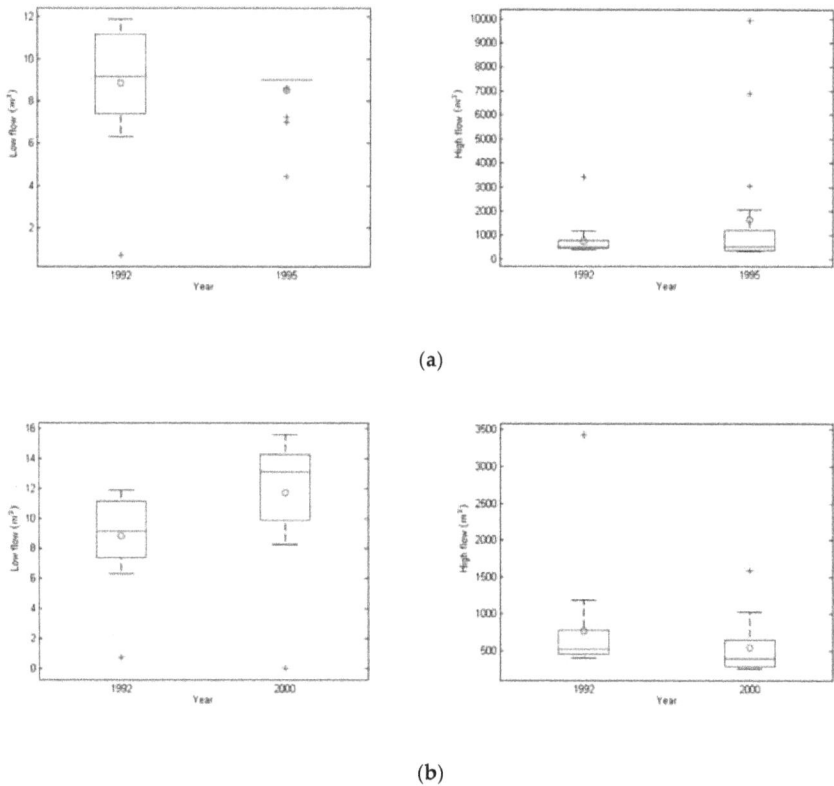

(a)

(b)

**Figure 10.** Comparison on the magnitude of low flows and high flows in selected pairs: (**a**) 1992 vs. 1995; and (**b**) 1992 vs. 2000.

**Table 7.** Statistical tests for the effect of forest disturbances on the low flows and high flows

| Pair | Year | Variables | Mann-Whitney U Test | |
|------|------|-----------|:---:|:---:|
| | | | **Z** | ***p*-Value** |
| #1 | 1992 vs. 1995 | Low flow | 0.53 | 0.65 |
| | | High flow | 0.61 | 0.50 |
| #2 | 1992 vs. 2000 | Low flow | −2.97 | <0.01 * |
| | | High flow | 0.58 | 0.11 |

\* Significant at α = 0.05.

## 4. Discussion

### 4.1. Forest Changes Due to Typhoon and Cold Wave

Sharp reductions of LAI saw in 1994 and 2000 in the No.1 experimental watershed (Figure 4). The sharp decline of LAI in 1994 was associated with the typhoon Lewis which happened in July 1993. After the pass of the storm, mean LAI during the dry season of 1994 (November 1993 to April 1994) was greatly decreased by 12.4% as compared to the mean for dry season of 1993. Storm, typhoon or hurricane associated damage on forests normally include uprooting, trunk breakage, branch snapping and defoliation, resulting in more forest gaps, lower canopy density, and even tree mortality and eventually lower LAI [27,30,31,65,66]. For example, in the 22-year-old community-managed secondary forest at Manobo near Tacloban on Leyte Island in the central Philippines, Zhang et al. (2015) found LAI dropped by 27.5% after Typhoon Haiyan on 8 November 2013 [67]. Similarly, the subtropical forests of Puerto Rico hit by Hurricane Hugo in September 1989 were found with about a quarter of the trees destroyed, and 9% tree mortality [68].

However, the impact of cold wave on forests can be more pronounced than typhoon in the study watershed. According to our analysis, mean LAI during the dry season of 2000 (November 1999 to April 2000) and the wet season of 2000 (May 2000 to October 2000) greatly decreased by 21.3% and 15.5% as compared to their means in 1999. Given the absence of large typhoons from 1997 to 1999, the sharp decline in LAI was due to the cold wave in December 1999, the strongest one in recent 50 years. This cold wave with long-lasting extreme low temperature below the long-term average inhibited the growth of vegetation and led to high mortality of trees, resulting in a sharp drop in LAI. As documented by many studies, the growth of tropical vegetation is very sensitive to winter temperature [69–72]. Extreme low temperature can slow down photosynthesis, transpiration, and translocation of the starch of tropical trees, leading to lower biomass accumulation [73–75]. A similar study by Hilliard and West (1970) also found the growth of *Digitaria decumbens* (Gramineae) (a tropical plant) was severely reduced when the temperature is 10 °C or below. Therefore, according to the above analysis, typhoon disturbances and cold waves yielded significant negative impact on forest growth [76].

### 4.2. Annual/Seasonal Streamflow Response to Forest Disturbances

According to the modified double mass curve and quantification analysis, forest disturbances produced significant positive effects on dry season, wet season, and annual streamflow in the No.1 experimental watershed from 1995 to 2005. Dry season, wet season, and annual streamflow were increased by 43.8% (127.0 mm), 10.8% (126.3 mm), and 17.2% (126.7 mm), respectively as a result of LAI reduction ascribed to forest disturbances including cold wave and typhoon. This is in accordance with some findings from tropical watershed studies that deforestation (e.g., harvesting, urbanization, and wildfire) can increase streamflow. For example, Costa et al. (2003) denoted that a 19% forest loss produced a significant increase in annual streamflow in Tocaintins River watershed in Southeast Asia [77]. On one hand, forest disturbance induced by typhoon, cold wave, and logging can lead to reduced canopy interception of rainfall and less transpiration, and consequently more water available for streamflow generation. On the other hand, these forest disturbances can lead to a decline in growth rate of tree due to less active photosynthesis, lower transpiration rate and less evaporation, and eventually with less water consumption and more streamflow [65,78].

It is well known that climatic conditions are crucial for hydrological processes in forest watersheds [59,79]. Our analysis showed that during the dry period (1995–1999), annual streamflow is more sensitive to forest disturbances than during the wet period (2000–2005). The average increment in annual streamflow was 140.3 mm (19.2%) from 1995–1999 (P = 2111.9 mm) while the increment was only 115.3 mm (15.8%) from 2000–2005 (P = 2814.1 mm) in this watershed. These results are similar to a global review, which indicates that the sensitivity of annual streamflow to forest change is closely related to dryness index [80]. In drier areas or drier years, forest change can produce a noticeable effect on streamflow. In dry season, water availability is the limiting factor for vegetation growth, while energy input in terms of temperature and radiation become the dominant factor given that saturated soils are prevalent in wet season [80]. Consequently, forest change tends to generate more pronounced hydrological impacts in dry season or drier years.

*4.3. The Effect of Forest Disturbances on High Flow and Low Flow*

The impact of typhoon-induced forest change on hydrological extremes tended to be less than that of cold wave-induced forest change. The differences in both high flows and low flows between 1995 (the year after the typhoon Liews) and 1992 (the reference year) were insignificant. Similarly, the response of high flows to cold wave-induced forest change in 2000 was insignificant. High flows in the study watershed often occur in the typhoon season, which are mainly caused by typhoon related heavy rain or storms. The effect of tropical forests in the generation of high flows is believed to be less than climate such as storms or typhoon associated heavy rain [81]. However, the magnitude of low flows was significantly increased by cold wave-induced forest disturbances. As mentioned before, the cold wave with long-lasting extreme low temperature can lead to a slow-down of photosynthesis and transpiration of tropical trees and losses of leaves, resulting in less water consumption and more water available for streamflow generation especially during the low flow season. In addition, this study watershed is classified as an energy-limited watershed (Budyko dryness index ($DI$) < 0.76), where forest growth tends to be less dependent on water availability but responds more strongly to temperature [59,79,80,82,83]. Therefore, lower temperature can lead to more pronounced hydrological responses in the study watershed.

*4.4. Implications for Watershed Management*

In forested watersheds, forest disturbances and climate variability are two major drivers for hydrological changes. Understanding their interactive, dynamic effects is important for sustainable water supply and flood control. Forest disturbances and climate variability can influence streamflow in the same or opposite (offsetting) directions with different strengths.

According to the analysis in the No.1 experimental watershed, forest disturbances and climate variability produced opposite impacts on dry season, wet season and annual streamflow during 1995 and 1999. During that dry period (1995–1999), forest disturbances increased dry/wet season and annual streamflow by 184.6 mm, 95.9 mm and 140.3 mm, respectively, while climate variability decreased dry/wet season and annual flow by 159.1 mm, 360.4 mm and 259.8 mm, respectively. The counteracting or cancelling effects of forest disturbances and climate variability eventually reduced annual variations of streamflow over the period of 1995 to 1999 in the study watershed, which can benefit water resources management by providing a stable water supply especially in dry years [55,57,84]. However, from 2000 to 2005 both forest disturbances and climate variability produced positive effects on streamflow. During the wet period (2000–2005), forest disturbances averagely yielded 151.7 mm and 115.3 mm increments in wet season and annual streamflow, respectively, and climate variability also contributed to 82.4 mm and 10.8 mm increments in the wet season and annual streamflow, respectively. The additive effects of forest disturbances and climate variability on wet season streamflow can lead to potential risks of larger floods, and greater challenges for flood control and eventually threatening downstream public safety [22]. As forests and climate continue to change in the future, their combined effects (offsetting or additive) on streamflow will have significant implications for watershed management and public safety.

In addition to the impact directions of climate variability and forest disturbances, their strength or relative contributions to streamflow variations are also important and meaningful. Over the whole disturbed period, forest disturbances accounted for a 48.6% change in annual streamflow while the relative contribution of climate variability is 43.0%, further suggesting that both forest disturbances and climate variability are important drivers for streamflow variations and forest disturbances are relatively more influential [55,85]. Contrarily, there are studies that found climate variability produced greater impact on streamflow than forest disturbances. In the Heihe River watershed of China, climate variability accounted for up to 95.8% of changes in streamflow while forest disturbance only explained 9.6% [86]. Another study in the central interior of British Columbia, Canada also found relative contributions from climate variability on annual streamflow (55%) was greater than forest harvesting (45%) in the Willow River Watershed [54]. These differences indicate the impact strength of different forests on water cycle. Tropical forests with higher evapotransipration may have greater impact on water cycle than temperate forests.

The relative contributions of forest disturbances and climate variability to streamflow variations may change over time especially in the context of global warming. During 1995–1999, climate variability was more influential on streamflow than forest disturbances. The relative contributions of climate variability on annual streamflow variation reaches 60.8% while the contribution from forest disturbances after the typhoon Lewis (1994) was only 32.8%. However, after the strong cold wave in December 1999, forest disturbances became a dominating factor on streamflow variation during 2000–2005. The relative contributions of forest disturbances on annual streamflow variation was 61.0%, while the contributions from climate variability were only 5.7%.

Forests play an important role in water balance in tropical regions to keep sustainable water supply. As is known, tropical forest ecosystem is very complex. Once the "sponge effect" of tropical forests is damaged, it will take a long time (more than 30 years) to reach a full hydrological recovery [5]. Moreover, Hainan Island is surrounded by ocean, frequently suffered from extreme weather events due to the monsoon from the Indian Ocean and Pacific Ocean. At the same time, tropical coastal ecosystems can experience frequent extreme weather events caused by El Niño southern oscillation [87]. As suggested by this study, forest disturbances especially cold wave-induced forest changes also generated greater influence on low flows, reducing the risks of drought. In addition, the offsetting effects of forest disturbances and climate variability on dry season streamflow can benefit water resources management especially in dry years, while the additive effects of forest disturbances and climate variability on wet season streamflow will increase the risk of floods. Obviously, the frequency of extreme weather events are increasing with climate change, which are expected to yield more significant impact on forest ecosystem and hydrological cycle even in the coastal headwaters such as the No.1 experimental watershed.

### 4.5. Uncertainties Assciated with LAI Data

The LAI data we used is GLASS product with a spatial resolution of 0.05°, but the drainage area of the study watershed is only 3.01 ha. The mismatched spatial resolution may lead to a concern about the representation of forest conditions by the LAI data used in the study watershed. In fact, LAI was used to indicate the average status of vegetation at a watershed scale rather than showing the spatial patterns or variations of disturbed patches in the forests. Since the study watershed and Jianfengling National Forest Park have similarities in topography, climate and forest conditions including age, species and structures, the changes in the average status of vegetation can be similar after typhoon or cold wave. In order to validate the LAI data in the study watershed, we also collected the LAI data of the Jianfengling National Forest Park (452.67 km$^2$) and compared the changes in the LAI of study watershed and the Jianfengling National Forest Park from 1990–2005 (Figure 4b). We found the change patterns in the dry, wet and annual LAI of the study watershed were in accordance with those in the Jianfengling National Forest Park. The linear correlation and Kendall tau correlation analysis confirmed that dry, wet and annual LAI of the study watershed were significantly correlated with those in the Jianfengling National Forest Park (Table 8). In addition, we actually used $\Delta LAI_a$ that indicated temporal variations in LAI as the regressor for ARIMAX modelling. Thus, we believe the bias associated with the use of GLASS product in the study watershed can be minor. However, ground measurements of LAI in the study watershed are suggested to perform in the future for a better validation of remote-sensed vegetation data.

**Table 8.** Statistical tests of LAI in the study watershed and Jianfengling National Forest Park.

|  | $R^2$ | Kendall Tau |
|---|---|---|
| Dry season | 0.70 ** | 0.53 ** |
| Wet season | 0.80 ** | 0.62 ** |
| Annual | 0.86 ** | 0.52 ** |

** Significant at $\alpha = 0.01$.

## 5. Conclusions

Forest disturbances were more influential on streamflow variation than climate variability in the No.1 experimental watershed from 1995 to 2005. Forest disturbances generally produced significant positive effects on dry season, wet season, and annual streamflow over the study period, while climate variability yielded negative or positive effects on streamflow in different periods. In addition, forest disturbances especially cold wave-induced forest change also generated greater influence on low flows, reducing the risks of drought. In addition, the offsetting effects of forest disturbances and climate variability on streamflow, in particular, on dry season streamflow can benefit water resources management, while the additive effects of forest disturbances and climate variability on wet season streamflow will lead to potential risks of larger floods and greater challenges for flood control. These findings are of great importance for water resource and forest management in the tropical forested watersheds in the context of global warming.

**Author Contributions:** Data interpretation, T.Y.; Data analysis, Y.H.; Manuscript drafting, Y.H. and M.Z.; Research design, M.Z.; Manuscript revision, S.L., P.S., Y.L., L.Y., Q.L. and X.W.

**Funding:** This research was supported by National Program on Key Research and Development Project of China (No. 2017YFC0505006), Open Funds of Key Laboratory for Groundwater and Ecology in Arid and Semi-Arid Areas, CGS (No.KLGEAS201801) and China National Science Foundation (No.31770759).

**Conflicts of Interest:** The authors declare no conflict of interest.

## References

1. Bosch, J.M.; Hewlett, J.D. A review of catchment experiments to determine the effect of vegetation changes on water yield and evapotranspiration. *J. Hydrol.* **1982**, *55*, 3–23. [CrossRef]
2. Sahin, V.; Hall, M.J. The effects of afforestation and deforestation on water yields. *J. Hydrol.* **1996**, *178*, 293–309. [CrossRef]
3. Stednick, J.D. Monitoring the effects of timber harvest on annual water yield. *J. Hydrol.* **1996**, *176*, 79–95. [CrossRef]
4. Andréassian, V. Waters and forests: From historical controversy to scientific debate. *J. Hydrol.* **2004**, *291*, 1–27. [CrossRef]
5. Bruijnzeel, L.A. Hydrological functions of tropical forests: Not seeing the soil for the trees? *Agric. Ecosyst. Environ.* **2004**, *104*, 185–228. [CrossRef]
6. Brown, A.E.; Zhang, L.; Mcmahon, T.A.; Western, A.W.; Vertessy, R.A. A review of paired catchment studies for determining changes in water yield resulting from alterations in vegetation. *J. Hydrol.* **2005**, *310*, 28–61. [CrossRef]
7. Moore, R.D.; Wondzell, S.M. Physical hydrology and the effects of forest harvesting in the Pacific Northwest: A review. *JAWRA J. Am. Water Resour. Assoc.* **2010**, *41*, 763–784. [CrossRef]
8. Van Dijk, A.I.J.M.; A-Arancibia, J.L.P.; Bruijnzeel, L.A. Land cover and water yield: Inference problems when comparing catchments with mixed land cover. *Hydrol. Earth Syst. Sci.* **2012**, *16*, 3461–3473. [CrossRef]
9. Floren, A.; Linsenmair, K.E. The influence of anthropogenic disturbances on the structure of arboreal arthropod communities. *Plant Ecol.* **2001**, *153*, 153–167. [CrossRef]
10. Hong, N.M.; Chu, H.J.; Lin, Y.P.; Deng, D.P. Effects of land cover changes induced by large physical disturbances on hydrological responses in Central Taiwan. *Environ. Monit. Assess.* **2010**, *166*, 503–520. [CrossRef] [PubMed]

11. Liu, M.; Tian, H.; Lu, C.; Xu, X.; Chen, G.; Ren, W. Effects of multiple environment stresses on evapotranspiration and runoff over Eastern China. *J. Hydrol.* **2012**, *426*, 39–54. [CrossRef]

12. Hallema, D.W.; Sun, G.; Caldwell, P.V.; Norman, S.P.; Cohen, E.C.; Liu, Y.; Bladon, K.D.; Mcnulty, S.G. Burned forests impact water supplies. *Nat. Commun.* **2018**, *9*, 1307. [CrossRef] [PubMed]

13. Schindler, D.W. The cumulative effects of climate warming and other human stresses on Canadian freshwaters in the new millennium. *Can. J. Fish. Aquat. Sci.* **2001**, *58*, 18–29. [CrossRef]

14. Stinson, G. Mountain pine beetle and forest carbon feedback to climate change. *Nature* **2008**, *452*, 987–990.

15. Anderegg, W.R.L.; Kane, J.M.; Anderegg, L.D.L. Consequences of widespread tree mortality triggered by drought and temperature stress. *Nat. Clim. Chang.* **2013**, *3*, 30–36. [CrossRef]

16. Pretzsch, H.; Biber, P.; Schütze, G.; Uhl, E.; Rötzer, T. Forest stand growth dynamics in Central Europe have accelerated since 1870. *Nat. Commun.* **2011**, *5*, 4967. [CrossRef] [PubMed]

17. Jolly, W.M.; Cochrane, M.A.; Freeborn, P.H.; Holden, Z.A.; Brown, T.J.; Williamson, G.J.; Bowman, D.M.J.S. Climate-induced variations in global wildfire danger from 1979 to 2013. *Nat. Commun.* **2015**, *6*, 7537. [CrossRef] [PubMed]

18. Bebi, P.; Seidl, R.; Motta, R.; Fuhr, M.; Firm, D.; Krumm, F.; Conedera, M.; Ginzler, C.; Wohlgemuth, T.; Kulakowski, D. Changes of forest cover and disturbance regimes in the mountain forests of the alps. *For. Ecol. Manag.* **2016**, *388*, 43–56. [CrossRef] [PubMed]

19. Woods, A. Is the health of British Columbia's forests being influenced by climate change? If so, was this predictable? *Can. J. Plant Pathol.* **2011**, *33*, 117–126. [CrossRef]

20. Stahl, K.; Moore, R.; Mckendry, I. Climatology of winter cold spells in relation to mountain pine beetle mortality in British Columbia, Canada. *Clim. Res.* **2006**, *32*, 13–23. [CrossRef]

21. Logan, J.A.; Powell, J.A. Ghost forest, global warming, and the mountain pine beetle (Coleoptera: Scolytidae). *Am. Entomol.* **2001**, *47*, 160–172. [CrossRef]

22. Zhang, M.; Wei, X. The effects of cumulative forest disturbance on streamflow in a large watershed in the central interior of British Bolumbia, Canada. *Hydrol. Earth Syst. Sci.* **2012**, *16*, 2021–2034. [CrossRef]

23. Lugo, A.E. Visible and invisible effects of hurricanes on forest ecosystems: An international review. *Austral Ecol.* **2010**, *33*, 368–398. [CrossRef]

24. Rodriguez-Iturbe, I. Ecohydrology: A hydrologic perspective of climate-soil-vegetation dynamies. *Water Resour. Res.* **2000**, *36*, 3–9. [CrossRef]

25. Schlyter, P.; Stjernquist, I.; Bärring, L.; Jönsson, A.; Nilsson, C. Assessment of the impacts of climate change and weather extremes on boreal forests in Northern Europe, focusing on Norway spruce. *Clim. Res.* **2006**, *31*, 75–84. [CrossRef]

26. Bernard, B.; Vincent, K.; Frank, M.; Anthony, E. Comparison of extreme weather events and streamflow from drought indices and a hydrological model in river Malaba, Eastern Uganda. *Int. J. Environ. Stud.* **2013**, *70*, 940–951. [CrossRef]

27. Jayakaran, A.D.; Williams, T.M.; Ssegane, H.; Amatya, D.M.; Song, B.; Trettin, C.C. Hurricane impacts on a pair of coastal forested watersheds: Implications of selective hurricane damage to forest structure and streamflow dynamics. *Hydrol. Earth Syst. Sci. Discuss.* **2013**, *10*, 11519–11557. [CrossRef]

28. Clarke, P.J.; Knox, K.J.E.; Bradstock, R.A.; Munoz-Robles, C.; Kumar, L. Vegetation, terrain and fire history shape the impact of extreme weather on fire severity and ecosystem response. *J. Veg. Sci.* **2014**, *25*, 1033–1044. [CrossRef]

29. Harder, P.; Pomeroy, J.W.; Westbrook, C.J. Hydrological resilience of a Canadian rockies headwaters basin subject to changing climate, extreme weather, and forest management. *Hydrol. Process.* **2015**, *29*, 3905–3924. [CrossRef]

30. Hogan, J.A.; Zimmerman, J.K.; Thompson, J.; Nytch, C.J.; Uriarte, M. The interaction of land-use legacies and hurricane disturbance in subtropical wet forest: Twenty-one years of change. *Ecosphere* **2016**, *7*. [CrossRef]

31. Yang, H.; Liu, S.; Cao, K.; Wang, J.; Li, Y.; Xu, H. Characteristics of typhoon disturbed gaps in an old-growth tropical montane rainforest in Hainan Island, China. *J. For. Res.* **2017**, *28*, 1231–1239. [CrossRef]

32. Stanturf, J.A.; Goodrick, S.L.; Outcalt, K.W. Disturbance and coastal forests: A strategic approach to forest management in hurricane impact zones. *For. Ecol. Manag.* **2007**, *250*, 119–135. [CrossRef]

33. Kupfer, J.A.; Myers, A.T.; Mclane, S.E.; Melton, G.N. Patterns of forest damage in a Southern Mississippi landscape caused by hurricane Katrina. *Ecosystems* **2008**, *11*, 45–60. [CrossRef]

34. Amatya, D.M.; Harrison, C.A.; Trettin, C.C. Water quality of two first order forested watersheds in coastal South Carolina. In Proceedings of the Watershed Management to Meet Water Quality Standards and TMDLS (Total Maximum Daily Load), San Antonio, TX, USA, 10–14 March 2007.

35. Rojas, R.; Feyen, L.; Dosio, A.; Bavera, D. Improving pan-european hydrological simulation of extreme events through statistical bias correction of rcm-driven climate simulations. *Hydrol. Earth Syst. Sci.* **2011**, *15*, 2599–2620. [CrossRef]

36. Stednick, J.D. *Long-Term Streamflow Changes Following Timber Harvesting*; Springer: New York, NY, USA, 2008; pp. 139–155.

37. Kirchner, J.W. Catchments as simple dynamical systems: Catchment characterization, rainfall-runoff modeling, and doing hydrology backward. *Water Resour. Res.* **2009**, *45*, 335–345. [CrossRef]

38. Chen, J.M.; Cihlar, J. Retrieving leaf area index of boreal conifer forests using landsat tm images. *Remote Sens. Environ.* **1996**, *55*, 153–162. [CrossRef]

39. Tian, Y.; Woodcock, C.E.; Wang, Y.; Privette, J.L.; Shabanov, N.V.; Zhou, L.; Zhang, Y.; Buermann, W.; Dong, J.; Veikkanen, B. Multiscale analysis and validation of the MODIS lai product: Ii. Sampling strategy. *Remote Sens. Environ.* **2002**, *83*, 431–441. [CrossRef]

40. Fu, G.-A.; Hong, X.J. Flora of vascular plants of Jianfengling, Hainan Island. *Guihaia* **2008**, *2*, 226–229.

41. Qian, W.C. Effects of deforestation on flood characteristics with particular reference to Hainan Island, China. *Int. Assoc. Sci. Hydrol. Publ.* **1983**, *140*, 249–257.

42. Xu, H.; Li, Y.; Luo, T.S.; Lin, M.X.; Chen, D.X.; Mo, J.H.; Luo, W.; Hong, X.J.; Jiang, Z.L. Community structure characteristics of tropical montane rain forests with different regeneration types in Jianfengling. *Sci. Silvae Sin.* **2009**, *45*, 14–20.

43. Zhou, G.; Chen, B.; Zeng, Q.; Wu, Z.; Li, Y.; Lin, M. Hydrological effects of typhoon and severe tropical storm on the regenerative tropical mountain forest at Jianfengling. *Acta Ecol. Sin.* **1996**, *16*, 555–558.

44. Li, Y. Biodiversity of tropical forest and its protection strategies in Hainan Island, China. *Rorest Res.* **1995**, *8*, 455–461.

45. Zhao, X.; Liang, S.; Liu, S.; Yuan, W.; Xiao, Z.; Liu, Q.; Cheng, J.; Zhang, X.; Tang, H.; Zhang, X. The Global Land Surface Satellite (glass) remote sensing data processing system and products. *Remote Sens.* **2013**, *5*, 2436–2450. [CrossRef]

46. Xiao, Z.; Liang, S.; Wang, J.; Xiang, Y.; Zhao, X.; Song, J. Long-time-series Global Land Surface Satellite leaf area index product derived from MODIS and AVHRR surface reflectance. *IEEE Trans. Geosci. Remote Sens.* **2016**, *54*, 5301–5318. [CrossRef]

47. Xiao, Z.; Liang, S.; Wang, J.; Chen, P.; Yin, X.; Zhang, L.; Song, J. Use of general regression neural networks for generating the glass leaf area index product from time-series MODIS surface reflectance. *IEEE Trans. Geosci. Remote Sens.* **2013**, *52*, 209–223. [CrossRef]

48. Vermote, E.F.; Vermeulen, A. Atmospheric Correction Algorithm: Spectral Reflectances (mod09). Available online: http://dratmos.geog.umd.edu/files/pdf/atbd_mod09.pdf (accessed on 23 October 2018).

49. Pedelty, J.; Devadiga, S.; Masuoka, E.; Brown, M.; Pinzon, J.; Tucker, C.; Vermote, E.; Prince, S.; Nagol, J.; Justice, C. Generating a long-term land data record from the AVHRR and MODIS instruments. In Proceedings of the IEEE International Geoscience and Remote Sensing Symposium (IGARSS 2007), Barcelona, Spain, 23–27 July 2007; pp. 1021–1025.

50. Morisette, J.T.; Baret, F.; Privette, J.L.; Myneni, R.B.; Nickeson, J.E.; Garrigues, S.; Shabanov, N.V.; Weiss, M.; Fernandes, R.A.; Leblanc, S.G. Validation of global moderate-resolution LAI products: A framework proposed within the CEOS land product validation subgroup. *IEEE Trans. Geosci. Remote Sens.* **2006**, *44*, 1804–1817. [CrossRef]

51. Dinpashoh, Y.; Jhajharia, D.; Fakheri-Fard, A.; Singh, V.P.; Kahya, E. Trends in reference evapotranspiration over Iran. *J. Hydrol.* **2011**, *399*, 422–433. [CrossRef]

52. Jhajharia, D.; Dinpashoh, Y.; Kahya, E.; Singh, V.P.; Fakheri-Fard, A. Trends in reference evapotranspiration in the humid region of Northeast India. *Hydrol. Process.* **2012**, *26*, 421–435. [CrossRef]

53. Hou, Y.; Zhang, M.; Meng, Z.; Liu, S.; Sun, P.; Yang, T. Assessing the impact of forest change and climate variability on dry season runoff by an improved single watershed approach: A comparative study in two large watersheds, China. *Forests* **2018**, *9*, 46. [CrossRef]

54. Wei, X.; Zhang, M. Quantifying streamflow change caused by forest disturbance at a large spatial scale: A single watershed study. *Water Resour. Res.* **2010**, *46*, W12525. [CrossRef]

55. Zhang, M.; Wei, X.; Sun, P.; Liu, S. The effect of forest harvesting and climatic variability on runoff in a large watershed: The case study in the Upper Minjiang River of Yangtze River Basin. *J. Hydrol.* **2012**, *464*, 1–11. [CrossRef]

56. Zhang, M.; Wei, X.; Li, Q. Do the hydrological responses to forest disturbances in large watersheds vary along climatic gradients in the interior of British Columbia, Canada? *Ecohydrology* **2017**, *10*, e1840. [CrossRef]

57. Liu, W.; Wei, X.; Liu, S.; Liu, Y.; Fan, H.; Zhang, M.; Yin, J.; Zhan, M. How do climate and forest changes affect long-term streamflow dynamics? A case study in the upper reach of Poyang River Basin. *Ecohydrology* **2015**, *8*, 46–57. [CrossRef]

58. Jassby, A.D.; Powell, T.M. Detecting changes in ecological time series. *Ecology* **1990**, *71*, 2044–2052. [CrossRef]

59. Rodriguez-Iturbe, I.; D'Odorico, P.; Laio, F.; Ridolfi, L.; Tamea, S. Challenges in humid land ecohydrology: Interactions of water table and unsaturated zone with climate, soil, and vegetation. *Water Resour. Res.* **2007**, *43*, W09301. [CrossRef]

60. Engle, R.; Watson, M. A one-factor multivariate time series model of metropolitan wage rates. *Publ. Am. Stat. Assoc.* **1981**, *76*, 774–781. [CrossRef]

61. Law, T.H.; Umar, R.S.; Zulkaurnain, S.; Kulanthayan, S. Impact of the effect of economic crisis and the targeted motorcycle safety programme on motorcycle-related accidents, injuries and fatalities in Malaysia. *Int. J. Inj. Contr. Saf. Promot.* **2005**, *12*, 9–21. [CrossRef] [PubMed]

62. Zhang, M.; Wei, X.; Li, Q. A quantitative assessment on the response of flow regimes to cumulative forest disturbances in large snow-dominated watersheds in the interior of British Columbia, Canada. *Ecohydrology* **2016**, *9*, 843–859. [CrossRef]

63. Poulin, R. Global warming and temperature-mediated increases in cercarial emergence in trematode parasites. *Parasitology* **2006**, *132*, 143–151. [CrossRef] [PubMed]

64. Atkin, O.K.; Edwards, E.J.; Loveys, B.R. Response of root respiration to changes in temperature and its relevance to global warming. *New Phytol.* **2010**, *147*, 141–154. [CrossRef]

65. Saloman, C.H.; Naughton, S.P. Effect of hurricane Eloise on the benthic fauna of Panama city beach, Florida, USA. *Mar. Biol.* **1977**, *42*, 357–363. [CrossRef]

66. Dai, Z.; Amatya, D.M.; Sun, G.; Trettin, C.C.; Li, C.; Li, H. Climate variability and its impact on forest hydrology on South Carolina coastal plain, USA. *Atmosphere* **2011**, *2*, 330–357. [CrossRef]

67. Zhang, J.; van Meerveld, I.H.J.; Waterloo, M.J.; Bruijnzeel, L.A., Sr. Typhoon haiyan's effects on interception loss from a secondary tropical forest near tacloban, leyte, the Philippines. In Proceedings of the AGU Fall Meeting, San Francisco, CA, USA, 14–18 December 2015.

68. Zimmerman, J.K.; Emiii, E.; Waide, R.B.; Lodge, D.J.; Taylor, C.M.; Nvl, B. Responses of tree species to hurricane winds in subtropical wet forest in Puerto Rico: Implications for tropical tree life histories. *J. Ecol.* **1994**, *82*, 911–922. [CrossRef]

69. Latham, R.E.; Ricklefs, R.E.; Ricklefs, R.E.; Schluter, D. *Continental Comparisons of Temperate-Zone Tree Species Diversity*; University of Chicago Press: Chicago, IL, USA, 1993; pp. 294–314.

70. Latham, R.E.; Ricklefs, R.E. Global patterns of tree species richness in moist forests: Energy-diversity theory does not account for variation in species richness. *Oikos* **1993**, *67*, 325–333. [CrossRef]

71. Wiens, J.J.; Donoghue, M.J. Historical biogeography, ecology and species richness. *Trends Ecol. Evol.* **2004**, *19*, 639–644. [CrossRef] [PubMed]

72. Wang, D.; Hejazi, M. Quantifying the relative contribution of the climate and direct human impacts on mean annual streamflow in the contiguous United States. *Water Resour. Res.* **2011**, *47*, 411. [CrossRef]

73. Mckinley, V.L.; Vestal, J.R. Biokinetic analysis of adaptation and succession: Microbial activity in composting municipal sewage sludge. *Appl. Environ. Microbiol.* **1984**, *47*, 933–941. [PubMed]

74. Dover, C.L.V.; Lutz, R.A. Experimental ecology at deep-sea hydrothermal vents: A perspective. *J. Exp. Mar. Biol. Ecol.* **2004**, *300*, 273–307. [CrossRef]

75. De, S.B.; Clavel, T.; Clerté, C.; Carlin, F.; Giniès, C.; Nguyenthe, C. Influence of anaerobiosis and low temperature on bacillus cereus growth, metabolism, and membrane properties. *Appl. Environ. Microbiol.* **2012**, *78*, 1715–1723.

76. Hilliard, J.H.; West, S.H. Starch accumulation associated with growth reduction at low temperatures in a tropical plant. *Science* **1970**, *168*, 494–496. [CrossRef] [PubMed]

77. Costa, M.H.; Botta, A.; Cardille, J.A. Effects of large-scale changes in land cover on the discharge of the Tocantins River, Southeastern Amazonia. *J. Hydrol.* **2003**, *283*, 206–217. [CrossRef]

78. Michot, T.C.; Burch, J.N.; Arrivillaga, A.; Rafferty, P.S.; Doyle, T.W.; Kemmerer, R.S. *Impacts of Hurricane Mitch on Seagrass Beds and Associated Shallow Reef Communities along the Caribbean Coast of Honduras and Guatemala*; U.S. Geological Survey: Reston, VA, USA, 2003.

79. Asbjornsen, H.; Goldsmith, G.R.; Alvaradobarrientos, M.S.; Rebel, K.; Osch, F.P.V.; Rietkerk, M.; Chen, J.; Gotsch, S.; Tobón, C.; Geissert, D.R. Ecohydrological advances and applications in plant–water relations research: A review. *J. Plant Ecol.* **2011**, *4*, 3–22. [CrossRef]

80. Zhang, M.; Liu, N.; Harper, R.; Li, Q.; Liu, K.; Wei, X.; Ning, D.; Hou, Y.; Liu, S. A global review on hydrological responses to forest change across multiple spatial scales: Importance of scale, climate, forest type and hydrological regime. *J. Hydrol.* **2017**, *546*, 44–59. [CrossRef]

81. Graff, J.V.D.; Ahmad, R.; Scatena, F.N. Recognizing the importance of tropical forests in limiting rainfall-induced debris flows. *Environ. Earth Sci.* **2012**, *67*, 1225–1235. [CrossRef]

82. Troch, P.A.; Martinez, G.F.; Pauwels, V.R.N.; Durcik, M.; Sivapalan, M.; Harman, C.; Brooks, P.D.; Gupta, H.; Huxman, T. Climate and vegetation water use efficiency at catchment scales. *Hydrol. Process.* **2010**, *23*, 2409–2414. [CrossRef]

83. Jackson, R.B.; Cook, C.W.; Pippen, J.S.; Palmer, S.M. Increased belowground biomass and soil CO2 fluxes after a decade of carbon dioxide enrichment in a warm-temperate forest. *Ecology* **2009**, *90*, 3352–3366. [CrossRef] [PubMed]

84. Jones, J.A.; Creed, I.F.; Hatcher, K.L.; Warren, R.J.; Adams, M.B.; Benson, M.H.; Boose, E.; Brown, W.A.; Campbell, J.L.; Covich, A. Ecosystem processes and human influences regulate streamflow response to climate change at long-term ecological research sites. *Bioscience* **2012**, *62*, 390–404. [CrossRef]

85. Jones, J.A.; Post, D.A. Seasonal and successional streamflow response to forest cutting and regrowth in the Northwest and Eastern United States. *Water Resour. Res.* **2004**, *40*, 191–201. [CrossRef]

86. Li, Z.; Liu, W.Z.; Zhang, X.C.; Zheng, F.L. Impacts of land use change and climate variability on hydrology in an agricultural catchment on the Loess Plateau of China. *J. Hydrol.* **2009**, *377*, 35–42. [CrossRef]

87. Wolter, K.; Timlin, M.S. El Niño/Southern oscillation behaviour since 1871 as diagnosed in an extended multivariate ENSO index (MEI.Ext). *Int. J. Climatol.* **2011**, *31*, 1074–1087. [CrossRef]

![forests logo] *forests*

MDPI

*Article*

# Attribution Analysis for Runoff Change on Multiple Scales in a Humid Subtropical Basin Dominated by Forest, East China

Qinli Yang [1,2,3], Shasha Luo [1], Hongcai Wu [1], Guoqing Wang [3], Dawei Han [4], Haishen Lü [5] and Junming Shao [2,6,*]

[1] School of Resources and Environment, University of Electronic Science and Technology of China, Chengdu 611731, China; qinli.yang@uestc.edu.cn (Q.Y.); 201621180125@std.uestc.edu.cn (S.L.); wuhongcai@std.uestc.edu.cn (H.W.)

[2] Big Data Research Center, University of Electronic Science and Technology of China, Chengdu 611731, China

[3] State Key Laboratory of Hydrology-Water Resources and Hydraulic Engineering, Nanjing Hydraulic Research Institute, Nanjing 210029, China; gqwang@nhri.cn

[4] Department of Civil Engineering, Faculty of Engineering, University of Bristol, Bristol BS8 1TR, UK; d.han@bristol.ac.uk

[5] State Key Laboratory of Hydrology-Water Resources and Hydraulic Engineering, College of Hydrology and Water Resources, Hohai University, Nanjing 210098, China; lvhaishen@hhu.edu.cn

[6] School of Computer Science and Engineering, University of Electronic Science and Technology of China, Chengdu 611731, China

[*] Correspondence: junmshao@uestc.edu.cn; Tel.: +86-028-6183-1677

Received: 12 January 2019; Accepted: 19 February 2019; Published: 20 February 2019

**Abstract:** Attributing runoff change to different drivers is vital in order to better understand how and why runoff varies, and to further support decision makers on water resources planning and management. Most previous works attributed runoff change in the arid or semi-arid areas to climate variability and human activity on an annual scale. However, attribution results may differ greatly according to different climatic zones, decades, temporal scales, and different contributors. This study aims to quantitatively attribute runoff change in a humid subtropical basin (the Qingliu River basin, East China) to climate variability, land-use change, and human activity on multiple scales over different periods by using the Soil and Water Assessment Tool (SWAT) model. The results show that runoff increased during 1960–2012 with an abrupt change occurring in 1984. Annual runoff in the post-change period (1985–2012) increased by 16.05% (38.05 mm) relative to the pre-change period (1960–1984), most of which occurred in the winter and early spring (March). On the annual scale, climate variability, human activity, and land-use change (mainly for forest cover decrease) contributed 95.36%, 4.64%, and 12.23% to runoff increase during 1985–2012, respectively. On the seasonal scale, human activity dominated runoff change (accounting for 72.11%) in the dry season during 1985–2012, while climate variability contributed the most to runoff change in the wet season. On the monthly scale, human activity was the dominant contributor to runoff variation in all of the months except for January, May, July, and August during 1985–2012. Impacts of climate variability and human activity on runoff during 2001–2012 both became stronger than those during 1985–2000, but counteracted each other. The findings should help understandings of runoff behavior in the Qingliu River and provide scientific support for local water resources management.

**Keywords:** climate variability; land-use change; human activities; SWAT

## 1. Introduction

Runoff is a key quantitative indicator for water resources availability and river regimes. Its variation may affect water-use patterns (in different sectors such as agriculture, domestic, and industry) and water resources management. Climate variability and human activity are suggested to be two primary driving factors for runoff variation [1]. Climate variability may alter runoff by changing the spatiotemporal distributions of temperature, evaporation, and precipitation [2]. Human activity includes land-use change, water withdraw, and hydraulic engineering operation, which exert impacts on runoff by changing the underlying properties and the runoff generation mechanisms of the basin [3]. Attributing the runoff change to different driving factors could help better understand how and why runoff varies, and could further support decision makers with water resources planning and management.

Numerous methods have been proposed for runoff attribution analysis, which can be broadly classified into three categories: diagrammatizing methods (i.e., Budyko curves [4], Tomer–Schilling framework [5], modified double-mass curves [6,7]), numerical calculation-based methods (e.g., elasticity [8], sensitivity [9]), hydrologic models [3,10], and field data-based methods [11]. Different methods have both advantages and shortcomings. Specifically, the field data-based method is often limited by the availability of long-term observations of paired experimental catchments. The diagrammatizing methods are easy for intuitive understanding, but the introduced parameters for actual evaporation estimation are difficult to be quantified [2]. Numerical calculation-based methods, particularly for hydrological model-based approaches, have been widely adopted in runoff attribution studies due to their consideration of basin hydrological processes. Among the diverse hydrological models, the Soil and Water Assessment Tool (SWAT) has demonstrated its effectiveness. For instance, Marhaento et al. [12] and Anand et al. [13] investigated the impact of land-use change on water balance using the SWAT model. Zuo et al. [14] assessed the effects of land-use and climate changes on runoff on the Loess Plateau of China by using the SWAT model.

There also have been extensive studies that separate the impacts of climate variability and human activity (including urbanization, deforestation) on runoff change across world [1,7,15]. Montenegro and Ragab [16] separated the contributions of climate and land-use changes in Northeast Brazil. Poelmans et al. [17] attributed runoff variation to climate change and human activity in central Belgium. Zhao et al. [18] estimated the effects of vegetation and climate changes on runoff in the paired catchments in Australia, New Zealand, and South Africa. Zhang et al. [19] investigated the attribution of runoff changes for 107 catchments in central and northern China. However, most existing studies more focus on runoff change in the arid and semi-arid areas, which are more likely to face water scarcity.

Nevertheless, rivers in the humid and semi-humid climate regions also suffer frequent and intensive flood and drought events, which result in huge losses in the riverine communities. For example, in the 1998 flood disaster in the Yangtze River, 223 million people were affected, 3004 people died, 15 million became homeless, and 15 million farmers lost their crops. By the end of August 1998, the direct economic damage was estimated at over USD $20 billion [20]. In addition, the effects of climate and human activity on runoff may vary greatly depending on the different geographic locations, climatic zones, decades, temporal scales, and different contributors considered [21,22]. Furthermore, different kinds of human activities exert different impacts on runoff. Land-use change, as one common and key type of human activity, and its impact on runoff is supposed to be quantified separately from that of the whole human activities. Therefore, it would be of great importance to comprehensively study the effects of climate variability, land-use change, and human activity on runoff variation at multiple scales in the river located in the humid and semi-humid areas.

In China, among seven major rivers, five large rivers (Yellow River [10,23], Haihe River [2,24], Huaihe River [25,26], Liao River [21], and Songhua River [27,28]) are situated in arid and semi-arid northern China, and have been widely investigated. In contrast, attribution for runoff change in the humid and semi-humid climate zone is insufficiently studied [29]. More recently, Shen et al. [21]

evaluated runoff variation and its causes within the Budyko framework across 224 catchments in China, including the humid climate regions. Zhai and Tao [3] studied the contributions of climate change and human activities to runoff change in seven typical catchments in China by using the hydrological model. However, separating contributions of land-use change from human activity and runoff attribution on multiple temporal scales still needs to be further studied.

In this study, the authors aim to quantitatively assess the contributions of climate variability, land-use change, and human activity to runoff change in the Qingliu River on multiple (monthly, seasonal, annual) scales over different periods (1985–2012, 1985–2000, and 2001–2012) by using a SWAT model. The objectives are to answer the following questions: (1) What changes have occurred in the runoff of the forest-dominated Qingliu River basin? (2) What are the dominant driving factors behind runoff change in the Qingliu River: climate variability, human activity, or land-use/cover change? (3) Will the dominant driving factor differ according to the different temporal scales and change over different periods?

## 2. Study Area and Data Acquisition

### 2.1. Study Area

The Qingliu River, a secondary order tributary of the lower Yangtze River, originates from the Huangfu and Mopan mountains, and flows southeast through Chuzhou city with a length of 84 km. The Qingliu River basin (36°21′–37°19′ N, 108°38′–110°29′ E; Figure 1) covers a drainage area of 1070 km². The basin was dominated by evergreen broad-leaved forest, but has experienced climate variability, forest-cover change, and dam constructions in recent years [30–32]. The mean altitude of the basin is 77 m, and the mean slope of river channel is 1.23‰ [33]. The study area belongs to a humid subtropical climate zone that is hot and rainy in summer while mild and dry in winter. The mean annual temperature ranges from 11.1 °C to 17.2 °C, and the mean annual precipitation was about 1053 mm between 1960 and 2012 [34]. Seven rain gauges and one hydrological station were set up in the study area. The Shaheji reservoir and the Chengxi reservoir started operations since 1962 and 1965 with storage of 52.1 million m³ and 44 million m³, respectively [31].

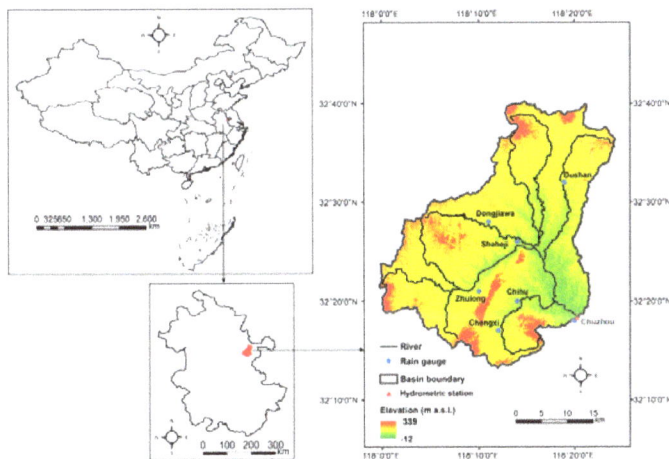

**Figure 1.** Location of the Qingliu River basin, East China.

Chuzhou city experienced rapid socio-economic development between 1960 and 2012. Specifically, statistical data indicate that the total population in Chuzhou has increased from 2.12 million in 1964 to 3.44 million in 1984 and 4.52 million in 2012 [35]. The mean annual growth rates of the population over the periods of 1964–1984, 1984–2000, and 2000–2012 were 24.40‰, 13.36‰, and

5.17‰, respectively. The gross domestic product (GDP) has grown from 2.45 billion CNY (China Yuan) in 1978 to 97.07 billion CNY in 2012. The structure of the primary, secondary, and tertiary industries has changed from 70.51:15.42:14.07 in 1974 to 19.8:52.3:27.9 in 2012 [35]. With the development of society and the economy, land use also has undergone changes, mainly transferring from forest and farmland to residential areas [32,36].

The basin also undergone extreme hydro-meteorological events such as floods in 1991, 2003, and 2008, and droughts in 1967–1969 and 2017. According to the hydrological record [33], the flood event from 2003 caused the flooding of an area of 2680 km$^2$, and an economic loss of USD $0.65 billion. Besides, 2.43 million people in 163 counties were affected, 79,000 houses collapsed, 139 embankments were destroyed, and the Beijing–Shanghai railway was interrupted twice.

### 2.2. Data Acquisition

Climatic data of daily precipitation, temperature, wind speed, pan evaporation, and solar radiation from 1960–2012 were collected by the China Meteorological Data Service Center (http://data.cma.cn). Monthly runoff data of the Chuzhou hydrometric station were provided by the Anhui Hydrology Bureau (http://www.ahsl.gov.cn). The digital elevation data (DEM) data with a spatial resolution of 30 m were obtained from the Geospatial Data Cloud (http://www.gscloud.cn). Land-use maps from 1981 and 2010 were interpreted from Landsat imagery downloaded from the United States Geological Survey (USGS) (https://www.usgs.gov). Here, the land-use classification results in 1981 and 2010 were used to represent land-use status in the baseline period and the disturbed period, respectively. Soil type data were collected from the Harmonized World Soil Database (HWSD) (https://daac.ornl.gov/SOILS/guides/HWSD.html). Necessary soil attributes (e.g., soil depth, soil texture, and soil grain composition) were extracted from the Soil Topography of China. Sectional attributes were calculated in the Soil–Plant–Atmosphere–Water Model (SPAW) according to organic matter, soil texture, organic matter, and gravel content.

## 3. Methodology

### 3.1. Mann–Kendall Test for Trend Analysis

The non-parametric rank-based Mann–Kendall test [37], which has been recommended by the World Meteorological Organization to identify trends in the hydrological and meteorological time series [14], has been adopted for the purpose of this study. In the Mann–Kendall test, there is a null hypothesis ($H_0$) that no trend exists in the data. For a data series $x = \{x_1, x_2, \ldots, x_n\}$ in which $n > 10$, the standard normal statistic $Z$ is estimated by following Equations (1) to (4)):

$$Z = \begin{cases} \frac{S-1}{\sqrt{var(S)}} & S > 0 \\ 0 & S = 0 \\ \frac{S+1}{\sqrt{var(S)}} & S < 0 \end{cases} \tag{1}$$

$$S = \sum_{i=1}^{n-1} \sum_{j=i+1}^{n} sgn(x_j - x_i) \tag{2}$$

$$var(S) = \frac{n(n-1)(2n+5)}{18} \tag{3}$$

$$sgn(x_j - x_i) = \begin{cases} +1 & x_j - x_i > 0 \\ 0 & x_j - x_i = 0 \\ -1 & x_j - x_i < 0 \end{cases} \tag{4}$$

where $n$ is the number of the data points. Given a certain level of confidence ($\alpha$), the null hypothesis of no trend is rejected if $|Z| \geq Z_{1-\alpha/2}$. When the significance levels are set at 0.01, 0.05, and 0.1, the values of $|Z_{1-\alpha/2}|$ are 2.58, 1.96, and 1.65, respectively. A positive value of $Z$ denotes an increasing trend, and the opposite corresponds to a decreasing one.

### 3.2. Mann–Kendall Method for Change Detection

A number of parametric and non-parametric methods have been widely used for abrupt change detection in time series [38]. Most parametric methods (e.g., the moving $t$-test technique [14] and Yamamoto method [39]) assume that the data series are independent and normally distributed. However, unknown probability distribution is frequently encountered in hydrological and meteorological records. Therefore, in this study, the authors selected the non-parametric Mann–Kendall method to detect the abrupt change point of the runoff series.

The Mann–Kendall method is based on the Mann–Kendall test (see Section 3.1) statistic, but it is calculated for sub-sets of the series. For the time series $x_i$ ($1 \leq i \leq k$), the numbers $a_i$ of elements $x_j$ preceding it ($j \leq i$) have been calculated (Equations (5) and (6)). Under the null hypothesis (i.e., no trend assumed), the rank series $S_k$ is normally distributed with mean and variance being $E(S_k)$ (Equation (7)) and $Var(S_k)$ (Equation (8)), respectively. Let $UF_k$ be the standard normal distribution (Equation (9)).

$$S_k = \sum_{i=1}^{k} a_i (k = 2, 3, 4, \ldots, n) \tag{5}$$

$$a_i = \begin{cases} 1 \ X_i \geq X_j \\ 0 \ X_i < X_j \end{cases} ; 1 \leq j \leq i \tag{6}$$

$$E(S_k) = \frac{k(k+1)}{4} \tag{7}$$

$$Var(S_k) = k(k-1)(2k+5)/72 \tag{8}$$

$$UF_k = \frac{[S_k - E(S_k)]}{\sqrt{Var(S_k)}} \ (k = 1, 2, \ldots, n) \tag{9}$$

For a specific level of significance $\alpha$, if $UF_k$ is greater than $U_{\alpha/2}$ then the sequence has a significant increasing or descending trend. $UB_k$ is the opposite of $UF_k$; namely, $UB_k = -UF_k$ and $k = n + 1 - k$ ($k = 1, 2, \ldots, n$). The intersection of the forward trend ($UF_k$) and backward trend lines ($UB_k$) within the confidence interval indicates an abrupt change point in the runoff series.

### 3.3. SWAT Model Construction, Calibration, and Validation

The SWAT model construction, calibration, and validation were all implemented via a visual user interface in the ArcSWAT version 2012. Firstly, the stream network was generated from the DEM of the Qingliu River catchment. A total of 23 sub-basins and 109 hydrological response units (HRU) were further created based on the land-use, soil, and slope layers. Afterwards, according to the abrupt change point (1984) of runoff, the entire study period was divided into two phases: the baseline period (natural phase, 1960–1984) without significant human activities, and the disturbed period (1985–2012) associated with intensive human activities. Considering the availability of clear Landsat imagery before 1984, the authors used a land-use map from 1981 to represent the land-use conditions over the baseline period. Regarding the land-use status over the disturbed period, the authors selected a land-use map in the almost middle year (2000) to represent. Subsequently, the SWAT model was calibrated over the period 1960–1971 and validated over the years 1972–1984 with the first three years (1957–1959) as the warm-up period. The partial parameters of the SWAT model that were determined in this study are listed in Table 1. CN2, ESCO, and GWQMN are the most sensitive parameters for runoff simulation. The SWAT model is configured to output monthly runoff.

**Table 1.** Ranges and final values of parameters used in the SWAT model calibration.

| Parameter | Parameter Description | Range of Values | Value Used |
|---|---|---|---|
| CN2 | SCS curve number for Moisture Condition II | 35–98 | 75 |
| ESCO | Soil Evaporation Compensation Factor | 0–1 | 0.142 |
| GWQMN | Threshold depth of water in the shallow aquifer required for return flow to occur | 0–5000 | 1.4 |
| GW_REVAP | Groundwater revap coefficient | 0.02–0.2 | 0.16 |
| ALPHA_BF | Base flow alpha factor (days) | 0–1 | 1 |
| SOL_AWC | Available water capacity of the soil layer (mm/mm soil) | 0–1 | 0.2 |
| RCHRG-DP | Osmosis ratio in deep aquifer | 0–1 | 0.28 |
| USLE-P | USLE equation support practice | 0–1 | 0.52 |
| SMTMP | Snow melt base temperature | −5–5 | 5 |
| BIOMIX | Biological mixing efficient | 0–1 | 1 |
| GW-DELAY | Groundwater delay | 0–500 | 75 |
| CH-N2 | Manning's "$n$" value for the main channel | −0.01–0.3 | 0.04 |
| REVAPMN | Threshold depth of water in the shallow aquifer for "revap" or percolation to the deep aquifer to occur (mm) | 0–500 | 0 |

Note: SWAT: Soil and Water Assessment Tool; SCS: Soil Conservation Service; USLE: universal soil loss equation.

To assess the performance of the SWAT model, the Nash–Sutcliffe coefficient (NSE) and the relative error (RE) were selected as evaluation indicators. The calculation formulas (Equations (10) and (11)) of NSE and RE are as follows:

$$NSE = 1 - \frac{\sum_{i=1}^{n}(Q_{obs}(i) - Q_{sim}(i))^2}{\sum_{i=1}^{n}(Q_{obs}(i) - \overline{Q}_{obs})^2} \tag{10}$$

$$RE = \frac{\overline{Q}_{sim} - \overline{Q}_{obs}}{\overline{Q}_{obs}} \times 100\% \tag{11}$$

where $Q_{obs}(i)$ is the observed runoff at time step $i$, $Q_{sim}(i)$ is the simulated runoff at time step $i$, $\overline{Q}_{obs}$ is the mean value of the observed runoff, $\overline{Q}_{sim}$ is the mean value of the simulated runoff, and $n$ is the number of the data sequence.

*3.4. Attribution Analysis*

Based on the calibrated SWAT model, the authors run the SWAT model to simulate runoff under three different scenarios. Specifically, Scenario 1 (S1) takes climatic data during 1960–1984 and the land-use map for 1981. Scenario 2 (S2) uses climatic data during 1989–2012 and the land-use map for 1981. Scenario 3 (S3) is based on climatic data during 1960–1984 and the land-use map for 2000.

Comparing S1 and S2 with each other, the only difference is their climatic inputs. Thereby, the simulated runoff under Scenario 2 can be regarded as the naturalized runoff in the disturbed period, and its difference from the simulated runoff under S1 in the baseline period is primarily caused by climate variability. In contrast to Scenario 1, the only discrepancy in Scenario 3 is the input of land-use data. Hence, the difference of runoff simulation between Scenario 3 and Scenario 1 is induced by land-use change.

Finally, runoff variations caused by climate variability, land-use change, and human activity are quantified (as shown in Equations (12)–(15)), and their relative contributions to runoff changes relative to the baseline period are formulated (as written in Equations (16)–(18)) as follows:

$$\Delta Q_{Total} = Q_{obs,d} - Q_{obs,b} \tag{12}$$

$$\Delta Q_{Climate} = Q_{S2} - Q_{S1} \tag{13}$$

$$\Delta Q_{Haman\ activity} = \Delta Q_{Total} - \Delta Q_{Climate} \tag{14}$$

$$\Delta Q_{Land\ use} = Q_{S3} - Q_{S1} \tag{15}$$

$$C_{Climate} = \frac{\Delta Q_{Climate}}{\Delta Q_{Total}} \times 100\% \tag{16}$$

$$C_{Human\ activity} = \frac{\Delta Q_{Human\ activity}}{\Delta Q_{Total}} \times 100\% \tag{17}$$

$$C_{Land\ use} = \frac{\Delta Q_{Land\ use}}{\Delta Q_{Total}} \times 100\% \tag{18}$$

where $\Delta Q_{Total}$ is the total change of runoff; $Q_{obs,b}$ and $Q_{obs,d}$ represent the mean observed runoff over the baseline period and the disturbed period, respectively; $Q_{S1}$, $Q_{S2}$, and $Q_{S3}$ are the mean simulated runoff under three different scenarios; $\Delta Q_{Climate}$, $\Delta Q_{Land\ use}$ and $\Delta Q_{Human\ activity}$ mean the runoff change induced by climate variability, land-use change and human activity, respectively; Page: 7 and $C_{Climate}$, $C_{Land\ use}$, and $C_{Human\ activity}$ indicate the contributions in percentage of climate variability, land-use change, and human activity to runoff change, respectively.

## 4. Results and Discussion

### 4.1. Change Trend of Hydro-Meteorological Variables

The trends of precipitation, temperature, pan evaporation, wind speed, solar radiation, and runoff depth over the period of 1960–2012 for the Qingliu River basin have been analyzed by using the simple linear regression method and the Mann–Kendall test method. Figure 2 illustrates the inter-annual change of these hydro-meteorological variables, and Table 2 shows the statistical information of the trend analysis on the annual and seasonal scales.

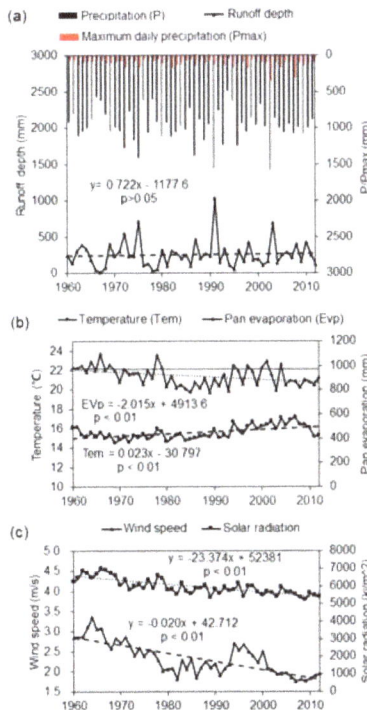

**Figure 2.** The inter-annual change of hydro-meteorological variables ((**a**) precipitation, maximum daily precipitation, and runoff depth; (**b**) temperature and pan evaporation; and (**c**) wind speed and solar radiation) between 1960 and 2012 for the Qingliu River basin.

**Table 2.** Statistical information of trend analysis for the hydro-meteorological variables on annual and seasonal scales between 1960 and 2012 for the Qingliu River basin, East China.

| Statistic | Annual | | Spring | | Summer | | Autumn | | Winter | |
|---|---|---|---|---|---|---|---|---|---|---|
| | Slope | Z | Slope | Z | Slope | Z | Slope | Z | Slope | Z |
| Precipitation (mm) | 1.68 | 0.71 | −0.69 | −0.69 | 1.74 | 1.00 | −0.47 | −0.39 | 1.19 * | 3.21 * |
| Temperature (°C) | 0.02 * | 3.63 * | 0.04 * | 3.86 * | 0.01 | 0.72 | 0.02 * | 3.51 * | 0.03 * | 3.06 * |
| Evp (mm) | −2.01 * | −2.79 * | 0.22 | 0.25 | −1.62 * | −3.90 * | −0.16 | −0.52 | −0.46 * | −2.22 # |
| Runoff depth (mm) | 0.72 | 0.48 | 0.06 | 0.58 | 0.09 | 0.59 | 0.17 | 1.64 | 0.37 * | 2.75 * |
| D (P > 0 mm) | 0.03 | 0.12 | −0.16 * | −2.51 # | 0.06 | 1.08 | −0.17 * | −2.22 # | 0.28 * | 4.48 * |
| D (P > 25 mm) | −0.002 | −0.15 | −0.02 | −0.65 | 0.003 | 0.15 | 0.01 | 0.41 | 0.01 # | 1.5 |
| D (P > 50 mm) | 0.001 | 0.18 | 0 | 0.71 | 0.01 | 0.48 | −0.004 | −0.57 | 0 | 0 |
| Pmax | 1.18 # | 2.148 # | 0.07 | 0.77 | 1.24 # | 2.10 # | −0.01 | −0.23 | 0.13 # | 1.77 |

Note: Z means the statistic value derived from the Mann–Kendall test method. * and # indicate that the variable has a significant change at the significance level of 0.01 and 0.05, respectively. Evp: pan evaporation; D (P > 0 mm), D (P > 25 mm), and D (P > 50 mm) represent the number of days with daily precipitation large than 0 mm, 25 mm, and 50 mm, respectively; Pmax: maximum daily precipitation.

The results indicate that the mean annual runoff increased insignificantly, while the winter runoff increased significantly with a magnitude of 0.37 mm/year (Figure 2a). The mean annual temperature shows an upward trend during 1960–2012 with a rate of 0.02 °C/year ($p < 0.01$, Figure 2b). All of the seasonal temperatures increased significantly ($p < 0.01$) except for the summer temperature. In contrast, the average annual pan evaporation declined significantly with a magnitude of 2.01 mm/year ($p < 0.01$, Figure 2b). It infers that there exists an evaporation paradox, which has been widely observed across China [40,41]. The phenomenon of the evaporation paradox can be attributed to the significant decrease of solar radiation and wind speed (Figure 2c) in the study area. The findings are consistent with the results published by Han et al. [41,42].

The mean annual precipitation exhibited a weak increase trend (1.68 mm/year, $p > 0.05$) during 1960–2012, which was associated with insignificant decrease in spring and autumn, and a significant increase in winter (1.19 mm/year, $p < 0.01$) (Table 2). Correspondingly, it can be noted that the number of rainy days has decreased in the spring ($p < 0.05$) and autumn ($p < 0.05$), and increased in winter ($p < 0.01$). The annual maximum daily precipitation indicated an increasing trend ($p = 0.018$) over the study period. Similar trends can also be identified in the summer maximum daily precipitation and winter maximum daily precipitation. The number of days with daily precipitation over 25 mm increased significantly in winter. The results imply that: (1) the winter became wetter and warmer, while the spring and autumn became drier and warmer; (2) the maximum daily precipitation intensity on the annual, summer, and winter scales all tended to be higher; and (3) precipitation in winter became more intensive.

### 4.2. Abrupt Change Detection of Runoff Series

Based on the Mann–Kendall method, an abrupt change of runoff in the Qingliu River basin was detected in 1984 (as shown in Figure 3). The *UF* value for each year indicates the trend from the starting year (1960) to that specific year. In addition, Figure 4 presents the double-mass curve of the accumulated runoff depth and precipitation. It can be observed that since 1988, the curve started to deviate from the original regression line, which implies that human activities had more intensive impacts on runoff.

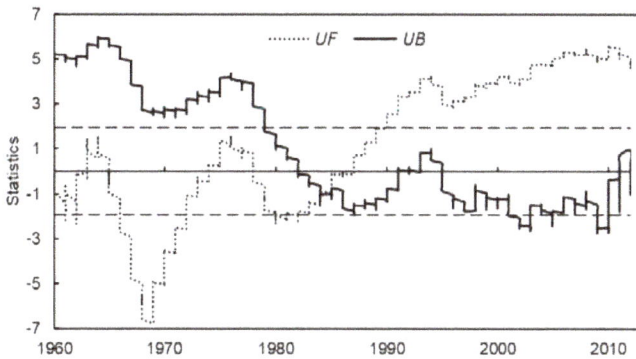

**Figure 3.** Change detection of runoff during 1960–2012 for the Qingliu River basin, China. *UF* and *UB* present the forward trend and backward trend line of the runoff series. *UF* >0 and *UF* <0 indicate the increasing trend and decreasing trend, respectively.

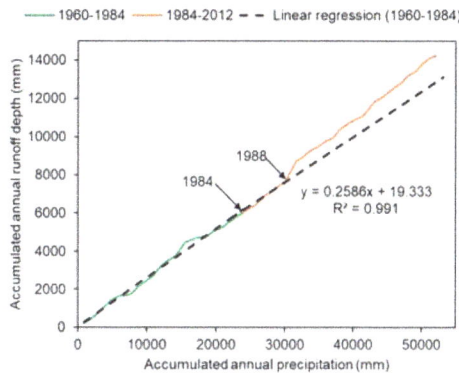

**Figure 4.** Double-mass curve of the accumulated annual runoff depth and the accumulated annual precipitation.

Furthermore, since 1984, the population has increased rapidly, with a mean annual growth rate of 13.36% during 1984–2000. Due to the reform of the economic system during 1978–1984, agriculture and partial industry have developed dramatically since 1984 [43]. Compared with the GDP in 1978 being 4.2 times that in 1949, the GDP in 1986 doubled from 1980, indicating a more rapid development [43]. With the growths of the population, agriculture, and industry, transportation and urbanization have developed, and water withdrawals have increased accordingly [43]. The Chengxi reservoir provided irrigation for an area of 70 km$^2$ during 1960–1980, and was mainly used for domestic and industrial water supply since the mid-1980s.

According to the results of the Mann–Kendall method and the double-mass curve as well as the socio-economic development, we selected 1984 as the abrupt change point of the runoff series, and divided the study period into the baseline period (1960–1984) and the disturbed period (1985–2012).

To investigate the variations of hydro-meteorological data before and after the abrupt change point, comparisons were made on the mean annual level and inner-annual level, respectively (illustrated in Table 3 and Figure 5, correspondingly). Table 3 indicates that during 1985–2012, the temperature rose by 0.67 °C, while precipitation, pan evaporation, and runoff changed by 4.65%, −6.67%, and 16.05% relative to 1960–1984, respectively. However, the variations were uneven over the year (Figure 5).

**Table 3.** Comparison of hydro-meteorological variables between the baseline period and the disturbed period.

|  | Baseline Period (1960–1984) | Disturbed Period (1985–2012) | Change | Change (%) |
|---|---|---|---|---|
| Temperature (°C) | 15.25 | 15.92 | 0.67 | / |
| Precipitation (mm) | 957.83 | 1002.40 | 44.56 | 4.65 |
| Pan evaporation (mm) | 947.22 | 884.06 | −63.16 | −6.67 |
| Runoff depth (mm) | 236.67 | 274.65 | 37.98 | 16.05 |

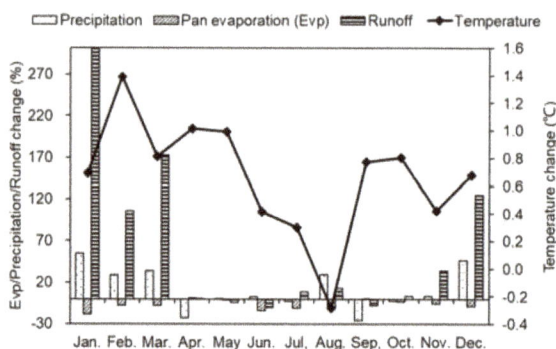

**Figure 5.** Intra-annual variations of meteorological variables during the disturbed period (1985–2012) relative to the baseline period (1960–1984) for the Qingliu River basin, East China.

In general, the variations in winter and early spring were much higher than those between April and November. The larger the change magnitudes of precipitation and pan evaporation, the large change in the runoff. Runoff increases were often associated with an increase of precipitation and decline of pan evaporation. Specifically, the temperature rose in all the months except for August, ranging from −0.28 °C to 1.39 °C, with the highest rise occurring in February. The mean precipitation increased by 12.27%, with the largest increase (54.61%) in January and the largest reduction (−25.67%) in September. Pan evaporation declined in all months except for April, May, and September, with the highest decrease (−17.89%) occurring in January. As an integrated result, runoff increased in most months (especially for January, February, March, and December), ranging from 0.69% to 301.20%. Runoff increased the most in January, which is consistent with the change features of precipitation and pan evaporation. The largest decrease (−9.73%) of runoff occurred in June. The variations in April, May, and June imply that there should be some other factors affecting runoff.

*4.3. Land-Use Change for the Qingliu River Basin*

Based on the Landsat imagery and random forest classifier, land use was classified into five categories, namely forest, farmland, residential area, water body, and barren land. Figure 6 shows the land-use classification results for 1981 and 2000, respectively. It can be observed that the forest dominates land use in the study area, and the residential area has expanded between 1981 and 2000. Table 4 displays the transmission matrix for land-use changes from 1981 to 2000. The results indicate that in total, 10.35% of the study area (110.79 km$^2$) experienced land-use changes, which mainly refer to transmission among the forest, farmland, and residential area. Specifically, the residential area expanded from 52.40 km$^2$ in 1981 to 73.64 km$^2$ in 2000, increasing by 40.53% relative to 1981. However, the total increase of residential area (21.24 km$^2$) only accounts for 1.99% of the whole research area. The forest decreased by 57.44 km$^2$, while the farmland increased by 41.59 km$^2$, accounting for 5.37% and 3.89% of the study region, respectively.

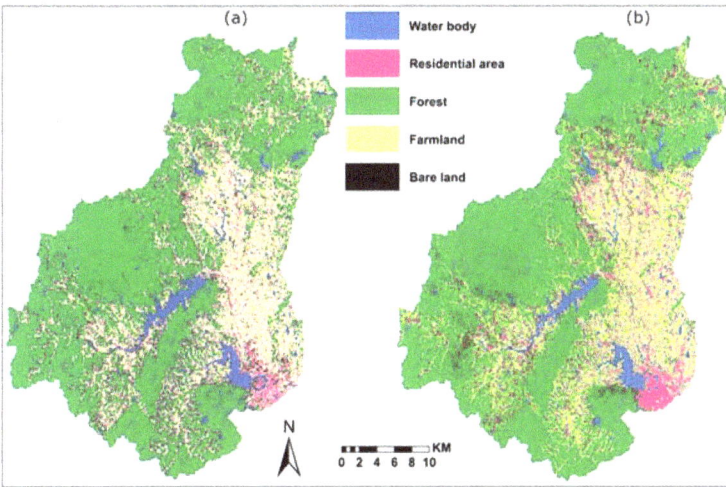

**Figure 6.** Land-use classification results in 1981 (**a**) and 2000 (**b**) for the Qinliu River basin, East China.

**Table 4.** Transition matrix of land-use changes from 1981 to 2000 for the Qingliu River basin (unit: km$^2$).

| 1981 | 2000 | | | | | | Percentage (%) |
|---|---|---|---|---|---|---|---|
| | Water Body | Residential Area | Forest | Farmland | Bare Land | Total | |
| Water body | 39.96 | 1.06 | 2.68 | 1.18 | 0 | 44.88 | 4.19 |
| Residential area | 0.87 | 43.75 | 3.33 | 4.42 | 0.03 | 52.4 | 4.90 |
| Forest | 0.40 | 11.97 | 556.56 | 58.54 | 0.41 | 627.88 | 58.68 |
| Farmland | 0.44 | 16.04 | 6.6 | 312.43 | 0.02 | 335.53 | 31.36 |
| Bare land | 0.16 | 0.82 | 1.27 | 0.55 | 6.51 | 9.31 | 0.87 |
| Total | 41.83 | 73.64 | 570.44 | 377.12 | 6.97 | 1070 | |
| Percentage (%) | 3.91 | 6.88 | 53.31 | 35.24 | 0.65 | | |

*4.4. SWAT Model Calibration and Validation for Monthly Runoff Simulation*

Taking the detected abrupt change in 1984 as a break point, the whole study period was divided into the baseline period (1960–1984) and the disturbed period (1985–2012). The SWAT model was calibrated and validated over the baseline period. Specifically, data during 1960–1971 were used for model calibration, while data during 1972–1984 were applied for model validation.

Figure 7 illustrates the simulated and observed monthly runoff between 1960 and 1984. The performance evaluation of the SWAT model is listed in Table 5. It indicates that on the monthly scale, the Nash–Shutcliffe efficiency (NSE) were 0.76 and 0.81 for the calibration and validation periods, respectively. The relative error values imply that the monthly runoff was underestimated in the calibration period, and overestimated in the validation period.

Over the validated period (1972–1984), in contrast to performance of the monthly runoff simulation, the SWAT model performed better on the annual runoff simulation (NSE = 0.83) and runoff simulation in the wet season (NSE = 0.87). The relative error of runoff simulation in the dry season (11.57%) was higher than that in wet season (−3.30%), which might be because the Qingliu River almost dried up in dry seasons between 1976 and 1980, but the SWAT model overestimated the runoff.

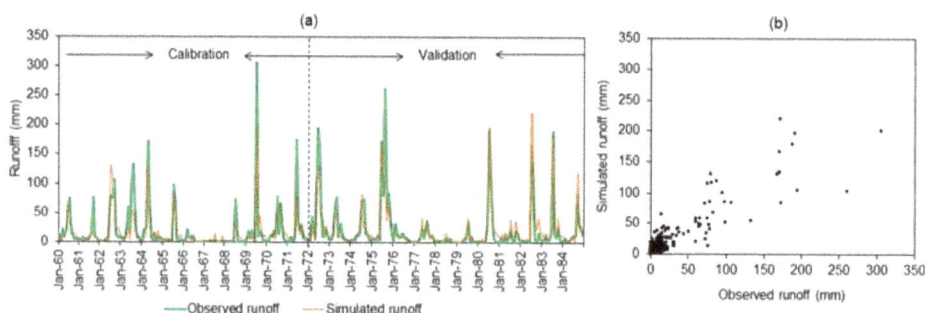

**Figure 7.** Comparison of the observed and simulated runoff in the Qingliu River for the calibration (1960–1971) and validation (1972–1984) periods (**a**), and the scatter plot of the observed versus simulated runoff (**b**). Jan means January.

**Table 5.** Performance of the SWAT model for runoff simulation for the Qingliu River basin, East China.

| Indicator | Calibration Period (1960–1971) | Validation Period (1972–1984) | | | |
|---|---|---|---|---|---|
| | Monthly | Monthly | Annual | Wet Season | Dry Season |
| NSE | 0.76 | 0.81 | 0.83 | 0.87 | 0.73 |
| Relative error (%) | −8.13 | 6.56 | 6.56 | −3.30 | 11.57 |

Note: SWAT: Soil and Water Assessment Tool; NSE: Nash–Sutcliffe coefficient.

### 4.5. Attribution Analysis of Runoff Change for the Qingliu River Basin

Based on the runoff observations and runoff simulations derived from the SWAT model under three scenarios (S1, S2, and S3), the contributions of individual factors (climate variability, land-use change, and human activity) to runoff change in the Qingliu River were quantified at multiple scales (annual, seasonal, and monthly) over three periods (1985–2012, 1985–2000, and 2001–2012). Specially, the attribution to land-use change was only analyzed on the annual and seasonal scales over the whole disturbed period (1985–2012). One reason is that the computation of land use-induced runoff change (i.e., the discrepancies between runoff simulations under Scenario 3 and Scenario 1) requires the comparison period having the same length as the baseline period. Furthermore, land-use change is mostly characterized by annual and seasonal scale changes instead of at a monthly scale.

Table 6 presents the attribution analysis results for the Qingliu River on the annual scale. It indicates that annual runoff during 1985–2012 increased by 38.05 mm in total relative to 1960–1984. Climate variability, land-use change, and human activity contribute 95.36%, 4.64%, and 12.26% to the total change of runoff, respectively. The results imply that climate variability dominates runoff variation in the Qingliu River on the annual scale. Land-use changes (mainly for deforestation and urbanization) resulted in a runoff increase, which may be derived from the reduction of evaporation and interception, and the increase of impermeable area. Compared with land-use change, human activity showed a smaller contribution to runoff change. It infers that some other human activities caused runoff decrease such as water withdraw by irrigation due to the increase of farmland from 1981 to 2000 (41.59 km$^2$).

**Table 6.** Results of attribution analysis for runoff change in the Qingliu River basin on annual scale.

| Period | Observed Runoff (mm) | Simulated Runoff—Land Use in 1981 (mm) | Simulated Runoff—Land Use in 2000 (mm) | Total Change (mm) | Climate-Induced Change | | Human-Induced Change | | Land Use-Induced Change | |
|---|---|---|---|---|---|---|---|---|---|---|
| | | | | | mm | % | mm | % | mm | % |
| 1960–1984 | 236.67 | 236.59 | 241.26 | / | / | / | / | / | / | / |
| 1985–2012 | 274.72 | 272.87 | 277.53 | 38.05 | 36.28 | 95.36 | 1.77 | 4.64 | 4.67 | 12.26 |
| 1985–2000 | 279.48 | 261.70 | 266.17 | 42.81 | 25.11 | 58.64 | 17.71 | 41.36 | | |
| 2001–2012 | 268.36 | 287.78 | 292.66 | 31.70 | 51.19 | 161.49 | −19.49 | −61.49 | | |

It is worth noting that runoff change and its attribution vary over different periods. For instance, the total increase of runoff during 1985–2000 (42.81 mm) was larger than that during 2001–2012 (31.70 mm). Over the periods of 1985–2000 and 2001–2012, climate variability and human activity accounted for 58.64% and 41.36%, and 161.49% and −61.49% of runoff change, respectively. The results imply that human activity after 2000 has become more intensive, and runoff subsequently decreased. It may partially due to more water withdrawals for different sectors to support the socio-economic development.

Here, considering the property of precipitation distribution, we divided the whole year into the wet season (from May to September) and the dry season (from January to April, and from October to December). Table 7 displays the results of attribution for runoff change in the Qingliu River on a seasonal scale during 1985–2012. The results indicate the total change of runoff in the wet season over the whole disturbed period increased by 6.26 mm relative to the baseline period. However, the runoff increased by 27.42 mm, which was induced by climate variability, and declined by 21.15 mm, which was induced by human activity, respectively. The finding implies that in the wet season, climate variability and human activity both have a major impact on the runoff, and their influences counteract with each other. In contrast, the total change of runoff in the dry season was 31.78 mm, accounting for 61.12% of that in the baseline period. The contributions of climate variability, land-use change, and human activity account for 27.89%, 0.18%, and 72.11% of runoff change in the dry season, respectively. It indicates that human activity is the primary contributor to runoff increase in the dry season, which might be because of the operations of reservoirs in the basin. However, due to the relative error of runoff simulation (−3.30% for the wet season, 11.57% for the dry season), uncertainties are associated with the attribution analysis results on the seasonal scale especially for the dry season.

**Table 7.** Results of attribution analysis for runoff change in the Qingliu River basin on a seasonal scale.

| Period | Season | Observed Runoff (mm) | Simulated Runoff—Land Use in 1981 (mm) | Simulated Runoff—Land Use in 2000 (mm) | Total Change (mm) | Climate-Induced Change | | Human-Induced Change | | Land Use-Induced Change | |
|---|---|---|---|---|---|---|---|---|---|---|---|
| | | | | | | mm | % | mm | % | mm | % |
| 1960–1984 | Dry season | 52.00 | 69.81 | 69.87 | / | / | / | / | / | / | / |
| | Wet season | 184.67 | 166.78 | 171.39 | / | / | / | / | / | / | / |
| 1985–2012 | Dry season | 83.78 | 78.68 | 78.70 | 31.78 | 8.87 | 27.89 | 22.92 | 72.11 | 0.06 | 0.18 |
| | Wet season | 190.94 | 194.19 | 198.83 | 6.26 | 27.42 | 437.68 | −21.15 | −337.68 | 4.61 | 73.60 |

Furthermore, Figure 8 shows the contributions of different drivers to runoff change in the dry season and the wet season over three periods (1985–2012, 1985–2000, and 2001–2012). Similar to 1985–2012, during the two sub-periods, human activity dominated runoff change in the dry season, while climate variability dominated runoff change in the wet season. In the dry season, both climate variability and human activity increased runoff. The impact of climate variability on runoff during 1985–2000 (10.36%) became weaker than that during 2001–2012 (35.80%), while the impact of human activity exhibited an opposite trend. Regarding the wet season, climate variability increased runoff, while human activity decreased runoff, and their effects became stronger during 2001–2012 compared with the period of 1985–2000. The impact of land use on runoff in the wet season was much higher than that in the dry season.

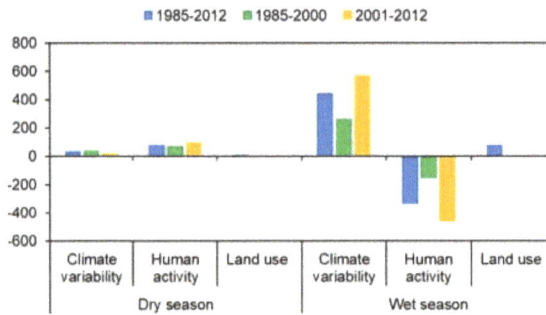

**Figure 8.** Contributions of climate variability, human activity, and land-use change to runoff variation on the seasonal scale over three different periods (1985–2012, 1985–2000, and 2001–2012) for the Qingliu River basin, East China.

Runoff change and its attribution in individual months for the Qingliu River are illustrated in Figure 9. The results show that the impacts of driving factors (i.e., climate variability and human activity) differ across different months and over different periods. During 1985–2012, the largest total change (14.30 mm) appeared in March. The highest changes caused by climate variability and human activity both occurred in August, where runoff increased by 19.64 mm due to climate variability and decreased by 14.37 mm due to human activity. Human activity dominated runoff change in all months except January, May, July, and August, most of which belong to the wet season.

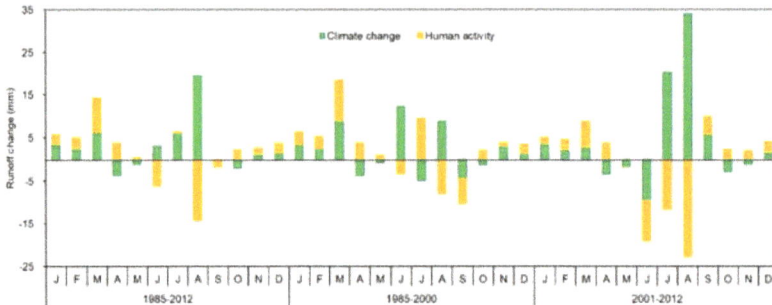

**Figure 9.** Runoff change (mm) induced by climate variability and human activity on the monthly scale over three different periods (1985–2012, 1985–2000, and 2001–2012) for the Qingliu River basin, East China. J–D are the abbreviations of January to December.

In contrast, during 1985–2000, the largest total change was 18.50 mm in March. Climate variability caused the largest increase in runoff in June (12.29 mm), while human activity mostly increased runoff in March. Over the period of 2001–2012, the largest changes induced by climate and human activity appeared in August as well, but with greater magnitudes. Compared with 1985–2000, human activity had a more intensive impact on runoff decline during 2001–2012.

In combination with Table 8, it can be noted that the largest total change of runoff and the largest runoff change induced by climate are highly related with the magnitude of precipitation change. For all three periods, the largest increase of runoff induced by human activity occurred in March, and the largest decrease of runoff induced by human activity occurred in August. The reason behind it might be related to reservoir operations, which also reflects the roles that human beings play in water resources management.

**Table 8.** Changes of precipitation (in mm) during different periods relative to the baseline period (1960–1984) for the Qingliu River basin, East China.

| | January | February | March | April | May | June | July | August | September | October | November | December |
|---|---|---|---|---|---|---|---|---|---|---|---|---|
| 1985–2000 | 11.40 | 8.64 | **29.14** | −25.13 | 11.47 | 26.90 | −31.80 | 24.12 | −31.23 | 7.82 | 1.79 | 2.21 |
| 2001–2012 | 13.51 | 11.12 | 4.91 | −8.96 | −14.29 | −27.98 | 31.14 | **57.76** | −16.29 | −13.39 | 1.10 | 18.26 |
| 1989–2012 | 12.30 | 9.70 | 18.75 | −18.20 | 0.43 | 3.38 | −4.83 | **38.54** | −24.83 | −1.27 | 1.49 | 9.09 |

Note: The bold numbers indicate the maximum values of runoff changes.

Comparing the findings on three different scales, people may find that attribution on finer (seasonal and monthly) scales would reveal more detailed information for runoff attribution analysis, which might be weakened or offset on a coarser (annual) scale. Investigating runoff change and its causes over different time periods can also benefit attribution analysis. It can deliver how the dominant drivers change along with time. Therefore, attributing runoff change to different driving forces on multiple scales over different periods would provide more detailed support for decision and policy makers on adaptive water resources management, sustainable utilization, and planning.

A previous study on runoff attribution in humid and semi-humid regions in southern China indicated that climate change dominated runoff change and contributed to runoff increase [21]. In this study, findings on the annual scale are consistent with the results in the previous work. Zhai et al. [3] reported that in the 2000s, climate variability contributed to runoff decrease, and human activity contributed more to runoff change than in the 1990s, and dominated runoff change in the Xixian basin and Changle basin. In this study, human activity during 2001–2012 exerted more impact on runoff change compared with that during 1985–2000, which is consistent with the findings in the literature published by Zhai et al. [3]. However, differently, we found that climate variability caused an increase in runoff and dominated runoff change in all periods.

### 4.6. Discussion on Uncertainty and Future Work

The attribution analysis in this study is based on comparison between the observed runoff and the simulated runoff derived from the SWAT model. Although the hydrological model performed well, there exist some differences between the simulations and the observations. Specifically, the relative errors of runoff simulations on annual and monthly scales (6.56%) were relatively lower than those in the dry season (11.57%) and higher than those in the wet season (−3.30%). Correspondingly, uncertainties are associated in the attribution results of runoff change and differ from different temporal scales. Briefly, attribution results of runoff variation in the wet season and on monthly and annual scales have lower uncertainty than those in the dry season. However, the conclusions drawn regarding the contributions trends of different driving factors (climate variability, land-use change, and human activity) to runoff change are sound.

In this study, the impact of land-use change on run variation was quantified separately from that of human activity. However, other human activities such as reservoir operations and water withdrawals were not analyzed separately due to the lack of detailed data. Nevertheless, different kinds of human activities may exert different impacts on runoff. Therefore, separation of the impacts of different kinds of human activities on runoff variation can be further investigated in the future.

It has been well-documented that climate variability and human activities are two main contributors to runoff changes. In this study, similar to many previous studies, the authors have assumed that the impacts of climate variability and human activities are independent. In reality, climate and human activities interact with each other, and it is difficult to separate their impacts on runoff. Therefore, more detailed assessments could be made on how to integrate the relationships among different driving factors into an attribution analysis of runoff change in future work.

Based on the Mann–Kendall test method, double-mass curve, and the facts of socio-economic development, the study period was divided into the baseline period and the disturbed period (which was further divided into two sub-periods), and the authors assumed that there was no human intervention in the baseline period. However, human activity existed over the whole period and exerted different magnitudes of impacts on runoff over different periods. Therefore, more dynamic attribution that considers changing attribution should be recommended in the future.

## 5. Conclusions

The mean annual runoff in the Qingliu River increased during 1960–2012 and changed abruptly in 1984. Relative to 1960–1984, the temperature rose by 0.67 °C; summer and winter got wetter and warmer, while spring and autumn became drier and warmer between 1985 and 2012. On the annual scale, climate variability, human activity, and land-use change contributed 95.36%, 4.64%, and 12.23% to runoff increase during 1985–2012, respectively. Impacts of climate variability and human activity on runoff during 2001–2012 became stronger than those during 1985–2000. On the seasonal scale, climate variability was the dominant contributor to runoff increase in the wet season, while human activity was the primary driving force to runoff increase in the dry season. Climate variability and human activity accounted for 27.89% and 72.11% of runoff change in the dry season during 1985–2012. Land-use change had more impact on runoff in the wet season than in the dry season. Compared with 1985–2000, the contribution of climate variability to runoff change decreased in the dry season and increased in the wet season, while the contribution of human activity increased over the year during 2001–2012. On the monthly scale, human activity was the dominant contributor to runoff change in all of the months except for January, May, July, and August. It can be concluded that the dominant driver of runoff change differs on different temporal scales and varies over different periods. The results on different scales imply that attribution analysis on finer (seasonal and monthly) scales would reveal more detailed information, which might be weakened or offset on a coarser (annual) scale. Uncertainties are associated with the attribution results of runoff change, especially for those in the dry season. The findings provide better understanding of runoff behavior in the Qingliu River and provide scientific support for decision and policy makers on local water resources management.

**Author Contributions:** Data interpretation, H.L. and G.W.; Data analysis, Q.Y., S.L., J.S. and H.W.; Manuscript drafting, Q.Y.; Research design, Q.Y. and J.S.; Manuscript revision, J.S., D.H., H.L. and G.W.

**Funding:** This research was funded by the National Key Research and Development Program of China [grant number 2016YFA0601501, 2016YFA0601601], National Natural Science Foundation of China [grant numbers 41601025, 41830863, and 41830752], State Key Laboratory of Hydrology-Water Resources and Hydraulic Engineering [grant number 2017490211], Science-Technology Foundation for Young Scientist of Sichuan Province [grant number 2016JQ0007], Fok Ying Tong Education Foundation for Young Teachers in the Higher Education Institutions of China [Grant number 161062], the International Exchange Grant from the UK Royal Society and China NSFC [grant number IEC\NSFC\170123], and the Fundamental Research Funds for the Central Universities (ZYGX2018J087).

**Acknowledgments:** We are grateful to the Editor and the three anonymous reviewers for their valuable advices and conscientious work.

**Conflicts of Interest:** The authors declare no conflict of interest.

## References

1. Dey, P.; Mishra, A. Separating the impacts of climate change and human activities on streamflow: A review of methodologies and critical assumptions. *J. Hydrol.* **2017**, *548*, 278–290.
2. Li, R.; Zheng, H.; Huang, B.; Xu, H.; Li, Y. Dynamic impacts of climate and land-use changes on surface runoff in the mountainous region of the Haihe river basin, China. *Adv. Meteorol.* **2018**. [CrossRef]
3. Zhai, R.; Tao, F. Contributions of climate change and human activities to runoff change in seven typical catchments across China. *Sci. Total Environ.* **2017**, *605*, 219–229. [PubMed]
4. Budyko, M.I. *Climate and Life*; Miller, D.H., Ed.; Academic Press: San Diego, CA, USA, 1974.
5. Tomer, M.D.; Schilling, K.E. A simple approach to distinguish land-use and climate-change effects on watershed hydrology. *J. Hydrol.* **2009**, *376*, 24–33.
6. Zhang, M.; Wei, X. The effects of cumulative forest disturbance on streamflow in a large watershed in the central interior of British Columbia, Canada. *Hydrol. Earth Syst. Sci.* **2012**, *16*, 2021–2034.

7. Li, Q.; Wei, X.; Zhang, M.; Liu, W.; Giles-Hansen, K.; Wang, Y. The cumulative effects of forest disturbance and climate variability on streamflow components in a large forest-dominated watershed. *J. Hydrol.* **2018**, *557*, 448–459.

8. Schaake, J.C. From climate to flow. In *Climate Change and U.S. Water Resources*; Waggoner, P.E., Ed.; John Wiley: New York, NY, USA, 1990.

9. Milly, P.C.D.; Dunne, K.A. Macroscale water fluxes 2. Water and energy supply control of their interannual variability. *Water Resour. Res.* **2002**, *38*, 24-1–24-9.

10. Wang, G.; Zhang, J.; Yang, Q. Attribution of runoff change for the Xinshui river catchment on the Loess Plateau of China in a changing environment. *Water* **2016**, *8*, 267–280.

11. Hou, Y.; Zhang, M.; Liu, S.; Sun, P.; Yin, L.; Yang, T.; Li, Q.; Wei, X. The hydrological impact of extreme weather-induced forest disturbances in a tropical experimental watershed in south China. *Forests* **2018**, *9*, 734.

12. Marhaento, H.; Booij, M.J.; Rientjes, T.H.M.; Hoekstra, A.Y. Attribution of changes in the water balance of a tropical catchment to land use change using the SWAT model. *Hydrol. Process.* **2017**, *31*, 2029–2040.

13. Anand, J.; Gosain, A.K.; Khosa, R. Prediction of land use changes based on Land Change Modeler and attribution of changes in the water balance of Ganga basin to land use change using the SWAT model. *Sci. Total Environ.* **2018**, *644*, 503–519. [PubMed]

14. Zuo, D.; Xu, Z.; Yao, W.; Jin, S.; Xiao, P.; Ran, D. Assessing the effects of changes in land use and climate on runoff and sediment yields from a watershed in the Loess Plateau of China. *Sci. Total Environ.* **2016**, *544*, 238–250. [PubMed]

15. Ahn, K.H.; Merwade, V. Quantifying the relative impact of climate and human activities on streamflow. *J. Hydrol.* **2014**, *515*, 257–266.

16. Montenegro, A.; Ragab, R. Hydrological response of a brazilian semi-arid catchment to different land use and climate change scenarios: A modelling study. *Hydrol. Process.* **2010**, *24*, 2705–2723.

17. Poelmans, L.; Rompaey, A.V.; Ntegeka, V.; Willems, P. The relative impact of climate change and urban expansion on peak flows: A case study in central Belgium. *Hydrol. Process.* **2011**, *25*, 2846–2858.

18. Zhao, F.; Zhang, L.; Xu, Z.; Scott, D.F. Evaluation of methods for estimating the effects of vegetation change and climate variability on streamflow. *Water Resour. Res.* **2010**, *46*, 742–750.

19. Zhang, S.L.; Yang, D.W.; Yang, H.B.; Lei, H.M. Analysis of the dominant causes for runoff reduction in five major basins over China during 1960–2010. *Adv. Water Sci.* **2015**, *26*, 605–613.

20. United Nations Disaster Assessment and Coordination Team (UNDAC)/United Nations Inter-Agency Mission). Final Report on 1998 Floods in the People's Republic of China. UN Office for the Coordination of Humanitarian Affairs, 7–25. 1998. Available online: https://reliefweb.int/report/china/final-report-1998-floods-peoples-republic-china (accessed on 31 December 2018).

21. Shen, Q.; Cong, Z.; Lei, H. Evaluating the impact of climate and underlying surface change on runoff within the Budyko framework: A study across 224 catchments in China. *J. Hydrol.* **2017**, *554*, 251–262.

22. Zhang, M.; Liu, N.; Harper, R.; Li, Q.; Liu, K.; Wei, X. A global review on hydrological responses to forest change across multiple spatial scales: Importance of scale, climate, forest type and hydrological regime. *J. Hydrol.* **2017**, *546*, 44–59.

23. Zheng, Y.; Huang, Y.; Zhou, S.; Wang, K.; Wang, G. Effect partition of climate and catchment changes on runoff variation at the headwater region of the Yellow River based on the Budyko complementary relationship. *Sci. Total Environ.* **2018**, *643*, 1166–1177.

24. Wang, W.G.; Shao, Q.X.; Yang, T.; Peng, S.Z.; Xing, W.Q.; Sun, F.C.; Luo, Y.F. Quantitative assessment of the impact of climate variability and human activities on runoff changes: A case study in four catchments of the Haihe River basin, China. *Hydrol. Process.* **2013**, *27*, 1158–1174.

25. Gao, C.; Tian, R. The influence of climate change and human activities on runoff in the middle reaches of the huaihe river basin, China. *J. Geogr. Sci.* **2018**, *28*, 79–92.

26. Zhang, J.Y.; Wang, G.Q.; Pagano, T.C.; Jin, J.L.; Liu, C.S.; He, R.M.; Li, Y.L. Using hydrologic simulation to explore the impacts of climate change on runoff in the Huaihe River Basin of China. *J. Hydrol. Eng.* **2013**, *18*, 1393–1399.

27. Meng, D.; Mo, X. Assessing the effect of climate change on mean annual runoff in the songhua river basin, China. *Hydrol. Process.* **2012**, *26*, 1050–1061.

28. Xi, Y.; Peng, S.; Ciais, P.; Guimberteau, M.; Li, Y.; Piao, S.; Zhou, F. Contributions of climate change, CO2, land-use change and human activities to changes in river flow across ten Chinese basins. *J. Hydrometeorol.* **2018**, *19*, 1899–1914.

29. Wu, L.; Wang, S.; Bai, X.; Luo, W.; Tian, Y.; Zeng, C.; He, S. Quantitative assessment of the impacts of climate change and human activities on runoff change in a typical karst watershed, SW China. *Sci. Total Environ.* **2017**, *601*, 1449–1465. [PubMed]

30. Liu, J.; Zhang, Q.; Singh, V.P.; Shi, P. Contribution of multiple climatic variables and human activities to streamflow changes across China. *J. Hydrol.* **2017**, *545*, 145–162.

31. Editorial Committee of Encyclopedia of Rivers and Lakes in China (EC-ERLC). Section of Yangtze River basin (Vol. 2). In *Encyclopedia of Rivers and Lakes in China*; China Water Power Press: Beijing, China, 2010.

32. Yang, Q.; Zhang, H.; Peng, W.; Lan, Y.; Luo, S.; Shao, J.; Chen, D.; Wang, G. Assessing climate impact on forest cover in areas undergoing substantial land cover change using Landsat imagery. *Sci. Total Environ.* **2019**, *659*, 732–745.

33. Cheng, J.; Xie, B. Trend analysis and countermeasures of flood peak flow trend of flood Peak water level in Qingliuhe County Station. *South China Agric.* **2017**, *11*, 118–120. (In Mandarin). Available online: http://www.cnki.com.cn/Article/CJFDTotal-NFNY201715064.htm (accessed on 30 July 2018).

34. Yang, Q.; Zhang, H.; Wang, G.; Luo, S.; Chen, D.; Peng, W.; Shao, J. Dynamic runoff simulation in a changing environment: A data stream approach. Environ. *Model. Softw.* **2019**, *112*, 157–165.

35. Zhang, J. A comparative study of population size prediction models in Chuzhou. *J. Anhui Agric. Univ.* **2007**, *34*, 405–409. Available online: http://www.airitilibrary.com/Publication/alDetailedMesh?docid=1672352x-200707-34-3-405-409-a (accessed on 30 July 2018).

36. Zhang, J.; Pu, L. On coordination between urbanization and farmland area of Chuzhou city in recent 30 years. *Soils* **2008**, *40*, 523–528. (In Mandarin)

37. Kendall, M.G. *Rank Correlation Measures*; Charles Griffin: London, UK, 1975.

38. Pohlert, T. Non-Parametric Trend Tests and Change-Point Detection. Available online: http://cran.stat.upd.edu.ph/web/packages/trend/vignettes/trend.pdf (accessed on 30 July 2018).

39. Yamamoto, R.; Iwashima, T.; Hoshiai, M. An analysis of climatic jump. *J. Meteor. Soc. Jpn.* **1986**, *64*, 273–281.

40. Cong, Z.T.; Yang, D.W.; Ni, G.H. Does evaporation paradox exist in china? *Hydrol. Earth Syst. Sci.* **2009**, *13*, 357–366.

41. Han, S.; Xu, D.; Wang, S. Decreasing potential evaporation trends in China from 1956 to 2005: Accelerated in regions with significant agricultural influence? *Agric. For. Meteorol.* **2012**, *154*, 44–56.

42. Han, S.; Tian, F.; Hu, H. Positive or negative correlation between actual and potential evaporation? Evaluating using a nonlinear complementary relationship model. *Water Resour. Res.* **2014**, *50*, 1322–1336.

43. Chuzhou Local Chronicles Compilation Committee. Chuzhou Chorography. Available online: http://dfz.chuzhou.gov.cn/4290494.html (accessed on 9 February 2019).

![forests logo] *forests*

MDPI

*Article*

# The Cumulative Effects of Forest Disturbance and Climate Variability on Streamflow in the Deadman River Watershed

**Krysta Giles-Hansen** [1], **Qiang Li** [2,*] **and Xiaohua Wei** [1]

1 Department of Earth, Environmental and Geographic Sciences, University of British Columbia Okanagan, 1177 Research Road, Kelowna, BC V1V 1V7, Canada; Krysta.Giles-Hansen@ubc.ca (K.G.-H.); adam.wei@ubc.ca (X.W.)

2 Department of Civil Engineering, University of Victoria, 3800 Finnerty Road, Victoria, BC V8P 5C2, Canada

* Correspondence: liqiang1205@gmail.com

Received: 11 January 2019; Accepted: 20 February 2019; Published: 22 February 2019

**Abstract:** Climatic variability and cumulative forest cover change are the two dominant factors affecting hydrological variability in forested watersheds. Separating the relative effects of each factor on streamflow is gaining increasing attention. This study adds to the body of literature by quantifying the relative contributions of those two drivers to the changes in annual mean flow, low flow, and high flow in a large forested snow dominated watershed, the Deadman River watershed (878 km$^2$) in the Southern Interior of British Columbia, Canada. Over the study period of 1962 to 2012, the cumulative effects of forest disturbance significantly affected the annual mean streamflow. The effects became statistically significant in 1989 at the cumulative forest disturbance level of 12.4% of the watershed area. The modified double mass curve and sensitivity-based methods consistently revealed that forest disturbance and climate variability both increased annual mean streamflow during the disturbance period (1989–2012), with an average increment of 14 mm and 6 mm, respectively. The paired-year approach was used to further investigate the relative contributions to low and high flows. Our analysis showed that low and high flow increased significantly by 19% and 58%, respectively over the disturbance period ($p < 0.05$). We conclude that forest disturbance and climate variability have significantly increased annual mean flow, low flow and high flow over the last 50 years in a cumulative and additive manner in the Deadman River watershed.

**Keywords:** cumulative effects; forest disturbance; climate change; annual streamflow; low flow; high flow

---

## 1. Introduction

Forested watersheds are important sources of streamflow, water regulation and generation for more than 30% of the world's population [1,2]. Changes in forest cover have important effects on the sustainability of water resources, aquatic habitat, and many other ecological functions. In forested watersheds, forest cover change and climate variability are regarded as the two main drivers of hydrological variation [3,4]. The compounding effects of climate variability and land cover change (e.g., forest harvesting, land use conversion) have driven scientific research to focus on how these drivers and their interactions affect the hydrological regime [5–9].

The effect of forest cover change on streamflow has been studied for many decades, with periodic summaries consolidating key findings [4,9–11]. However, most research has been carried out using paired watershed experiments (PWE) in small watersheds (<100 km$^2$) since the 1910s with long term land cover and streamflow data. The inability to simply extrapolate results from the PWE to watersheds of different sizes, topography, climate and land cover types, suggests an important need to study hydrological responses to forest changes at other spatial scales. This gap is recently and clearly recognized in a recent global assessment report on forests and water [12]. Additionally, advances have been made in hydrological modeling (e.g., the Soil and Water Assessment Tool) and statistical methods such as time-trend analysis, sensitivity-based method, double mass curves (DMC), and those based on the Budyko theory [3,13]. These advances make it possible to evaluate the forest-water relationship in large watersheds.

Numerous forest indicators have been used to characterise forest change through time in hydrological studies including: percentage forested [14], the area disturbed, and remotely sensed vegetation indices such as the Normalized Difference Vegetation Index (NDVI) [2,15], Leaf Area Index (LAI) [16–18], and normalized burn ratio (NBR) [19]. Although these indices have been widely implemented in assessing spatial and temporal dynamics of forest cover, they do not explicitly account for hydrological recovery due to regeneration following forest disturbance, and consequently cumulative effects on hydrology. Cumulative effects are defined as the combined results from actions that are individually minor but collectively significant in the past, present, and foreseeable future [20,21]. Therefore, a comprehensive index is required, which can account for spatial and temporal cumulative forest cover change. ECA (equivalent clear-cut area) is calculated as the cumulative area that is clear-cut with a reduction factor implemented over time to account for hydrological recovery as the forest regenerates [22]. ECA has been widely used in scientific research and operational practices in the Pacific Northwest for decades [23–27], and can therefore, be a good indicator to quantify cumulative forest cover change in a watershed.

Previous studies generally conclude that deforestation increases annual mean flow ($Q_{mean}$),while reforestation decreases it across multiple spatial scales [4,11], with less consistent results in the literature on low and high flows. For instance, it has been found that forest logging could increase the frequency and magnitude of high flows but is unlikely to affect large flooding events [23,28]. In snow dominated watersheds, the effect is less significant and forest disturbance has been found to advance the timing of snow-melt generated high flows [29]. The effect on low flow is tightly coupled to soil disturbance, the history of land use, and climatic regime [30,31]. Numerous studies have found that disturbance can increase low flow, however some show inconsistent results with negative and insignificant changes in low flows [26,31]. Our collective understanding of how forest cover change might affect streamflow has progressed, however, there is an increasing demand for more case studies, shown by the limited number of studies and lack of consistency or explanation of differences between results from various regions characterised with different climate regimes, forest structure, soil property, and topography [9].

The Deadman River watershed (878 km$^2$) is located in the Southern Interior of British Columbia, Canada. It is characterised by a snow dominated hydrological regime and has experienced dramatic forest disturbance from the mountain pine beetle (MPB) infestation and forest harvesting. The significant level of forest disturbance in the watershed has raised serious concerns over alteration of the hydrological regime. To address these hydrological concerns, this study answers two research questions: 1) how have forest disturbance and climate cumulatively affected annual streamflow in Deadman River watershed, and 2) how have forest disturbance and climate cumulatively affected high and low flows?

## 2. Materials and Methods

### 2.1. Watershed Description

The Deadman River watershed is located in the Southern Interior of British Columbia, Canada. (Figure 1). The drainage area is 878 km$^2$ of which 91.3% is forested, with most of the remaining area being grassland in the valley bottoms. Deadman River is an important Salmonoid-bearing river that is a tributary of Thompson River. Communities and First Nations rely directly on the water to drink, irrigate, and sustain a healthy fish population, ecological functioning and aquatic resources. The elevation ranges from 527 m above sea level at the southern outlet and main branch of the Deadman River up to 1779 m in the upper reaches to the east and west of the watershed (Figure 1). From 1962–2012, the mean daily temperature in the winter (December–February) was −6 °C and 12.9 °C in the summer (June–August). Approximately 27% of the annual precipitation (P) accumulates as snow in the winter, with the snow-melt event producing the spring freshet (Figure S1 in the Supplementary Materials (SM)). Lower elevation slopes support Douglas-fir (*Pseudotsuga menziesii* (Mirb.) Franco) and Lodgepole pine (*Pinus contorta* Douglas ex Loudon) dominated forests with smaller amounts of Spruce (*Picea glauca* (Moench) Voss) and Balsam *(Abies lasiocarpa* (Hook.) Nutt.). A small amount of agriculture and urban development dominate the valley bottom (around 1% of the total watershed area), the total area of which has not changed over the course of this study; while forestry is the main land use across the watershed [32]. The recent Province-wide mountain pine beetle (MPB) epidemic [33] has brought widespread Lodgepole pine mortality and salvage harvesting throughout the watershed (Figure 1).

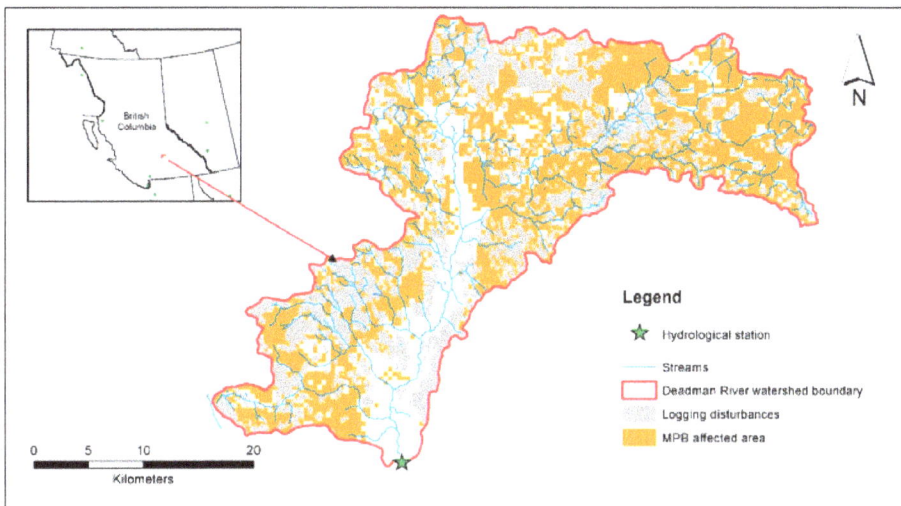

**Figure 1.** The location of watershed boundary, hydrometric station, stream network, forest logging, and mountain pine beetle (MPB) infestation of the Deadman River watershed, located in British Columbia, Canada.

## 2.2. Watershed Data

The age, height, density, and species composition of the forest was sourced from the 2013 provincial vegetation resources inventory (VRI). Disturbance indicators were calculated using three complementary spatial databases that are maintained by the BC Ministry of Forests, Lands and Natural Resource Operations (FLNRO). The spatial and temporal location of historical harvesting was accounted for using the FLNRO consolidated cutblocks layer (2013), wildfires from the BC wildfire database, and mountain pine beetle (MPB) disturbance from the British Columbia Mountain Pine Beetle (BCMPB) projections version 11 [33].

Daily discharge data ($m^3s^{-1}$) from 1960 to 2012 was acquired from Environment Canada (Station ID: 08LF027), which was standardized to millimeters (mm year$^{-1}$) using watershed area (Figure S2). The watershed monthly P, mean, maximum, and minimum temperatures ($T_{mean}$, $T_{max}$, $T_{min}$) were derived from the ClimateBC model at the spatial resolution of 500 × 500 m [34] (Figures S3 and S4 in the SM). Elevation (m) for each VRI polygon was calculated from the FLNRO digital elevation model at a spatial resolution of 30 meters.

*2.3. Cumulative Equivalent Clear-Cut Area*

Forest logging, MPB infestation, and wildfire are the three major disturbance types and cumulate over space and time in the study watershed (Figure 1). Cumulative equivalent clear-cut area (CECA) was used to account for cumulative forest disturbance at the watershed level [22,35]. At the stand level, equivalent clear-cut area (ECA) is defined as the area that has been clear-cut, killed by fire, or infested by MPB, with a reduction factor to account for hydrological recovery as the forest regenerates. All forest logging in the area is clear-cut. Following the watershed assessment procedure in British Columbia [27], the ECA is set at 100% after harvesting, reflecting changes in hydrological processes such as infiltration, evapotranspiration, and run-off. Stand height was used to represent the relationship between forest growth and hydrological recovery [22,27] (Table 1). Hydrological recovery after wildfire was assumed to follow the same relationship as harvesting [36]. The hydrological recovery of a forest stand is related to its growth rate over time and is therefore determined by many factors including disturbance type, climate, and tree species. Accounting for this, stand height is projected through time using standard models that are calibrated in British Columbia. The Variable Density Yield Prediction model version 7 (VDYP7) is used for stands of natural origin (wildfire) and the Table Interpolation Program for Stand Yields model version 4.3 (TIPSYv4.3) is used for stands that regenerate after harvesting. In MPB affected forest, the ECA shown in Figure 2 [29,37] is applied to the affected portion of a stand. It follows an asymmetrical bell-shaped curve through time as tree death and needle drop occur gradually over years [38,39] and regeneration and hydrological recovery starts around 20 years. Stand-level ECA values are summed annually to give the watershed–level CECA timeseries.

**Table 1.** Hydrological recovery and equivalent clear-cut area (ECA) for stands disturbed by wildfire or harvesting according to height in meters (m) of the leading tree species.

| Height (m) | Hydrologic Recovery (%) | ECA (%) |
|:---:|:---:|:---:|
| $0 \leq 3$ | 0 | 100 |
| $3 \leq 5$ | 25 | 75 |
| $5 \leq 7$ | 50 | 50 |
| $7 \leq 9$ | 75 | 25 |
| >9 | 90 | 10 |

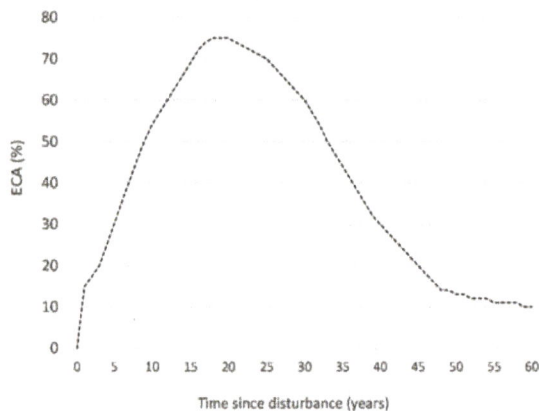

**Figure 2.** Equivalent clear-cut area (ECA) (%) since year of disturbance by mountain pine beetle (MPB).

In 2012, the CECA was 41.3% of the watershed area (Figure 3). Logging was the dominant disturbance type through the study period. The average annual harvest rate in the Deadman River watershed was 4 km$^2$ year$^{-1}$ from 1960 till 1999. From the year 2000 onwards, this rate increased to an average of 11 km$^2$ year$^{-1}$. In 2012, the CECA from logging was 27.2% of the watershed area. MPB affected a total of 370 km$^2$ since 2000, a rate of 46 km$^2$ year$^{-1}$ (Figure 3). The CECA of MPB in 2012 was 13.6%. Wildfire is an insignificant disturbance agent during the study period with only 3 km$^2$ in total. The total CECA rose from 0% in 1960 to 41% in 2012. Overall, the Deadman River watershed has experienced significant forest disturbance.

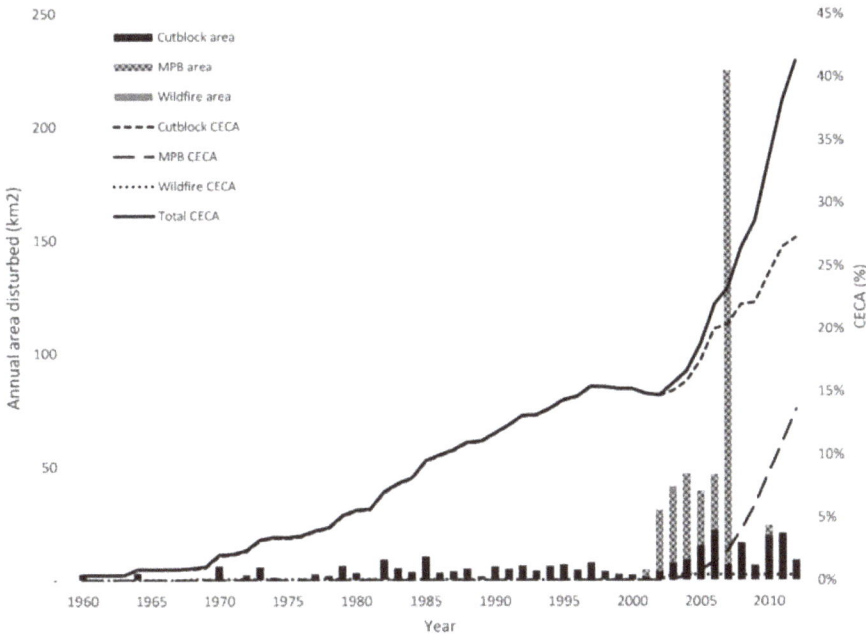

**Figure 3.** Annual area disturbed (km$^2$) and Cumulative Equivalent clear-cut area (CECA) in percent (% of watershed area) by disturbance type and total in the Deadman River watershed from 1960 to 2012.

*2.4. Cross-Correlation Analysis*

Cross-correlation analysis is an effective method to examine whether there are significant relationships between two time series variables. The advantage of cross-correlation is that it can remove autocorrelations existing in data series and identify lagged causality between them [29]. In this study, cross-correlation tests were used to detect relationships and lagged effects between cumulative forest change (CECA) and three flow variables, including: annual mean streamflow ($Q_{mean}$), a low flow parameter ($Q_{95\%}$) which is defined as the daily flow equalled or exceeded for 95% of days in a year, and high flow parameter ($Q_{5\%}$) which is the daily flow equalled or exceeded 5% of days in year.

All variables were pre-whitened to remove autocorrelation by fitting Autoregressive Integrated Moving Average (ARIMA) models in R software [40]. The *ndiffs* function from the forecast package [41] was first used to find the level of differencing needed to achieve stationarity and then the ARIMA model with the best performancewas selected for cross-correlation analysis [42]. Cross-correlation analysis of ARIMA model residuals was carried out in R using the *ccf* function from the forecast package [41]. The correlation coefficient and significant lags are used to assess if there is a correlation between tested variables.

*2.5. Quantifying the Effects of Forest Disturbance and Climate Variability on Streamflow Components*

Three complementary statistical methods were used to make a robust assessment of the separate effects of forest disturbance and climate variation on streamflow components. The modified double mass curve (MDMC) method and sensitivity-based method were applied to $Q_{mean}$, while the paired-year approach can be used for $Q_{mean}$, $Q_{5\%}$ and $Q_{95\%}$.

### 2.5.1. Timeseries Trend Analysis

As a first step, the Mann–Kendall test [43,44] was used to assess what trends exist in the climatic and hydrological data over the study period, helping with the interpretation of results. The Mann–Kendall non-parametric test is widely used in hydrology for the trend detection [23]. Prior to implementing the Mann–Kendall test, each timeseries was pre-whitened to remove the influence of serial correlations [45], following the process recommended in Yue et al. [46]. However, Razavi and Vogel [47] recently demonstrated how pre-whitening can underestimate extreme conditions in hydrological timeseries analysis, leading to a potentially conservative correlation in this analysis. All statistical tests are at a significance level of 5%.

### 2.5.2. Modified Double Mass Curve

A modified double mass curve (MDMC) is a graph of cumulative annual $Q_{mean}$ versus cumulative effective precipitation ($P_e$) which is the difference between annual P and actual evapotranspiration (AET) [3,22]. Annual potential evapotranspiration (PET) was estimated as the average of the Priestley–Taylor [48] and Hamon method [49–51] and then AET was calculated as the average of the Budyko [52] and Zhang's [53] equations. The basic assumptions of the MDMC are 1) the linear relationship holds between the $P_e$ and $Q_{mean}$; and 2) all climate variability is lumped into P and ET [22,54]. In a period with no forest disturbance (reference period), the slope of the MDMC should be straight, and a break point in this line suggests a regime shift in annual mean flow caused by forest disturbance (disturbance period) [3]. The Pettitt break point test was introduced to detect the statistically significant break point between the reference and disturbance periods [55]. Before carrying out the Pettitt test, autocorrelations in the MDMC slope were removed following the methodology found in Yue et al. [46]. To validate the statistical significance of the break point, the nonparametric Mann–Whitney U test Z statistic was adopted [8,30,56]. The linear relationship in the reference period was used to predict the cumulative annual streamflow for the disturbance period, with the difference between this and the observed line regarded as the cumulative impact of forest change on streamflow ($\Delta Q_f$). The deviations caused by climate variability on each annual streamflow component can be determined using Equation (1).

$$\Delta Q_c = \Delta Q - \Delta Q_f \tag{1}$$

where, $\Delta Q$, $\Delta Q_f$, and $\Delta Q_c$ are the deviations of each annual streamflow between disturbance and reference periods, annual flow deviations caused by forest disturbance, and annual flow deviations caused by climate variability, respectively. The relative contributions of forest disturbance and climate variability to $Q_{mean}$ is calculated from their respective proportion of $\Delta Q$ [22].

### 2.5.3. Sensitivity-Based Method

The sensitivity-based approach assumes that the change in $Q_{mean}$ can be determined using Equation (2) [57,58].

$$\Delta Q_c = \beta \Delta P + \gamma \Delta PET \tag{2}$$

where, $\Delta P$, $\Delta PET$, and $\Delta Q_c$ are the difference in annual P, PET, and Q due to climate between reference and disturbance periods, respectively. $\beta$ and $\gamma$ are the sensitivity coefficients of streamflow to P and PET [36]. $\beta$ and $\gamma$ can be derived from Equations (3) and (4) below, where $\omega$ is assumed to be 2 for a predominantly forested watershed.

$$\beta = \frac{1 + \frac{2PET}{P} + 3\omega\left(\frac{PET}{P}\right)^2}{\left[1 + \frac{PET}{P} + \omega\left(\frac{PET}{P}\right)^2\right]^2} \tag{3}$$

$$\gamma = -\frac{1 + 2\omega \frac{PET}{P}}{\left[1 + \frac{PET}{P} + \omega \left(\frac{PET}{P}\right)^2\right]^2} \quad (4)$$

As a result, the effects of forest change on $Q_{mean}$ can be derived using Equation (1).

### 2.5.4. Paired-Year Approach

Similar to the approaches for assessing the cumulative effects of forest disturbance on $Q_{mean}$, the effects of climate variability on $Q_{5\%}$ and $Q_{95\%}$ should be removed first. The paired-year approach has been effectively used to quantify the cumulative effects of forest disturbance on flow regimes across various climatic regions [35,42] and was selected for this study. The paired-year approach assumes that flow changes are mainly attributed to cumulative forest disturbance when climate conditions between the reference and disturbance year are similar. To identify climatic variables that can be used as proxies for equivalent climate conditions, the following steps are used: (1) Kendall tau correlation analyses are used to select the statistical relationship between flow ($Q_{5\%}$ and $Q_{95\%}$) and seasonal or annual climate variables, respectively (Table S1). (2) Once key climate variables were determined, several combinations or sets of the selected climate variables were composed. (3) Each set of the climate variables and the set of $Q_{5\%}$ and $Q_{95\%}$ serve as inputs for the canonical correlation analyses [35], the approach used to examine correlations between two sets of variables. To ensure that all combinations of climate variables were tested thoroughly, we also selected climate variables that were not statistically related to the $Q_{5\%}$ and $Q_{95\%}$ for the canonical correlation analyses. A total of 30 sets were tested and finally the $T_{mean}$, $T_{max}$, minimum summer temperature, winter P of the antecedent year and spring P in the current year, were determined as the controlling climate variables. To ensure a reliable pairing, a threshold of 10% biases were allowed in each climatic variable (Table S2). (4) As a result, ten pairs of years were chosen to compare $Q_{5\%}$ and $Q_{95\%}$ with similar climate (Table S3). The differences between each pair of years are denoted as the effects of cumulative forest disturbances on $Q_{5\%}$ and $Q_{95\%}$. (5) The Mann–Whitney U test is further used to confirm whether $Q_{5\%}$ and $Q_{95\%}$ between the reference and disturbance are statistically significant [35]. As such, the cumulative forest disturbance on $Q_{5\%}$ and $Q_{95\%}$ were quantified.

## 3. Results

### 3.1. Time-Trend Analysis of the Hydrometeorological Variables

Mann–Kendall trend analysis was used to study the annual and seasonal trends in hydrometeorological variables for the Deadman River watershed across the whole study period. Table 2 presents the results of this analysis after pre-whitening following Yue et al. [46]. Although only annual temperatures experienced significant upwards trends (at a significance level ($p$-value) < 0.05), spring and summer $T_{min}$ were significant. Related to increasing $T_{mean}$, annual PET also exhibited a significant upward trend. Spring P was the only season that showed a significant increasing trend for P. For streamflow, only autumn $Q_{mean}$ and $Q_{95\%}$ showed an increasing trend with no corresponding significant climate trend. The increasing trends in temperatures and PET may, therefore, play a role in reducing streamflow, while increased spring P may increase spring streamflow. While, time-trend analysis is useful to help understand hydrologic behaviour, it does not have explanatory power on its own. In addition, recent reviews by Razavi and Vogel [47] and Serinaldi et al. [59] have exposed shortcomings when used in hydrological timeseries analysis. Therefore, we use the results from time-trend analysis with caution and as one piece to inform the overall picture.

**Table 2.** Time-trend analysis of hydro-meteorological variables in the Deadman River watershed from 1962 to 2012 (spring: March–May; summer: June–August; autumn: September–November; winter: December–February; $T_{max}$ is maximum daily annual temperature (°C); $T_{min}$ is minimum daily annual temperature (°C); $T_{mean}$ is average daily annual temperature (°C); P is annual precipitation (mm); PET is potential evapotranspiration (mm); AET is actual evapotranspiration (mm); $Q_{mean}$ is annual mean flow (mm); $Q_{95\%}$ is the low flow parameter; $Q_{5\%}$ is the high flow parameter; tau is the z-statistic from the Mann–Kendall test indicating the direction of change of the variable; p-value is the level of significance from the Mann–Kendall test; and bolded italics indicate significant trends at a significance level of 0.05).

| Season | | $T_{max}$ | $T_{min}$ | $T_{mean}$ | P | PET | AET | $Q_{5\%}$ | $Q_{95\%}$ | $Q_{mean}$ |
|---|---|---|---|---|---|---|---|---|---|---|
| Annual | tau | *0.19* | *0.25* | *0.23* | −0.02 | *0.25* | 0.06 | −0.04 | *0.24* | 0.01 |
| | p-value | *0.05* | *0.01* | *0.01* | 0.87 | *0.01* | 0.55 | 0.73 | *0.02* | 0.95 |
| Spring | tau | 0.08 | *0.18* | 0.15 | *0.21* | 0.17 | *0.25* | −0.01 | 0.05 | −0.06 |
| | p-value | 0.41 | *0.05* | 0.11 | *0.02* | 0.08 | *0.01* | 0.91 | 0.59 | 0.55 |
| Summer | tau | 0.09 | *0.25* | 0.13 | 0.04 | 0.16 | 0.08 | −0.08 | 0.11 | −0.01 |
| | p-value | 0.35 | *0.01* | 0.17 | 0.68 | 0.09 | 0.42 | 0.45 | 0.29 | 0.96 |
| Autumn | tau | 0.16 | 0.04 | 0.06 | 0.05 | 0.10 | 0.05 | 0.06 | *0.26* | *0.23* |
| | p-value | 0.08 | 0.71 | 0.50 | 0.60 | 0.27 | 0.57 | 0.54 | *0.01* | *0.02* |
| Winter | tau | 0.13 | 0.15 | 0.12 | −0.07 | 0.11 | 0.10 | 0.13 | *0.21* | 0.14 |
| | p-value | 0.15 | 0.10 | 0.20 | 0.47 | 0.24 | 0.27 | 0.20 | *0.04* | 0.15 |

### 3.2. Cross-Correlation between Forest Disturbance and Streamflow Regime Components

Annual hydrological and CECA timeseries were first pre-whitened to remove serial correlations using ARIMA models as listed in Table 3. The pre-whitened variables from this process were used in the cross-correlation analysis. Cross-correlation analysis cannot prove causality but is used to calculate the statistical relationship between two data series considering the displacement ("lag") of one relative to the other. Cross-correlation analysis revealed that CECA is significantly related to $Q_{mean}$, $Q_{5\%}$ and $Q_{95\%}$ as shown by significant correlation coefficients for all hydrologic variables. The lags indicate that the response of hydrological variables are 8, 8, and 4 years after the change in CECA. The lag detected by cross-correlation reflects that the observed response of streamflow to forest cover change only occurs when disturbance accumulates to certain amount in a watershed. Additionally, changes caused by MPB mortality such as the cessation of root functioning, needle drop and decomposition occur gradually over a number of years [60]. The positive correlation coefficient implies that a higher level of cumulative disturbance, approximated by CECA, likely results in higher annual mean, low flows, and high flows (Table 3). The magnitude of the cross-correlation coefficient calculates the strength of the statistical relationship, all coefficients are around 0.3, which is considered a weak relationship, albeit statistically significant. Forest disturbance can explain some but not the majority of variation in the hydrological variables, reflecting the direct influence of climate variability on streamflow.

**Table 3.** Cross-correlation results between the residuals of logged Autoregressive Integrated Moving Average (ARIMA) time series models for cumulative equivalent clear-cut area (CECA) and annual streamflow components in the Deadman River watershed from 1960 to 2012. The bold italicised coefficient value indicates that the model is significant at p-value = 0.05. Streamflow components include annual mean flow (Q), low flow ($Q_{5\%}$), and high flow ($Q_{95\%}$). The ARIMA model used is denoted by ln(p,d,q), where ln represents that the log of the data was taken prior to running the ARIMA model, p is the order of the auto-regressive model, d is the degree of differencing, and q is the order of moving-average chosen for the ARIMA model.

| Hydrological Variables | Cross-Correlation with CECA (Pre-Whitened Using ARIMA (2, 2, 1)) | | |
|---|---|---|---|
| | ARIMA Model Used to Pre-Whiten | Coefficients | Lag |
| Annual mean flow (Q) | (1, 3, 0) | *0.32* | 8 |
| High flow ($Q_{5\%}$) | (0, 1, 1) | *0.33* | 8 |
| Low flow ($Q_{95\%}$) | (0, 1, 2) | *0.32* | 4 |

### 3.3. Separation of the Effects Of Forest Disturbance and Climate Variability on Annual Mean Flow

#### 3.3.1. Modified Double Mass Curve

A breakpoint was identified on the MDMC, which plots cumulative effective precipitation ($P_{ae}$) versus cumulative annual mean flow ($Q_a$) (Figure 4). The observed cumulative $Q_a$ is greater than predicted after the break point, indicating that forest disturbance has led to an increase in $Q_a$. The Pettitt break point test indicates that there is a statically significant regime shift in 1989, leading to the pre- and post-disturbance periods being defined as 1960 to 1989 and 1990 to 2012, respectively. Mann–Whitney U tests confirmed that the slopes of MDMCs before and after the break point were statistically different ($p$-value < 0.001). The break point coincides with the history of forest disturbance as at the end of 1989, the cumulative area disturbed is 80.4 km² and the CECA is 12.4% (Figure 3).

Further calculations on the difference between observed and predicted $Q_a$ in the post-disturbance period show that forest change increased Q ($\Delta Q_f$) by an average of 16.6 mm annually. In contrast, climate variability played a more minor role in streamflow alteration, i.e. the $Q_{mean}$ change attributed to climate ($\Delta Q_c$) is 2.7 mm in the disturbance period (Table 4). As a result, the relative contribution of forest change to the total change in Q ($R_f$) was 86.2% while the relative contribution of climate variability ($R_c$) was 18.8%. As indicated, the breaking point was not determined visually from the MDMC, but rather statistically using the Pettitt break point test. As a result, the calculated $\Delta Q_f$ and $\Delta Q_c$ for an individual year fluctuates with large variations in climatic inputs (primarily P). There are some dry years, when the observed line dips below the predicted in Figure 3 and the $\Delta Q_c$ is negative. However, over the entire disturbance period, the calculated $\Delta Q_f$ and $\Delta Q_c$ are both positive.

Overall, the effects of climate variability on streamflow were much lower than those from forest disturbance, indicating streamflow variations in the Deadman River watershed were mainly caused by cumulative forest disturbance.

The disturbance period was further divided into five sub-periods to examine the temporal role of forest disturbance and climate variability in streamflow. As shown in Table 4, the $R_f$ and $R_c$ to streamflow showed temporal variations. For example, with less forest disturbance, the $R_f$ was lower in 1995–1999 than in the other periods.

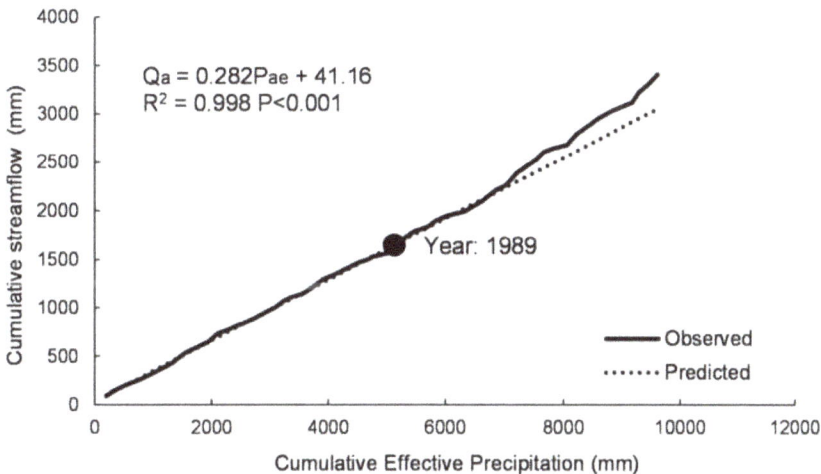

**Figure 4.** Modified Double Mass Curve (MDMC) of cumulative effective precipitation ($P_{ae}$) versus cumulative annual mean flow ($Q_a$) in the Deadman River watershed from 1960 to 2012. The 'Predicted' line is from the linear equation shown on the graph. The breakpoint is in 1989.

### 3.3.2. Sensitivity-Based Method

Similar to MDMC, results from the sensitivity-based method show an overall increase in $Q_{mean}$ due to cumulative forest change. Here, the increases in $Q_{mean}$ attributed to cumulative forest disturbance and climate variability in the period 1990–2012 were an average of 10.8 and 8.9 mm respectively (Table 4). The $R_f$ derived from the sensitivity-based method (86.2%) is lower than that from the MDMC (54.7%). As with MDMC, the $\Delta Q_f$ and $\Delta Q_c$ for an individual year (or selected period) will fluctuate depending on the climatic inputs for that time. For example, the MDMC method calculated $\Delta Q_c$ −1.61 mm in the period 2000 – 2004, reflecting the lower than average P in four out of five years in that period. The effect of forest disturbance is much more consistent with $\Delta Q_f$ greater than zero in all selected periods. So although the results for an individual year may fluctuate, on average across the whole disturbance period, the two methods clearly indicate that the cumulative forest disturbance played a dominant role in $Q_{mean}$ variation.

**Table 4.** Results from Modified Double Mass Curve (MDMC) and sensitivity-based analyses. Where $\Delta Q$ is the change in annual mean streamflow between the observed and predicted, $\Delta Q_f$ is the difference in streamflow attributed to forest change, $\Delta Q_c$ is the difference in streamflow attributed to climate variability, $R_f$ and $R_c$ are the relative contribution of forest change and climate variability to the total change in Q in that period expressed as a percentage. CECA is the average cumulative equivalent clear-cut area for the selected periods.

| Method | Selected Periods | $\Delta Q$ (mm) | $\Delta Q_f$ (mm) | $\Delta Q_c$ (mm) | $R_f$ (%) | $R_c$ (%) | CECA (%) |
|---|---|---|---|---|---|---|---|
| MDMC | 1990–1994 | 17.53 | 13.71 | 3.82 | 78.23 | 21.77 | 14.56 |
| | 1995–1999 | 30.53 | 14.23 | 16.30 | 46.62 | 53.38 | 17.32 |
| | 2000–2004 | 3.02 | 4.63 | −1.61 | 74.18 | 25.82 | 18.21 |
| | 2005–2009 | 21.57 | 19.71 | 1.85 | 91.40 | 8.60 | 32.43 |
| | 2010–2012 | 27.12 | 34.20 | −7.08 | 82.85 | 17.15 | 51.67 |
| | 1990–2012 | 19.95 | 16.59 | 2.66 | 86.20 | 13.80 | 24.68 |
| Sensitivity-based method | 1990–2012 | 19.65 | 10.75 | 8.90 | 54.70 | 45.30 | 24.68 |

### 3.4. Separation of Effects Of Forest Disturbance and Climate Variability on High and Low Flows

The paired-year approach revealed that the cumulative forest disturbance consistently increased $Q_{5\%}$ and $Q_{95\%}$. The average high flow in the reference period (1962–1989) is 8.07 m$^3$s$^{-1}$ increasing by 58% to 12.74 m$^3$s$^{-1}$ in the disturbance period (1990–2012) (Table 5), with a statistically significant *p*-value. Similarly, the cumulative forest disturbance increased the average low flow by 19% from 0.26 to 0.31 m$^3$s$^{-1}$. These results are consistent with those from the cross-correlation analysis that also showed a positive relationship between forest disturbance and high and low flows. So although there is some uncertainty introduced in the paired-year approach by inexact matching of climate variables in paired reference and disturbance years, the consistency with cross-correlation analysis strengthens the result.

**Table 5.** Average high flow variable ($Q_{5\%}$) and low flow variable ($Q_{95\%}$) in the reference and disturbance periods using the paired-year approach.

| Flow (m$^3$s$^{-1}$) | $Q_{5\%}$ | $Q_{95\%}$ |
|---|---|---|
| Reference | 8.07 | 0.26 |
| Disturbance | 12.74 | 0.31 |
| Change (m$^3$s$^{-1}$) | 4.67 | 0.05 |
| Change (%) | 58% | 19% |
| *p*-value | <0.001 | <0.001 |

## 4. Discussion

### 4.1. The Cumulative Effects of Forest Disturbance on Annual Mean Flow

In this study, cumulative forest disturbance has increased $Q_{mean}$ in the Deadman River watershed. Both the MDMC and sensitivity-based methods derived similar results, giving a more robust answer than one individual method. While the increase in $Q_{mean}$ is consistent with most other studies in the region with a similar climate [22,36,61], some have found no significant change with forest disturbance [23]. Zhang et al. [61] found no relationship between low CECA (<15%) and $Q_{mean}$, indicating that there may be no effect on $Q_{mean}$ with lower levels of disturbance. In two large neighboring watersheds in British Columbia, Zhang and Wei [23] found contrasting responses after high levels of forest disturbance, with the Willow watershed showing a significant increase in $Q_{mean}$ while the Bowron did not. They attributed this to differences in topography, climate and the spatial arrangement of harvesting. This variability highlights that the cumulative effects of forest disturbance are likely watershed specific. Two recent global review papers investigated 162 large watersheds (>1000 km²) [9] and 252 small (<1000 km²) [4] watersheds. These reviews identified that watershed area, climate regime, and forest type are key factors that affect the hydrological response to cumulative forest disturbance. As such, caution must be used when extrapolating results from one location to another and from one time period to another.

### 4.2. Additive Effects of Cumulative Forest Disturbance and Climate Variability on Annual Mean Flow

Both cumulative forest disturbance and climate variability were found to increase $Q_{mean}$ in the disturbance period, meaning that their effects are additive as opposed to offsetting. Here, the term 'additive' signifies that both forest change and climate variability affect streamflow in the same positive direction, it does not imply a simple additive relationship between the two drivers. Complex interactions do exist as studies have shown the effect of disturbance on streamflow is less pronounced at very high levels of rainfall [9,62]. In the Deadman River watershed since 1989, the combined effect of cumulative forest disturbance and climate variability has increased $Q_{mean}$ by an average of 19.6 mm year$^{-1}$. Of the other studies that have investigated the influence of both forest disturbance and climate in this region, all have found that forest disturbance has increased $Q_{mean}$ while the climate decreased $Q_{mean}$ [22,36,63,64]. For example, Wei and Zhang [22] found offsetting effect on $Q_{mean}$ of climate and forest disturbance in the Willow River watershed, as did Li et al. [36] in the Upper Similkameen River watershed. Zhang and Wei [65] found that forest disturbance increased $Q_{mean}$ by 48.4 mm year$^{-1}$, while climatic variability offset this by decreasing $Q_{mean}$ by 35.5 mm year$^{-1}$ in the Baker Creek watershed. Zhang [63] found that the Willow, Baker, Moffat, and Tulameen watersheds all had offsetting effects on $Q_{mean}$ of climate and forest disturbance.

This study is different, in that the MDMC and sensitivity analysis both found that climate has increased $Q_{mean}$, resulting in an additive rather than offsetting effect on $Q_{mean}$. Although most studies have found the climate in the interior of British Columbia to be getting drier, this depends on the period chosen, as illustrated in Zhang and Wei [65] where the influence of the Pacific Decadal Oscillation (PDO) affected the calculated climatic trend. Winkler et al. [8] found that the disturbance period in their study of the Upper Penticton Creek was wetter and cooler than the calibration period, and attributed this to the PDO. In the Deadman River watershed, time-trend analysis indicated that there was an increasing trend in annual $T_{mean}$ and PET, leading to the expectation that climate variability would have decreased $Q_{mean}$ over the study period. Surprisingly, both the MDMC and sensitivity-based method showed that climate variability increased $Q_{mean}$ over the disturbance period. The nearby meteorological station 'Kamloops A' was used to confirm annual trends in P, $T_{min}$, $T_{max}$, and $T_{mean}$ derived from the ClimateBC data. Additionally, more detailed time-trend analyses were conducted for the reference and disturbance periods separately (Figure S5). Results showed that $T_{mean}$ increased significantly over the entire study period, but in contrast, $T_{mean}$ has a decreasing trend over the disturbance period, with statistical insignificance. Unlike $T_{mean}$, P has an insignificant trend for all tested periods (Tables S4 and S5). Although the climate was getting drier over the entire period, there was no significant trend during the disturbance period. As a result, climate variability may have played a positive role in streamflow in the disturbance period. These findings also highlight the need to understand the temporal climate variability in the specific study watershed.

### 4.3. The Cumulative Effects of Forest Disturbance on High and Low Flows

Cross-correlation analysis showed a positive relationship between disturbance and both low and high flows. Similarly, the paired-year approach found that low and high flows increased significantly by 20% and 58% respectively in the disturbance period. The positive relationship of low and high flows to forest disturbance is consistent with most other studies of forest disturbance in neighboring areas [29,35,36]. In the Upper Penticton watershed, Winkler et al. [8] found that while $Q_{mean}$ increased by a small amount (5%) with disturbance, spring run-off (high flow) increased dramatically between 19% and 29% and summer water yield (low flow) decreased by a similar magnitude. In a snow dominated regime such as the Deadman River watershed, forest disturbance likely results in more snow accumulation and earlier snow melt driven spring freshet [8,25], leading to an increase in high flow [29,36]. Synchronisation of snow pack melt [23] or increased quick flow [66] may also contribute to the increase in high flows. The increase in low flow with disturbance is likely due to a different mechanism than high flow. Summer low flows may be increased by the reduced ET side of the water balance equation, with lower growth removing less water from the soil, and consequently increasing recharge [67]. Increased throughfall, infiltration and snowpack accumulation associated with canopy removal can also increase soil moisture and recharge [68]. Winter harvesting practices in the area reduce the soil compaction and associated loss of recharge, however diversion associated with roads and skid trails can modify run-off pathways [26].

However, some studies in large watersheds show no or insignificant change to low flow with forest disturbance. For example, Zhang and Wei [23] investigated low and high flows in two large highly disturbed watersheds, finding that forest harvesting increased peak flows in one but not the other and found no significant changes on low flow. A study in the Canadian Boreal forest found that forest disturbance did not impact peak flow, but increased low flow [69]. Liu et al. [14] found contrasting responses to reforestation in two large watersheds in China with one showing significant and positive effects on low flows and the other not. Another study in subtropical China, found that deforestation increased both low and high flows [42]. Wilk et al. [70] found no changes in hydrology in the 12,100 km$^2$ Nam Pong catchment in Thailand as forest cover was reduced from 80% to 27%. The influence of forest change on low flow depends upon factors such as soils, previous land uses, and climatic regime [30,31]. Soil properties such as porosity and organic matter content affect hydrological processes such as infiltration, surface and subsurface run-off, and can amplify or mute the effects of land use change [66]. For example, soils with a large holding capacity can dampen the effect of vegetation removal, while soil compaction and degradation may limit hydrological recovery with regeneration [31]. These contrary results highlight the watershed and region-specific nature of the effect of forest disturbance on high and low flows.

### 4.4. Management Implications

Our results show that $Q_{5\%}$ increases with greater CECA, implying that forest harvesting increases high flows in this area. Consequently, if harvest rates are not limited, changes to the peak flow regime could result in undesirable alterations to riparian ecosystems and aquatic habitat [29,71]. Much of the BC interior is managed to a maximum logged CECA threshold of 20% to 30% [24] of the watershed area, which is intended to serve as a coarse filter to identify watersheds that may have impacts from harvesting [27]. Stednick [72] suggested that more than 20% of forest harvest in a watershed could lead to significant annual streamflow change. However, in our study, the cumulative effects of forest disturbance became apparent in the MDMC slope in 1989 at a CECA of 12.4%, indicating that the effects of forest disturbance and climate may begin to affect some watersheds at a much lower CECA than is managed for currently in BC. Water use allocations, environmental flow needs, and forest management all need to consider the long term trajectory of forest condition and climate, and their interactions.

## 5. Conclusions

This study assessed how two main drivers, cumulative forest cover change and climate variability, have influenced streamflow in the Deadman River watershed. In the period 1990–2012, overall the effects of cumulative forest disturbance and climate variability are additive, —both have increased $Q_{mean}$. Forest disturbance increased $Q_{mean}$ by an average of 14 mm and climate variability increased it by 6 mm. Additionally, we found that forest disturbance increased low and high flows by 19% and 58% respectively. These insights provide an important contribution to the knowledge required for effective forest and watershed management under a changing climate.

**Supplementary Materials:** The following are available online at http://www.mdpi.com/1999-4907/10/2/196/s1. Figure S1: Average daily flow hydrograph of Deadman River watershed. Figure S2: Annual mean, low and high flows in the Deadman River watershed. Figure S3: Average annual daily average, maximum and miminum temperature in the Deadman River watershed. Figure S4: Annual precipitation, potential evapotranspiration and evapotranspiration in the Deadman River watershed. Figure S5: Average annual daily average, maximum and miminum temperature in the Deadman River watershed and the Kamloops A climate station. Table S1: Kendall tau tests between the hydrological variables and climate variables by season. Table S2: Canonical correlation analysis between hydrological variables and the set of climate variables. Table S3: Paired climate variables and hydrological variables in the paired-year analysis. Table S4: Trend analysis results for Kamloops A station data for seasonal and annual climatic variables over the reference and disturbance periods.Table S5: Trend analysis results for Deadman river watershed—level ClimateBC data for seasonal and annual climatic variables over the reference and disturbance periods.

**Author Contributions:** Conceptualization, K.G.-H., Q.L. and X.W.; Formal analysis, K.G.-H. and Q.L.; Funding acquisition, X.W.; Methodology, K.G.-H., Q.L. and X.W.; Supervision, X.W.; Validation, Q.L.; Writing—original draft, K.G.-H.; Writing—review & editing, Q.L. and X.W.

**Funding:** This research was funded by the Natural Sciences and Engineering Research Council of Canada (NSERC) Collaborative Research and Development Grant-Project (CRDPJ) entitled "The effects of reforestation on forest carbon and water coupling at multiple spatial scales", grant number CRDPJ 485176-15.

**Conflicts of Interest:** The authors declare no conflicts of interest.

## References

1. Ellison, D.; Morris, C.E.; Locatelli, B.; Sheil, D.; Cohen, J.; Murdiyarso, D.; Gutierrez, V.; Noordwijk, M.V.; Creed, I.F.; Pokorny, J.; et al. Trees, forests and water: Cool insights for a hot world. *Glob. Environ. Change* **2017**, *43*, 51–61. [CrossRef]
2. Wei, X.; Li, Q.; Zhang, M.; Giles-Hansen, K.; Liu, W.; Fan, H.; Wang, Y.; Zhou, G.; Piao, S.; Liu, S. Vegetation cover—Another dominant factor in determining global water resources in forested regions. *Glob. Change Biol.* **2018**, *24*, 786–795. [CrossRef] [PubMed]
3. Wei, X.; Liu, W.; Zhou, P. Quantifying the Relative Contributions of Forest Change and Climatic Variability to Hydrology in Large Watersheds: A Critical Review of Research Methods. *Water* **2013**, *5*, 728–746. [CrossRef]
4. Zhang, M.; Liu, N.; Harper, R.; Li, Q.; Liu, K.; Wei, X.; Ning, D.; Hou, Y.; Liu, S. A global review on hydrological responses to forest change across multiple spatial scales: Importance of scale, climate, forest type and hydrological regime. *J. Hydrol.* **2017**, *546*, 44–59. [CrossRef]
5. Zhang, L.; Hickel, K.; Shao, Q. Predicting afforestation impacts on monthly streamflow using the DWBM model. *Ecohydrology* **2016**. [CrossRef]
6. Ford, C.R.; Laseter, S.H.; Swank, W.T.; Vose, J.M. Can forest management be used to sustain water-based ecosystem services in the face of climate change? *Ecol. Appl.* **2011**, *21*, 2049–2067. [CrossRef] [PubMed]
7. Kelly, C.N.; McGuire, K.J.; Miniat, C.F.; Vose, J.M. Streamflow response to increasing precipitation extremes altered by forest management. *Geophys. Res. Lett.* **2016**, *43*, 3727–3736. [CrossRef]
8. Winkler, R.; Spittlehouse, D.; Boon, S. Streamflow Response to Clearcut Logging on British Columbia's Okanagan Plateau. *Ecohydrology* **2017**. [CrossRef]

9.    Li, Q.; Wei, X.; Zhang, M.; Liu, W.; Fan, H.; Zhou, G.; Giles-Hansen, K.; Liu, S.; Wang, Y. Forest cover change and water yield in large forested watersheds: A global synthetic assessment. *Ecohydrology* **2017**, *10*. [CrossRef]

10.   Bosch, J.M.; Hewlett, J.D. A review of catchment experiments to determine the effect of vegetation changes on water yield and evapotranspiration. *J. Hydrol.* **1982**, *55*, 3–23. [CrossRef]

11.   Brown, A.E.; Zhang, L.; McMahon, T.A.; Wes tern, A.W.; Vertessy, R.A. A review of paired catchment studies for determining changes in water yield resulting from alterations in vegetation. *J. Hydrol.* **2005**, *310*, 28–61. [CrossRef]

12.   Forest and Water on a Changing Planet:Vulnerability, Adaptation and Governance Opportunities. A Global Assessment Report. In Proceedings of the IUFRO World Series, Vienna, Austria, 10 July 2018; p. 192.

13.   Dey, P.; Mishra, A. Separating the impacts of climate change and human activities on streamflow: A review of methodologies and critical assumptions. *J. Hydrol.* **2017**, *548*, 278–290. [CrossRef]

14.   Liu, W.; Wei, X.; Li, Q.; Fan, H.; Duan, H.; Wu, J.; Giles-Hansen, K.; Zhang, H. Hydrological recovery in two large forested watersheds of southeastern China: The importance of watershed properties in determining hydrological responses to reforestation. *Hydrol. Earth Syst. Sci.* **2016**, *20*, 4747–4756. [CrossRef]

15.   Ruiz-Perez, G.; Koch, J.; Manfreda, S.; Caylor, K.; Frances, F. Calibration of a parsimonious distributed ecohydrological daily model in a data scarce basin using exclusively the spatio-temporal variation of NDVI. *Hydrol. Earth System Sci.* **2017**, *21*, 6235–6251. [CrossRef]

16.   Mikkelson, K.M.; Maxwell, R.M.; Ferguson, I.; Stednick, J.D.; McCray, J.E.; Sharp, J.O. Mountain pine beetle infestation impacts: modeling water and energy budgets at the hill-slope scale. *Ecohydrology* **2013**, *6*, 64–72. [CrossRef]

17.   Penn, C.A.; Bearup, L.A.; Maxwell, R.M.; Clow, D.W. Numerical experiments to explain multiscale hydrological responses to mountain pine beetle tree mortality in a headwater watershed. *Water Resour. Res.* **2016**, *52*, 3143–3161. [CrossRef]

18.   Schnorbus, M.; Bennett, K.; Werner, A. Quantifying the water resource impacts of mountain pine beetle and associated salvage harvest operations across a range of watershed scales: Hydrologic modelling of the Fraser River Basin. In *Information Report BC-X-423*; Natural Resources Canada, Canadian Forest Service, Pacific Forestry Centre: Victoria, BC, Canada, 2010.

19.   White, J.C.; Wulder, M.A.; Hermosilla, T.; Coops, N.C.; Hobart, G.W. A nationwide annual characterization of 25years of forest disturbance and recovery for Canada using Landsat time series. *Remote Sens. Environ.* **2017**, *194*, 303–321. [CrossRef]

20.   Reid, L.M. Cumulative watershed effects and watershed analysis. In *River Ecology and Management: Lessons from the Pacific Coastal Ecoregion*; Naiman, R.J., Bilby, R.E., Eds.; Springer: New York, NY, USA, 1998; pp. 476–501.

21.   Schindler, D.W.; Donahue, W.F. An impending water crisis in Canada's western prairie provinces. *Proc. Natl. Acad. Sci. USA* **2006**, *103*, 7210. [CrossRef] [PubMed]

22.   Wei, X.; Zhang, M. Quantifying streamflow change caused by forest disturbance at a large spatial scale: A single watershed study. *Water Resour. Res.* **2010**, *46*. [CrossRef]

23.   Zhang, M.; Wei, X. Contrasted hydrological responses to forest harvesting in two large neighbouring watersheds in snow hydrology dominant environment: implications for forest management and future forest hydrology studies. *Hydrol. Process.* **2014**, *28*, 6183–6195. [CrossRef]

24.   Winkler, R.; Boon, S. *Revised Snow Recovery Estimates for Pine-dominated Forests in Interior British Columbia*; Extension note 116; Ministry of Forests, Lands and Natural Resource Operations: Kamloops, BC, Canada, 2015.

25.   Winkler, R.D.; Moore, R.D.; Redding, T.E.; Spittlehouse, D.L.; Smerdon, B.D.; Carlyle-Moses, D.E. The Effects of Forest Disturbance on Hydrologic Processes and Watershed Response (Chapter 7). In *Compendium of Forest Hydrology and Geomorphology in British Columbia*; British Columbia: Victoria, BC, Canada, 2010.

26.   Moore, D.R.; Wondzell, S.M. Physical Hydrology and the Effects of Forest Harvesting in the Pacific Northwest: A Review. *JAWRA* **2005**, *41*, 763–784. [CrossRef]

27.   British Columbia Ministry of Forests. Forest Practices Code of British Columbia. In *Coastal Watershed Assessment Procedure Guidebook (CWAP) and Interior Watershed Assessment Procedure Guidebook (IWAP)*; British Columbia Ministry of Forests: Victoria, BC, Canada, 1999.

28. Jones, J.A.; Perkins, R.M. Extreme flood sensitivity to snow and forest harvest, western Cascades, Oregon, United States. *Water Resour. Res.* **2010**, *46*. [CrossRef]

29. Zhang, M.; Wei, X. Alteration of flow regimes caused by large-scale forest disturbance: a case study from a large watershed in the interior of British Columbia, Canada. *Ecohydrology* **2014**, *7*, 544–556. [CrossRef]

30. Liu, W.; Wei, X.; Liu, S.; Liu, Y.; Fan, H.; Zhang, M.; Yin, J.; Zhan, M. How do climate and forest changes affect long-term streamflow dynamics? A case study in the upper reach of Poyang River basin. *Ecohydrology* **2015**, *8*, 46–57. [CrossRef]

31. Bruijnzeel, L.A. Hydrological functions of tropical forests: not seeing the soil for the trees? *Agric. Ecosyst. Environ.* **2004**, *104*, 185–228. [CrossRef]

32. Ecoscape Environmental Consultants Ltd. *Deadman River Sensitive Habitat Inventory and Mapping (SHIM)-2009–2011*; Inventory Summary Report; Skeetchestn Indian Band: Kelowna, BC, Canada, 2012; p. 28.

33. Walton, A. Provincial-Level Projection of the Current Mountain Pine Beetle Outbreak: Update of the Infestation Projection Based on the Provincial Aerial Overview Surveys of Forest Health Conducted from 1999 through 2012 and the BCMPB Model (Year 10). BC Forest Service, 2013. Available online: https://www.google.com.sg/url?sa=t&rct=j&q=&esrc=s&source=web&cd=1&ved=2ahUKEwjPw-Tvn8zgAhWDLqYKHdg5AnlQFjAAegQIARAC&url=https%3A%2F%2Fwww.for.gov.bc.ca%2Fftp%2Fhre%2Fexternal%2F!publish%2Fweb%2Fbcmpb%2Fyear10%2FBCMPB.v10.BeetleProjection.Update.pdf&usg=AOvVaw1l11YYPl_TOX7qhgj15tq1 (accessed on 11 January 2019).

34. Wang, T.; Hamann, A.; Spittlehouse, D.; Carroll, C. Locally Downscaled and Spatially Customizable Climate Data for Historical and Future Periods for North America. *PLoS ONE* **2016**, *11*, e0156720. [CrossRef] [PubMed]

35. Zhang, M.; Wei, X.; Li, Q. A quantitative assessment on the response of flow regimes to cumulative forest disturbances in large snow-dominated watersheds in the interior of British Columbia, Canada. *Ecohydrology* **2016**, *9*, 843–859. [CrossRef]

36. Li, Q.; Wei, X.; Zhang, M.; Liu, W.; Giles-Hansen, K.; Wang, Y. The cumulative effects of forest disturbance and climate variability on streamflow components in a large forest-dominated watershed. *J. Hydrol.* **2018**, *557*, 448–459. [CrossRef]

37. Lewis, D.; Huggard, D. A Model to Quantify Effects of Mountain Pine Beetle on Equivalent Clearcut Area. *Streamline Watershed Manag. Bull.* **2010**, *13*, 42–51.

38. Axelson, J.N.; Alfaro, R.I.; Hawkes, B.C. Influence of fire and mountain pine beetle on the dynamics of lodgepole pine stands in British Columbia, Canada. *For. Ecol. Manage.* **2009**, *257*, 1874–1882. [CrossRef]

39. Baker, E.H.; Painter, T.H.; Schneider, D.; Meddens, A.J.H.; Hicke, J.A.; Molotch, N.P. Quantifying insect-related forest mortality with the remote sensing of snow. *Remote Sens. Environ.* **2017**, *188*, 26–36. [CrossRef]

40. R Core Team. R: A Language and Environment for Statistical Computing. R Foundation for Statistical Computing: Vienna, Austria, 2016.

41. Hyndman, R.J.; Khandakar, Y. Automatic time series forecasting: the forecast package for R. *J. Stat. Softw.* **2008**, *27*, 1–22. [CrossRef]

42. Liu, W.; Wei, X.; Fan, H.; Guo, X.; Liu, Y.; Zhang, M.; Li, Q. Response of flow regimes to deforestation and reforestation in a rain-dominated large watershed of subtropical China. *Hydrol. Process.* **2015**, *29*, 5003–5015. [CrossRef]

43. Mann, H.B. Nonparametric Tests Against Trend. *Econometrica* **1945**, *13*, 245–259. [CrossRef]

44. Kendall, M.G. *Rank Correlation Methods*; Oxford University Press: New York, NY, USA, 1975; p. 202.

45. Déry, S.J.; Wood, E.F. Decreasing river discharge in northern Canada. Geophys. *Res. Lett.* **2005**, *32*. [CrossRef]

46. Yue, S.; Pilon, P.; Phinney, B.; Cavadias, G. The influence of autocorrelation on the ability to detect trend in hydrological series. *Hydrol. Process.* **2002**, *16*, 1807–1829. [CrossRef]

47. Razavi, S.; Vogel, R. Prewhitening of hydroclimatic time series? Implications for inferred change and variability across time scales. *J. Hydrol.* **2018**, *557*, 109–115. [CrossRef]

48. Priestley, C.H.B.; Taylor, R.J. On the assessment of surface heat flux and evaporation using large-scale parameters. Mon. *Weather Rev.* **1972**, *100*, 81–82. [CrossRef]

49. Hamon, W.R. Computation of Direct Runoff Amounts From Storm Rainfall. *IAHS* **1963**, *63*, 52–62.

50. Lu, J.; Sun, G.; McNulty, S.G.; Amatya, D. A comparison of six potential evapotranspiration methods for regional use in the Southeastern United States. *JAWRA* **2005**, *41*, 621–633. [CrossRef]

51. McMahon, T.A.; Peel, M.C.; Lowe, L.; Srikanthan, R.; McVicar, T.R. Estimating actual, potential, reference crop and pan evaporation using standard meteorological data: a pragmatic synthesis. *Hydrol. Earth Syst. Sci.* **2013**, *17*, 1331–1363. [CrossRef]

52. Budyko, M.I. *Climate and Life*; Academic Press: London, UK, 1974; Volume 18.

53. Zhang, L.; Hickel, K.; Dawes, W.R.; Chiew, F.H.S.; Western, A.W.; Briggs, P.R. A rational function approach for estimating mean annual evapotranspiration. *Water Resour. Res.* **2004**, *40*. [CrossRef]

54. Yao, Y.; Cai, T.; Wei, X.; Zhang, M.; Ju, C. Effect of forest recovery on summer streamflow in small forested watersheds, Northeastern China. *Hydrol. Process.* **2012**, *26*, 1208–1214. [CrossRef]

55. Pettitt, A.N. A Non-Parametric Approach to the Change-Point Problem. *J. R. Stat. Soc. Ser. C Appl. Stat.* **1979**, *28*, 126–135. [CrossRef]

56. Li, Y.; Piao, S.; Li, L.Z.X.; Chen, A.; Wang, X.; Ciais, P.; Huang, L.; Lian, X.; Peng, S.; Zeng, Z.; et al. Divergent hydrological response to large-scale afforestation and vegetation greening in China. *Sci. Adv.* **2018**, *4*. [CrossRef] [PubMed]

57. Koster, R.D.; Suarez, M.J. A Simple Framework for Examining the Interannual Variability of Land Surface Moisture Fluxes. *J. Climate* **1999**, *12*, 1911–1917. [CrossRef]

58. Milly, P.C.D.; Dunne, K.A. Macroscale water fluxes 1. Quantifying errors in the estimation of basin mean precipitation. *Water Resour. Res.* **2002**, *38*, 1–14. [CrossRef]

59. Serinaldi, F.; Kilsby, C.G.; Lombardo, F. Untenable nonstationarity: An assessment of the fitness for purpose of trend tests in hydrology. *Adv. Water Resour.* **2018**, *111*, 132–155. [CrossRef]

60. Mikkelson, K.M.; Bearup, L.A.; Maxwell, R.M.; Stednick, J.D.; McCray, J.E.; Sharp, J.O. Bark beetle infestation impacts on nutrient cycling, water quality and interdependent hydrological effects. *Biogeochemistry* **2013**, *115*, 1–21. [CrossRef]

61. Zhang, M.; Wei, X.; Li, Q. Do the hydrological responses to forest disturbances in large watersheds vary along climatic gradients in the interior of British Columbia, Canada? *Ecohydrology* **2017**, *10*, e1840. [CrossRef]

62. Berghuijs Wouter, R.; Larsen Joshua, R.; van Emmerik Tim, H.M.; Woods Ross, A. A Global Assessment of Runoff Sensitivity to Changes in Precipitation, Potential Evaporation, and Other Factors. *Water Resour. Res.* **2017**, *53*, 8475–8486. [CrossRef]

63. Zhang, M. The effects of cumulative forest disturbances on hydrology in the interior of British Columbia, Canada. Ph.D. Thesis, The University of British Columbia (Okanagan), Kelowna, BC, Canada, June 2013.

64. Zhang, M.; Wei, X.; Sun, P.; Liu, S. The effect of forest harvesting and climatic variability on runoff in a large watershed: The case study in the Upper Minjiang River of Yangtze River basin. *J. Hydrol.* **2012**, *464–465*, 1–11. [CrossRef]

65. Zhang, M.; Wei, X. The effects of cumulative forest disturbance on streamflow in a large watershed in the central interior of British Columbia, Canada. *HESS* **2012**, *16*, 2021–2034. [CrossRef]

66. Roa-García, M.C.; Brown, S.; Schreier, H.; Lavkulich, L.M. The role of land use and soils in regulating water flow in small headwater catchments of the Andes. *Water Resour. Res.* **2011**, *47*. [CrossRef]

67. Clark, K.L.; Skowronski, N.; Gallagher, M.; Renninger, H.; Schäfer, K. Effects of invasive insects and fire on forest energy exchange and evapotranspiration in the New Jersey pinelands. *Agric. For. Meteorol.* **2012**, *166–167*, 50–61. [CrossRef]

68. Du, E.; Link, T.E.; Wei, L.; Marshall, J.D. Evaluating hydrologic effects of spatial and temporal patterns of forest canopy change using numerical modelling. *Hydrol. Process.* **2016**, *30*, 217–231. [CrossRef]

69. Buttle, J.M.; Metcalfe, R.A. Boreal forest disturbance and streamflow response, northeastern Ontario. *Can. J. Fish. Aquat. Sci.* **2000**, *57*, 5–18. [CrossRef]

70. Wilk, J.; Andersson, L.; Plermkamon, V. Hydrological impacts of forest conversion to agriculture in a large river basin in northeast Thailand. *Hydrol. Process.* **2001**, *15*, 2729–2748. [CrossRef]

71. Chen, W.; Wei, X. Assessing the relations between aquatic habitat indicators and forest harvesting at watershed scale in the interior of British Columbia. *For. Ecol. Manage.* **2008**, *256*, 152–160. [CrossRef]

72. Stednick, J.D. Monitoring the effects of timber harvest on annual water yield. *J. Hydrol.* **1996**, *176*, 79–95. [CrossRef]

*forests*

MDPI

*Article*

# Two Centuries-Long Streamflow Reconstruction Inferred from Tree Rings for the Middle Reaches of the Weihe River in Central China

**Na Liu** [1,2], **Guang Bao** [1,2,*], **Yu Liu** [2] and **Hans W. Linderholm** [3]

[1] Shaanxi Key Laboratory of Disaster Monitoring and Mechanism Simulating,
   College of Geography and Environment, Baoji University of Arts and Sciences, Baoji 721013, China;
   liuna_2000@163.com
[2] State Key Laboratory of Loess and Quaternary Geology, Institute of Earth Environment,
   Chinese Academy of Sciences, Xi'an 710061, China; liuyu@loess.llqg.ac.cn
[3] Regional Climate Group, Department of Earth Sciences, University of Gothenburg, Box 460,
   405 30 Gothenburg, Sweden; hans.linderholm@gvc.gu.se
*   Correspondence: baoguang@bjwlxy.edu.cn

Received: 20 January 2019; Accepted: 25 February 2019; Published: 26 February 2019

**Abstract:** Water source is one of the most important concerns for regional society and economy development, especially in the Weihe River basin which is located in the marginal zone of the Asian summer monsoon. Due to the weakness of short instrumental records, the variations of streamflow during the long-term natural background are difficult to access. Herein, the average June–July streamflow variability in the middle reaches of the Weihe River was identified based on tree-ring width indices of Chines pine (*Pinus tabulaeformis* Carr.) from the northern slope of the Qinling Mountains in central China. Our model could explain the variance of 39.3% in the observed streamflow period from 1940 to 1970 AD. There were 30 extremely low years and 26 high years which occurred in our reconstruction for the effective span of 1820 to 2005. Several common dryness and wetness periods appeared in this reconstructed streamflow, and other tree-ring precipitation series suggested the coherence of hydroclimate fluctuation over the Weihe River basin. Some significant peaks in cycles implied the linkages of natural forcing on the average June–July streamflow of the Weihe River, such as the Pacific Decadal Oscillation (PDO) and El Niño-Southern Oscillation (ENSO) activities. Spatial correlation results between streamflow and sea surface temperature in the northern Pacific Ocean, as well as extremely low/high years responding to the El Niño/La Nina events, supported the teleconnections. The current 186-year streamflow reconstruction placed regional twentieth-century drought and moisture events in a long-term perspective in the Weihe River basin, and provided useful information for regional water resource safety and forest management, particularly under climate warming conditions.

**Keywords:** tree rings; Weihe River; streamflow variability; reconstruction

## 1. Introduction

Weihe River is the largest tributary of the Yellow River, China, with a basin area of 134,800 km². The length of the Weihe River is 818 km, crossing three provinces of Gansu, Ningxia, and Shaanxi in the eastern part of the Northwest China [1,2]. The upper reaches locate in the semi-arid region of eastern Gansu and southern Ningxia, and the middle and down reaches locate in the semi-humid region of Shaanxi. Hydroclimate in the Weihe River basin is mainly influenced by the Asian summer monsoon, and is sensitive to climate anomalies [3,4]. Total annual precipitation over the basin is about 573 mm, and most precipitation occurs in the summer rainy season. The mean natural discharge is

$10.4 \times 10^9$ m$^3$, accounting for 17.3% of total annual runoff of the Yellow River. As the major water source utilized for the middle and down reaches, the most important region of industrial–agricultural production and ecosystem protection in Northwest China, the Weihe River plays an important role in the sustainable development of Shaanxi province. The average water resources of the Weihe River basin accounts for 17.4% of the total amount in Shaanxi, but it supports 63.6% of the population, 53% of the cultivated land area, 63.4% of the grain output and 64.8% of the gross domestic product (GDP) of Shaanxi [5]. Compared to industrial water and domestic water consumption, the rate of agricultural water is more than 60% of the total economic water consumption in the Weihe River in Shaanxi Province based on the period of 1997 to 2013 [6]. However, due to the influences of human activity and climate warming, water source availability has become a major limitation for regional sustainability during the last decades [2]. The Weihe River runoff is mainly dominated by precipitation which is strongly affected by the variations of the East Asian summer monsoon. The fluctuation of runoff is consistent with variability in precipitation during dry and wet seasons on the annual scale. Therefore, 415 reservoirs have been built in the middle and down reaches of the Weihe River to effectively utilize water resources, with a total storage capacity of $2.202 \times 10^9$ m$^3$ [7]. There are 274 irrigation divisions, including the areas of facilities irrigation and water-saving irrigation which are $11.32 \times 10^9$ m$^2$ and $71.95 \times 10^9$ m$^2$, respectively [7]. In addition, previous study shows that the synergistic effect of decreasing precipitation, increasing temperature, and increasing evaporation in the Weihe River basin during the past decades has a negative influence on the runoff variations [8]. Therefore, human activities and climatic conditions mentioned above have significantly changed the characteristics of runoff in the Weihe River.

Enhancing the understanding of regional hydroclimatical change mechanisms and runoff evolution characteristics is an urgent requirement for assessing current water resources security and future planning. However, relative shorter hydrological observations limit us in identifying hydrological variability during long historical periods. Tree-rings have proved to be one of the effective proxies for hydroclimatical changes study prior to the measured period, due to its wide spatial distribution, high resolution with an exact calendar, and sensitivity to hydroclimate [9].

Several hydroclimatical studies based on tree rings have been performed in the Weihe River basin, including seasonal precipitation reconstructions for the Huashan Mountain in the down reaches [10,11] and Tianshui in the upper reaches of the river [3], as well as the drought variability estimation for the Guiqing Mountain in Gansu province [12].

However, runoff or streamflow tree-ring reconstructions conducted in the Weihe River have not been reported. In this research, the variations of streamflow in the middle reaches were studied using tree-ring samples of Chinese pine collected from the Nanwutai area (NWT), northern slope of the Qinling Mountains. The relationships between the radial growth of pines, streamflow, and climatic factors are discussed. The driving mechanisms of hydroclimatical variability in the Weihe River basin need to be explored in the context of climatic circulation systems on the regional or global scale. Spatial correlation patterns between streamflow and sea surface temperature anomalies in key ocean areas is an effective method to establish remote linkages. Knowledge of atmospheric–ocean–land teleconnections could be of benefit in the assessment and prediction of hydrological and climate change over the Weihe River basin. Therefore, the features of the extreme streamflow event and teleconnection with remote climate forcing were also analyzed.

Our results provide the first hydrological reconstruction inferred from tree rings for the Weihe River, which will be valuable for water resources management and planning in the future.

## 2. Materials and Methods

### 2.1. Tree-ring Data

Increment cores from Chinese pine (*Pinus tabulaeformis*) were collected at the Nanwutai area (NWT) (33°58′–34°02′ N, 108°57′–108°59′ E, elevation 1500 m–1600 m), northern slope of the Qinling

Mountains [13]. To reveal the representativeness of the tree growth–climate response, we collected the cores at three sampling sites where old trees concentrated. A total of 94 samples were from three groups, i.e., NWTa, NWTb, NWTc, 30 cores/17trees, 31 cores/18trees, and 33 cores/18trees, respectively (Figure 1). All the samples were treated following the standard dendrochronological procedures [14]. The ring-width was measured using the Lintab system with a precision of 0.01 mm (www.rinntech.de), then each tree-ring series were given an accurate calendar year after cross-dating process performed by the output of COFECHA software [15]. Then each dated ring-width series was detrended and standardized to tree-ring width indices utilizing the ARSTAN software [16]. Negative exponential or linear regressions were applied to remove the age-related growth trends for 91 cores, and 3 cores from two trees were treated by cubic smoothing splines. Standard tree-ring width chronologies were established for NWTa, NWTb, NWTc and the variation and the sample depth are shown in Figure 2. Similar variability features were displayed by three groups, and significant correlation coefficients existed between three chronologies, i.e., 0.657 ($p < 0.001$, 1816–2005 for NWTa and NWTb), 0.666 ($p < 0.001$, 1765–2005 for NWTa and NWTc) and 0.724 ($p < 0.001$, 1816–2005 for NWTb and NWTc). Therefore, the regional standard chronology NWTabc were built based on 94 cores from 53 trees for the study area (Figure 2). The full length of NWTabc covered the period of 1760 to 2005AD. The mean sensitivity and all series correlation were 0.193 and 0.448, respectively. Statistical characteristics of the common span from 1880 to 2005 as follows, expressed population signal (EPS) was 0.948, signal-to-noise ratio (SNR) was 18.146, and variance in the first eigenvector was 30.6%. To make sure the reliability of the reconstruction, subsample signal strength (SSS) was used to identify the adequacy of replication of NWTabc chronology [17]. We restricted the further analysis to the period from 1820 to 2005 with the SSS value greater than 0.80, and the first year included 9 cores from 5 trees.

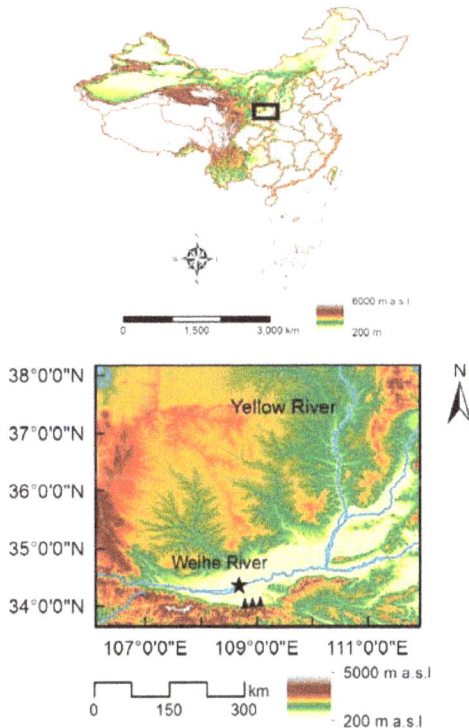

**Figure 1.** Map of the study area (black rectangle) and locations of the sample site (black tree) and Xianyang hydrological station (black star).

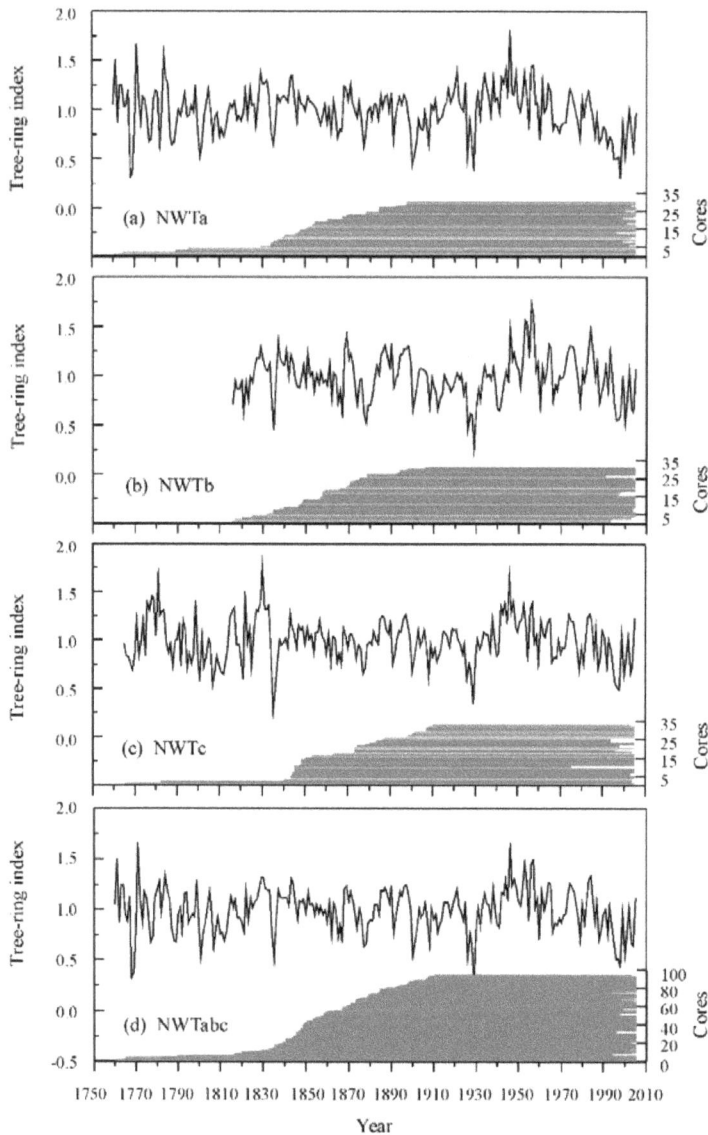

**Figure 2.** Single groups and composited regional tree-ring width standard chronologies (NWTa, NWTb, NWTc, and NWTabc), and numbers of sample cores.

## 2.2. Climatic and Hydrological Data

Regional monthly average temperature and precipitation from CRU (Climate Reach Unit) TS4.01 grid data within the range of 34°–35°N and 107°–109°E [18], and monthly streamflow data of the Xianyang hydrological station in the middle reaches of the Weihe River (108°42′ E, 34°19′ N, with a catchment area 46,827 km$^2$) provided by Hydrology Bureau of the Yellow River Conservancy Commission were selected to explore the response relationships between trees' radial growth and climatic hydrological elements (Figure 1). The calculation results showed that the study area was cold and dry in winter while hot and wet in summer, indicating East Asian monsoon climate characteristics

(Figure 3). The average annual precipitation in the region was 642.59 mm, the maximum precipitation 121.97 mm occurred in July, and the minimum amount of precipitation in January was 4.25 mm over the period 1940 to 2005. The annual average temperature was 11.63 °C, the highest and lowest monthly average temperatures appeared in July (23.95 °C) and January (−1.38 °C), respectively. Many water conservancy projects have been developed since the 1970s in the upper reaches of the Weihe River, and natural runoff changes have been significantly affected by human activities [2,19]. Therefore, monthly streamflow during the period from 1940 to 1970 was selected for subsequent analysis. The high-value period of the streamflow appeared from July to September during the monsoon rain season (Figure 3). The maximum flow occurred in September, but not in July, indicated that there was a lag effect in the river basin convergence process. It should be pointed out that the streamflow in June was the smallest one during the growing period from May to October. The main possible reason could be due to more evaporation loss relating to higher temperatures in June, resulting in less runoff.

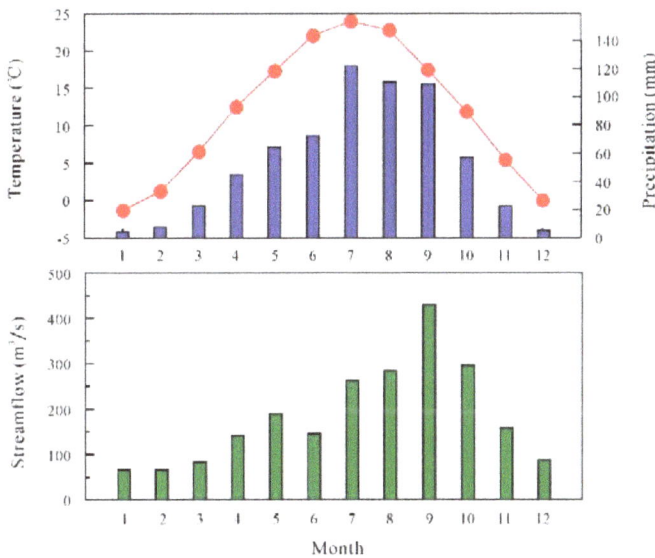

**Figure 3.** Monthly total precipitation (blue bar) and mean temperature (red circle) from CRU TS4.01 grid data within the range of 34°–35°N and 107°–109°E (1940–2005), and mean streamflow (green bar) of Xianyang hydrological station (1940–1970).

## 2.3. Statistical Methods

The relationships between climate/streamflow factors and tree-ring width index were performed using Pearson correlation analysis in this study [14]. Considering tree growth is affected by environmental conditions in the present growing season, as well as by factors in the previous year [14], previous October to current September data were selected for growth response analysis. The linear regression model and split calibration–verification were employed to identify the reliability of streamflow reconstruction [16]. Verification statistics of reduction of error (RE), coefficient of efficiency (CE), and the sign test (ST) were given [16]. Positive values of RE and CE suggested the regression model was valid and skillful for reconstruction. The ST demonstrated the numbers of agreements and disagreements between the estimated and measured streamflow data [14]. All the statistical results were performed using the software of Statistical Program for Social Sciences 19. Cycles of the dominant oscillation signals in streamflow series were done by spectral analysis of the multi-taper method (MTM) [20]. The MTM analysis could exactly reveal the signals of oscillation modes in reconstructed streamflow series based on a couple of tapers reducing the variance of spectral estimates. Parameters including red noise background estimation with the resolution of 2 and taper numbers of 3 were performed in this study [20]. To explore regional hydroclimatical variations and teleconnections reflected by current streamflow reconstruction, spatial correlations of our reconstructed streamflow with the gridded data including Standard precipitation evapotranspiration Index (SPEI) on the four-month scale [21], self-calibrating Palmer Drought Severity Index (scPDSI) [22] and sea surface temperature (SST) dataset of ERSST (Extended Reconstructed Sea Surface Temperature) v5 [23] were conducted for the period 1901 to 2005 through the online tool of KNMI (The Royal Netherlands Meteorological Institute) climate explorer (http://climexp.knmi.nl).

## 3. Results and Discussion

### 3.1. Climate/Streamflow-Growth Response

The responses of tree radial growth to climatic factors showed that the NWTabc chronology was significantly positively correlated with precipitation in June ($r = 0.572$, $n = 31$, 1940–1970, $p < 0.01$), significantly negatively correlated with temperature in June ($r = -0.401$, $n = 31$, 1940–1970, $p < 0.05$), and not correlated with precipitation and temperature in July, indicating that the growth of *Pinus tabulaeformis* was very sensitive to humidity conditions before the East Asian monsoon season started in July. Due to the river basin convergence process which mainly includes precipitation, surface infiltration, soil saturation, slope convergence, and river network convergence, a lag effect existed in the response of river runoff forming to precipitation. Therefore, the regional tree-ring chronology significantly correlated with the average June–July streamflow of the Weihe River, which inevitably reflected the hysteresis effect. Significant correlations occurred between June–July streamflow, June precipitation ($r = 0.753$, $n = 31$, $p < 0.01$) and June temperature ($r = -0.568$, $n = 31$, $p < 0.05$), while no responses existed in July precipitation and temperature. Runoff reflected the combined effects of climatic factors, such as temperature and precipitation, in the basin. Therefore, using the tree-ring chronology to reconstruct the average June–July streamflow of the Weihe River had a reliable physiological significance and physical basis (Figure 4).

### 3.2. Streamflow Reconstruction

Based on the analyses results mentioned above, the average June–July streamflow was reconstructed following the transfer function: Qs = 309.208 × NWTabc − 140.02 ($r = 0.627$, $n = 31$, F = 18.779, $p < 0.0001$), where Qs means the average June–July streamflow in the Weihe River. The Durbin–Watson value 1.341 ($p < 0.05$) suggests no significant first-order autocorrelation in the residuals of the regression model [14]. This function could explain the variance of 39.3% over the observed streamflow period 1940 to 1970 (37.2% considering the loss of degrees of freedom). The variations between reconstructed and observed streamflow agree quite well (Figure 5).

The reconstructed sequence reveals hydrological variability during the last 196 years in the Weihe River. Reduction of error (RE) and Coefficient of efficiency (CE) were used to identify shared variance between observation and reconstruction series, and both statistics had a theoretical range from $-\infty$ to +1. Positive RE and CE values demonstrated that the model was skillful and acceptable for streamflow reconstruction. Statistics results indicate that RE (0.191, 0.271) and CE (0.142, 0.243) were positive in both verification periods of 1960 to 1970 and 1940 to 1959, confirming our model is acceptable and skillful for streamflow reconstruction (Table 1). The ST was applied to check the numbers of agreement or disagreement signs between the paired observed and estimated departures from the series mean. The ST in the calibration spans 1940 to 1959 (15+/5−), and 1960 to 1970 (9+/1−) were at the 0.05 significant level, and, particularly, the ST result (24+/7−) in the full span 1940 to 1970 was at the 0.01 significant level.

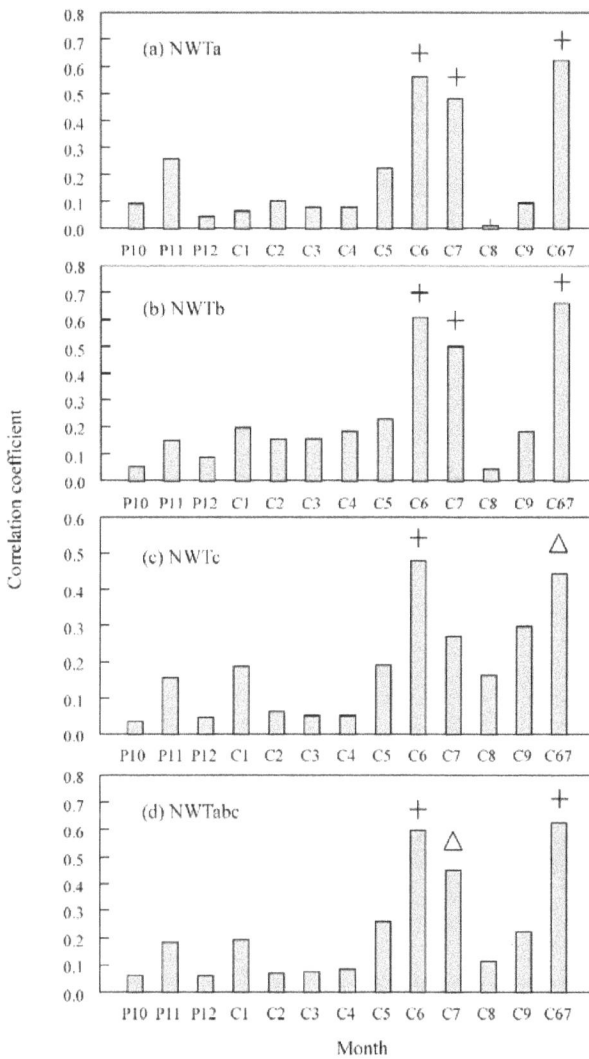

**Figure 4.** Correlations of tree-ring standard chronology with monthly average streamflow (grey bar) of Xianyang hydrological station (1940–1970) from the previous October (P10) to current September (C9) at the 95% confidence level (triangle) and 99% confidence level (cross). P indicates previous year, C indicates current year and C67 indicates the average June–July streamflow in the current year.

**Table 1.** Statistics of split calibration-verification for June–July streamflow reconstruction model.

| Calibration | | | | Verification | | | | | |
|---|---|---|---|---|---|---|---|---|---|
| Period | $r$ | $R^2$ | ST | Period | $r$ | $R^2$ | RE | CE | ST |
| 1940–1959 | 0.647 ** | 0.419 | 15 * | 1960–1970 | 0.672 * | 0.451 | 0.191 | 0.142 | 9 * |
| 1960–1970 | 0.672 * | 0.451 | 9 * | 1940–1959 | 0.647 ** | 0.419 | 0.271 | 0.243 | 11 |
| 1940–1970 | 0.627 ** | 0.393 | 24 ** | | | | | | |

* means $p < 0.05$, ** $p < 0.01$. RE means reduction of error; CE, coefficient of efficiency, and ST, sign test.

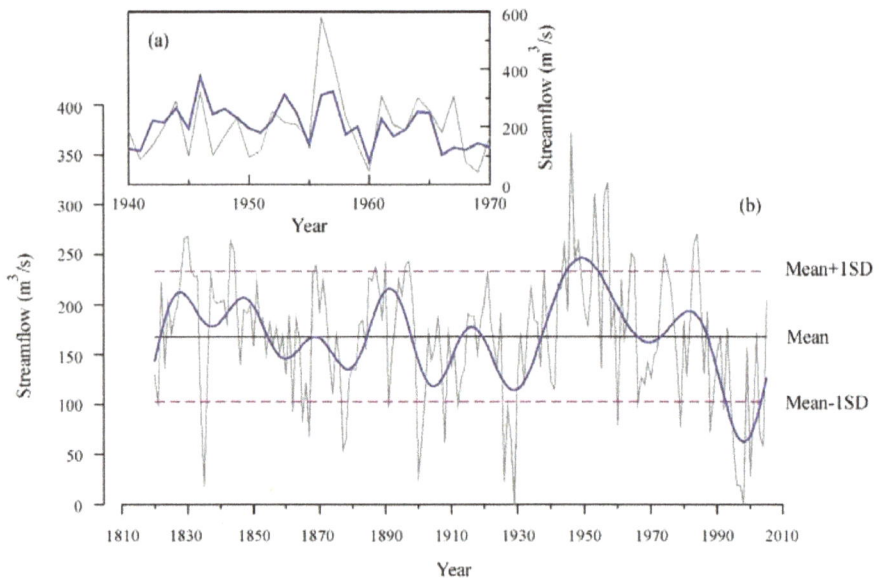

**Figure 5.** Comparisons (**a**) between reconstructed (bold line) and observed (thin line) average June–July streamflow during the period of 1940 to 1970, and (**b**) variations of the streamflow reconstruction during the entire span of 1820 to 2005 for the middle reaches of the Weihe River (the bold line indicates the 20-year low pass data, the horizontal line indicates the mean value and extremely high and low values with one standard deviation).

*3.3. Variation Features of Streamflow of the Weihe River*

The estimated mean June–July streamflow was 168.01 m$^3$/s, and the value of standard deviation (SD) was 65.22 m$^3$/s over the period of 1820 to 2005 AD. Extreme high and low years were determined when the reconstructed values were higher or lower one SD than the long-term mean streamflow. Based on these criteria, 30 extremely low years and 26 high years occurred in our reconstruction, and they accounted for 16.13% and 13.98% of the entire period of the 186 years, respectively (Figure 5). It should be noted that negative values were found in the two extremely lowest years of 1929 and 1998, suggesting the extremely dry conditions beyond the range of the calibration variations. A similar situation occurred in the reconstruction of seasonal streamflow inferred from tree rings in the Mongolian plateau [24]. We modified the negative values with zero in Table 2. The lowest values appeared in eight years, including 1929 (<mean-2SD), 1998 (<2SD), 1835 (<2SD), 1996 (<2SD), 1997 (<2SD), 1926 (<2SD), 2000 (<2SD), and 1900 (<2SD). In particular, the highest year appeared in 1946 with a value 372.65 m3/s (>mean+2SD), the other three higher years were 1953 (>2SD), 1956 (>2SD), and 1967 (>2SD). Low flow events lasting two-year or more occurred in 1834–1835, 1877–1878, 1900–1901,1926–1929, 1995–1998, 2000–2001, and 2003–2004, while high flow events appeared in 1829–1831, 1843–1844, 1896–1897,1946–1948, 1953–1954, 1956–1967, 1964–1965, 1974–1975, and 1983–1984. The low-frequency changes in reconstructed streamflow of the Weihe River showed obvious fluctuations on the inter-annual to multi-decadal scales during the last two centuries (Figure 5). Overall, the higher streamflow spans happened in the 1820s–1840s, 1890s, 1940s–1960s, and the lower spans existed in 1860s–1870s, 1900s, 1920s–1930s, and 1990s–2000s. The significant magnitude of streamflow variations presented from a moisture epoch in the 1950s to drought epoch in the 1990s.

**Table 2.** Extreme events of reconstructed June–July streamflow of the Weihe River.

| Year | Extreme Low Value (m³/s) | Year | Extreme High Value (m³/s) |
|------|--------------------------|------|---------------------------|
| 1821 | 98.07 | 1829 | 265.97 |
| 1834 | 101.78 | 1830 | 268.13 |
| 1835 | 17.68 | 1831 | 233.50 |
| 1862 | 91.89 | 1843 | 264.11 [b] |
| 1865 | 86.01[a] | 1844 | 253.29 |
| 1867 | 67.46 | 1869 | 240.31 |
| 1877 | 55.40 [a] | 1887 | 237.52 [b] |
| 1878 | 64.68 [a] | 1890 | 242.78 [b] |
| 1891 | 96.21 [a] | 1896 | 239.07 [b] |
| 1900 | 29.12 [a] | 1897 | 243.71 |
| 1901 | 63.44 [a] | 1921 | 233.81 [b] |
| 1908 | 61.89 | 1938 | 235.67 |
| 1912 | 98.38 [a] | 1944 | 264.11 |
| 1926 | 22.00 [a] | 1946 | 372.65 [b] |
| 1927 | 96.83 | 1947 | 242.47 |
| 1928 | 65.91 | 1948 | 262.26 |
| 1929 | 0.00 | 1953 | 311.42 [b] |
| 1960 | 78.59 | 1954 | 247.11 |
| 1966 | 100.85 [a] | 1956 | 309.26 [b] |
| 1979 | 77.04 [a] | 1957 | 322.86 [b] |
| 1988 | 76.43 | 1964 | 251.13 |
| 1992 | 95.29 [a] | 1965 | 246.80 |
| 1995 | 58.49 | 1974 | 249.58 [b] |
| 1996 | 18.91 | 1975 | 237.83 [b] |
| 1997 | 19.22 [a] | 1983 | 259.79 |
| 1998 | 0.00 | 1984 | 271.23 [b] |
| 2000 | 28.19 | | |
| 2001 | 84.77 | | |
| 2003 | 69.00 | | |
| 2004 | 57.56 | | |

Note: [a] indicates the El Niño year; [b] indicates La Nina year.

Standard precipitation evapotranspiration Index (SPEI) and self-calibrating Palmer Drought Severity Index (scPDSI) are widely used for dry and wet change studies in different regions and time-scales around the world [25,26]. Considering the time lag effect of runoff to precipitation and the multi-time scale properties of SPEI, four-month scale SPEI data were selected for analysis. The spatial correlation results showed that the reconstructed June–July streamflow was significantly positively correlated with both drought indices in the middle reaches of the Weihe River, indicating that different indicators had a similar ability to capturing hydroclimatical variations (Figure 6). Therefore, the reconstructed streamflow sequence could be proved to reflect regional hydrological variability in the study area within the Weihe River basin. Less precipitation tended to result in lower streamflow. Several extremely low values years (<1SD) were consistent with severe low precipitation events occurring in the upper and down reaches of the Weihe River [3,10], including 1834, 1862, 1867, 1891, 1900, 1908, 1926, and 1928. The drought periods, such as 1851–1867, 1877–1883, 1899–1905, 1925–1941, 1995–2004, inferred from a tree-ring based April–July precipitation reconstruction for Tianshui, the upper reaches of the Weihe River [3] almost coincided with the lower intervals 1860s–1870s, 1900s, 1920s–1930s, and 1990s–2000s existing in our June–July streamflow reconstruction. These results supported the common hydroclimatical variations in the whole basin of the Weihe River.

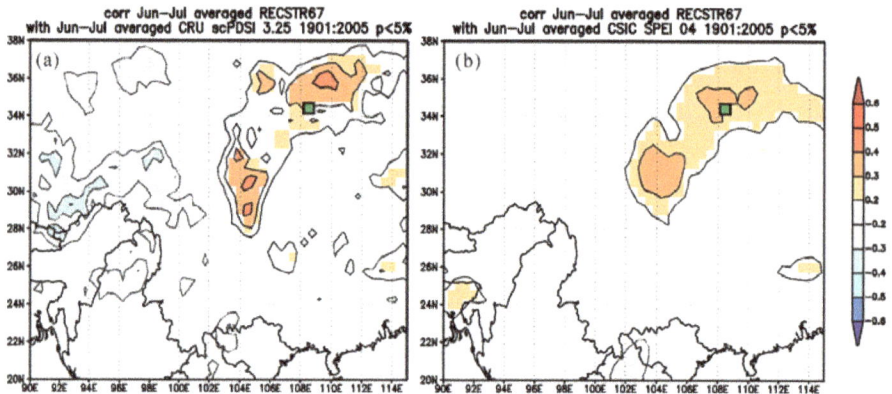

**Figure 6.** Spatial correlations between reconstructed average June–July streamflow (RECSTR67) for the Weihe River and the drought indices of (**a**) scPDSI and (**b**) SPEI on the four-month scale during the period 1901 to 2005 ($p < 0.05$). The sampling site marked by a green rectangle.

Previous studies had reported that the dryness and wetness conditions in northern China including our Weihe River basin were related to the climate mode forcing, such as the Pacific Decadal Oscillation (PDO) and El Niño-Southern Oscillation (ENSO) [27–30]. Spatial correlation of our reconstructed streamflow with SST and MTM spectral analysis results demonstrated the connections between June–July streamflow of the Weihe River and remote oceans (Figures 7 and 8).

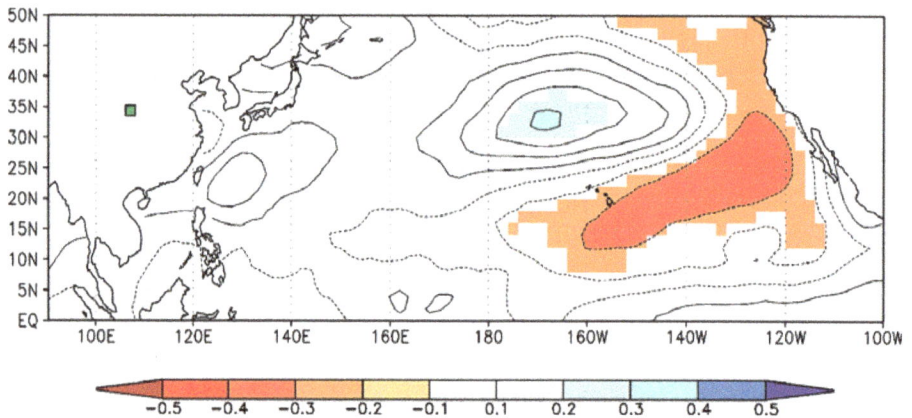

**Figure 7.** Spatial correlations between reconstructed June–July streamflow for the Weihe River and the average May–July sea surface temperature of NCDC ERSSTv5 during the period 1940 to 2005 ($p < 0.1$). All trends in the data were removed. The sampling site marked by a green rectangle.

Significantly negative correlation areas occurred in the SST over the eastern Pacific Ocean along the North America west coast. Meanwhile, positive correlation occurred in the central-north Pacific Ocean (Figure 7) suggesting the streamflow variability may be modulated by the PDO. In the PDO warm phase, negative precipitation abnormality causing more droughts occurred in North China due to the weak summer monsoon and the strong subtropical high with its position locating far to the south and west [31], whereas positive precipitation abnormality occurred in the middle and down reaches of the Yangtze River and South China associating with more flood conditions [27,32,33]. During the cold phase, the situations were opposite [34].

Significant spectrum peaks of June–July streamflow at 78.74a ($p < 0.1$), 72.99a ($p < 0.1$), 68.49a ($p < 0.1$), 64.10a ($p < 0.1$), 60.24a ($p < 0.1$), 35.34a ($p < 0.1$), 34.13a ($p < 0.1$), 27.70a ($p < 0.1$), 26.95a ($p < 0.1$), 26.25a ($p < 0.1$), 25.58a ($p < 0.1$), falling in the range of 50 to 70a as well as 20 to 30a variability of the PDO, supported the close linkages between of the Weihe River and north Pacific ocean on the multi-decadal scales.

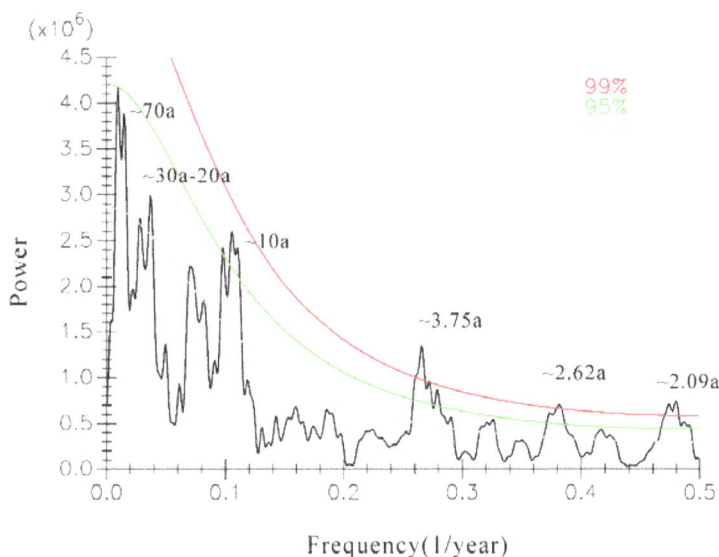

**Figure 8.** Cycles results of multi-taper spectrum (MTM) analyses for reconstructed June–July streamflow for the period 1820 to 2005. The confidence interval at 99%, 95%, and 90% for peaks in the power spectrum was indicated by the red, green, and blue lines.

Significant short cycles were also identified at 10.24a ($p < 0.05$), 10.14a ($p < 0.05$), ~8.98–9.66a ($p < 0.01$), ~3.75a ($p < 0.01$), ~2.62 a ($p < 0.01$), and ~2.09a ($p < 0.01$). Particularly, the ~2–3a periods suggested the strong teleconnections between streamflow variations in the Weihe River and ENSO. The similar frequency cycles had been revealed in other precipitation reconstructions obtained from tree rings in Tianshui [3] and Huashan [10], two studies conducted in the upper reaches and down reaches of the Weihe River, respectively. The PDSI variations over central and southern parts of Shaanxi province and the southeastern Gansu province, i.e., the Weihe River basin, showed an inverse relationship with the Niño 3.4 index in the period 1960 to 2009 [35]. Reconstructed events of the El Niño and La Nina for the period from 1525 to 2002AD provide another evidence that the extreme dryness (lower streamflow value) or wetness (higher streamflow value) conditions reflected the positive or negative phase of ENSO (Rable 2) [36]. Twelve El Niño and 12 La Nina events occurred in the estimated June–July streamflow series, accounting for 40% of the extreme dryness years and 46.15% of the extreme wetness years, respectively. Several lower streamflows in our reconstruction also were also consistent with the El Niño events identified for eastern Northwest China based on the Niño 3.4 index mentioned above, including 1966, 1992, 1997, and 2003. Two higher streamflow years coincided with the La Nina events in 1974 and 1975 [35].

The close relationships between hydroclimatical fluctuations in proximity to the Weihe River basin and remote oceans driving, such as the PDO and ENSO, have been demonstrated from several dendrohydrologial studies in the main tributaries of the Jinghe River, i.e., Kongtong Moutain [37] and Luohe River, i.e., Huanglong [38] within the Weihe River basin.

## 4. Conclusions

A robust regional tree-ring width chronology with the period from 1760 to 2005 AD was developed based on three groups samples of Chinese pine for the middle reaches of the Weihe River, in Central China. The highest significant correlation existed between regional chronology and average June–July streamflow. Therefore, a simple regression model was designed, and 39.3% of the actual variance for the calibration 1940 to 1970 AD was explained. Verification statistics proved the regression model was reliable and skillful to hydrology study for the confidence span from 1820 to 2005 AD. During the past 186 years, extremely low and high flow events occurred in 30 years and 26 years, respectively. The higher streamflow periods of 1820s–1840s, 1890s, 1940s–1960s, and the lower periods of 1860s–1870s, 1900s, 1920s–1930s, 1990s–2000s were identified. The significant decreasing trend occurred in the form moisture epoch in 1950s to the drought epoch in 1990s during the second half of the 20th century. Commonly, regional drought and moisture intervals captured in our streamflow reconstruction suggested it is representative of regional hydroclimate conditions over the Weihe River basin area. Significant spectral peaks were found on the multi-decadal and inter-annual scales, in the range of the bandwidths for natural climate oscillations, such as the PDO and ENSO. Spatial correlation patterns between streamflow and northern Pacific sea surface temperature, in addition to extreme streamflow events coinciding with the phase of ENSO activity, demonstrated the opposite relationships of regional streamflow variability with large-scale circulation systems mentioned above. This is the first hydrological reconstruction obtained from tree rings for the Weihe River. Our results demonstrate that there is great potential for recovering the characteristics and mechanisms of long-term hydrological changes in the Weihe River basin based on the relationships between the radial growth of trees and hydrological climatic factors. The reconstructed June–July streamflow results provide a new perspective for regional water resource assessment and forest protection in the Weihe River basin, which is useful to improve the adverse impacts of regional water cycles caused by global warming.

**Author Contributions:** Conceptualization, G.B. and N.L.; Data curation, G.B. and N.L.; Formal analysis, G.B. and N.L.; Funding acquisition, G.B., N.L. and Y.L.; Investigation, G.B. and N.L.; Methodology, G.B., N.L. and Y.L.; Project administration, G.B. and N.L.; Supervision, Y.L. and H.W.L; Writing-original draft, G.B. and N.L.; Writing-review and editing, G.B., N.L., Y.L. and H.W.L.

**Funding:** This work is supported by the Natural Science Basic Research Plan in Shaanxi Province of China (2018JQ4022); Shaanxi Key Laboratory of Disaster Monitoring and Mechanism Modeling (17JS005), Key program of the Baoji University of Arts and Sciences (ZK2018047), Second Outstanding Young Talents of Shaanxi Universities (2018), State Key Laboratory of Loess and Quaternary Geology (SKLLQG1711, SKLLQG1801) and the Young Scientist Project of Shaanxi Province (2016KJXX-41).

**Acknowledgments:** We thank B.F. Shen, R.Y. Wang, and W.P. Wang for their great assistance in the fieldwork. We also acknowledge the reviewers for their constructive comments to improve the manuscript.

**Conflicts of Interest:** The authors declare no conflict of interest.

## References

1. Song, J.X.; Wang, L.P.; Dou, X.Y.; Wang, F.J.; Guo, H.T.; Zhang, J.L.; Zhang, G.T.; Liu, Q.; Zhang, B. Spatial and depth variability of streambed vertical hydraulic conductivity under the regional flow regimes. *Hydrol. Process.* **2018**, *32*, 3006–3018. [CrossRef]
2. Du, J.; Shi, C.X. Effects of climatic factors and human activities on runoff of the Weihe River in recent decades. *Quat. Int.* **2012**, *282*, 58–65. [CrossRef]
3. Chen, F.; Yuan, Y.J.; Wei, W.S.; Fan, Z.A.; Yu, S.L.; Zhang, T.W.; Zhang, R.B.; Shang, H.M.; Qin, L. Reconstructed precipitation for the north-central China over the past 380 years and its linkages to East Asian summer monsoon variability. *Quat. Int.* **2013**, *283*, 36–45. [CrossRef]
4. Fang, K.; Guo, Z.; Chen, D.; Linderholm, H.W.; Li, J.; Zhou, F.; Gou, G.; Dong, Z.; Li, Y. Drought variation of western Chinese Loess Plateau since 1568 and its linkages with droughts in western North America. *Clim. Dyn.* **2017**, *49*, 3839–3850. [CrossRef]
5. Luo, W.G.; Guo, Z.Z.; Kou, X.M. Status Assessment and Improvement Measures Study on Ecological Water Volume in Wei River. *Northwest Hydropower* **2018**, *6*, 99–103.

6. Li, X.J.; Zhang, J.L.; Song, J.X.; Yang, X.G. Response of runoff to economic water consumptions of the Weihe River in Shaanxi Province. *Arid Land Geogr.* **2016**, *39*, 265–274.

7. Li, S.W.; Jin, l.P.; Zhang, J.; Jiang, T. Several advices on developments of water resources in Guanzhong region. *Shaanxi Water Resour.* **2018**, *2*, 197–200.

8. Sun, Y.; Li, D.L.; Zhu, Y.J. Advances in study about runoff variation of the Weihe River and its response to climate change and human activities. *J. Arid Meteor.* **2013**, *31*, 396–405.

9. Zhang, Z. Tree-rings, a key ecological indicator of environment and climate change. *Ecol. Indic.* **2015**, *51*, 107–116. [CrossRef]

10. Hughes, M.K.; Wu, X.; Shao, X.; Garfin, G. A preliminary reconstruction of rainfall in north-central China since A.D. 1600 from tree-ring density and width. *Quat. Res.* **1994**, *42*, 88–99. [CrossRef]

11. Chen, F.; Zhang, R.; Wang, H.; Qin, L.; Yuan, Y. Updated precipitation reconstruction (AD 1482–2012) for Huashan, north-central China. *Theor. Appl. Climatol.* **2016**, *123*, 723–732. [CrossRef]

12. Fang, K.; Gou, X.; Chen, F.; D'Arrigo, R.; Li, J. Tree-ring based drought reconstruction for the Guiqing Mountain (China): linkages to the Indian and Pacific Oceans. *Int. J. Climatol.* **2010**, *30*, 1137–1145. [CrossRef]

13. Liu, N. The dendroclimatology study on the northern slope of the Qinling Mountains. Master's Thesis, Xi'an Jiaotong University, Xi'an, China, May 2009.

14. Fritts, H.C. *Tree-Rings and Climate*; Academic Press: London, UK, 1976.

15. Holmes, R.L. Computer-assisted quality control in tree-ring dating and measurement. *Tree-Ring Bull.* **1983**, *43*, 69–95.

16. Cook, E.R.; Kairiukstis, L.A. *Methods of Dendrochronology: Applications in the Environmental Sciences*; Kluwer Academic Publishers: Boston, MA, USA, 1990.

17. Wigley, T.; Briffa, K.R.; Jones, P.D. On the average value of correlated time series, with applications in dendroclimatology and hydrometeorology. *J. Appl. Meteorol. Climatol.* **1984**, *23*, 201–213. [CrossRef]

18. Mitchell, T.D.; Jones, P.D. An improved method of constructing a database of monthly climate observations and associated high-resolution grids. *Int. J. Climatol.* **2005**, *25*, 693–712. [CrossRef]

19. Li, B.; Xie, J.C.; Hu, Y.H.; Jiang, R.G. Analysis of variation and abruption of annual runoff in middle and lower Weihe River. *Hydro-Sci. Eng.* **2016**, *3*, 61–69.

20. Mann, M.E.; Lees, J. Robust estimation of background noise and signal detection in climatic time series. *Clim. Chang.* **1996**, *33*, 409–445. [CrossRef]

21. Vicente-Serrano, S.M.; Beguería, S.; López-Moreno, J.I. A multiscalar drought index sensitive to global warming: the standardized precipitation evapotranspiration index-SPEI. *J. Clim.* **2010**, *23*, 1696–1718. [CrossRef]

22. Wells, N.; Goddard, S.; Hayes, M.J. A self–calibrating Palmer Drought Severity Index. *J. Clim.* **2004**, *17*, 2335–2351. [CrossRef]

23. Smith, T.M.; Reynolds, R.W.; Peterson, T.C.; Lawrimore, J. Improvements to NOAA's historical merged land-ocean surface temperature analysis (1880–2006). *J. Clim.* **2008**, *21*, 2283–2296. [CrossRef]

24. Pederson, N.; Lealand, C.; Nachin, B.; Hessl, A.E.; Bell, A.R.; Martin-Benito, D.; Saladyga, T.; Suran, B.; Brown, P.M.; Davi, N. Three centuries of shifting hydroclimatic regimes across the Mongolian Breadbasket. *Agric. For. Meteorol.* **2013**, *178–179*, 10–20. [CrossRef]

25. Dai, A.G. Drought under global warming: a review. *Interdiscip. Rev. Clim. Chang.* **2011**, *2*, 45–65. [CrossRef]

26. Vicente-Serrano, S.M.; Beguería, S.; López-Moreno, J.I.; Angulo, M.; Kenawy, A.E. A new global 0.5° gridded dataset (1901–2006) of a multiscalar drought index: comparison with current drought index datasets based on the Palmer Drought Severity Index. *J. Hydrometeorol.* **2010**, *11*, 1033–1043. [CrossRef]

27. Qian, C.; Zhou, T. Multidecadal Variability of North China Aridity and Its Relationship to PDO during 1900–2010. *J. Clim.* **2014**, *27*, 1210–1222. [CrossRef]

28. Bao, G.; Liu, Y.; Liu, N.; Linderholm, H.W. Drought variability in eastern Mongolian Plateau and its linkages to the large-scale climate forcing. *Clim. Dyn.* **2015**, *44*, 717–733. [CrossRef]

29. Zhang, Y.; Tian, Q.; Guillet, S.; Stoffel, M. 500-yr. precipitation variability in Southern Taihang Mountains, China, and its linkages to ENSO and PDO. *Clim. Chang.* **2017**, *144*, 419–432. [CrossRef]

30. Yang, Q.; Ma, Z.G.; Fan, X.G.; Yang, Z.L.; Xu, Z.F.; Wu, P.L. Decadal modulation of precipitation patterns over eastern China by sea surface temperature anomalies. *J. Clim.* **2017**, *30*, 7017–7033. [CrossRef]

31. Shen, C.; Wang, W.C.; Gong, W.; Hao, Z. A Pacific decadal oscillation record since 1470 AD reconstructed from proxy data of summer rainfall over eastern China. *Geophys. Res. Lett.* **2006**, *33*, L03702. [CrossRef]

32. Ma, Z.; Fu, C. Some evidence of drying trend over Northern China from 1951 to 2004. *Chin. Sci. Bull.* **2006**, *51*, 2913–2925. [CrossRef]

33. Zhou, T.; Song, F.; Lin, R.; Chen, X.; Chen, X. The 2012 North China floods: explaining an extreme rainfall event in the context of a longer-term drying tendency. *B. Am. Meteorol. Soc.* **2013**, *94*, S49–S51.

34. Gu, W.; Li, C.; Yang, H. Analysis on interdecadal variations of summer rainfall and its trend in East China. *Acta. Meteorol. Sin.* **2005**, *63*, 728–739.

35. Liu, Z.; Menzel, L.; Dong, C.; Fang, R. Temporal dynamics and spatial patterns of drought and the relation to ENSO: a case study in Northwest China. *Int. J. Climatol.* **2016**, *36*, 2886–2898. [CrossRef]

36. Gergis, J.L.; Fowler, A.M. A history of ENSO events since A.D. 1525: implications for future climate change. *Clim. Chang.* **2009**, *92*, 343–387. [CrossRef]

37. Fang, K.; Gou, X.; Chen, F.; Liu, C.; Davi, N.; Li, J.; Zhao, Z.; Li, Y. Tree-ring based reconstruction of drought variability (1615–2009) in the Kongtong Mountain area, northern China. *Glob. Planet Chang.* **2012**, *80–81*, 190–197. [CrossRef]

38. Chen, F.; Yuan, Y.; Zhang, R.; Qin, L. A tree-ring based drought reconstruction (AD 1760–2010) for the Loess Plateau and its possible driving mechanisms. *Glob. Planet Chang.* **2014**, *122*, 82–88. [CrossRef]

*forests*

MDPI

Article

# Contrasting Differences in Responses of Streamflow Regimes between Reforestation and Fruit Tree Planting in a Subtropical Watershed of China

Zhipeng Xu [1], Wenfei Liu [1,*], Xiaohua Wei [2], Houbao Fan [1], Yizao Ge [1], Guanpeng Chen [1] and Jin Xu [1]

[1] Jiangxi Province Key Laboratory for Restoration of Degraded Ecosystems & Watershed Ecohydrology, Nanchang Institute of Technology, Nanchang 330099, China; xuzhipeng34@163.com (Z.X.); hbfan@nit.edu.cn (H.F.); geyizao@163.com (Y.G.); 15279196557@163.com (G.C.); 18296150346@163.com (J.X.)

[2] Earth and Environmental Science Department, University of British Columbia (Okanagan), 1177 Research Road, Kelowna, BC V1V 1V7, Canada; adam.wei@ubc.ca

* Correspondence: 2007992987@nit.edu.cn; Tel.: +86-0791-82085311

Received: 13 January 2019; Accepted: 25 February 2019; Published: 27 February 2019

**Abstract:** Fruit tree planting is a common practice for alleviating poverty and restoring degraded environment in developing countries. Yet, its environmental effects are rarely assessed. The Jiujushui watershed (261.4 km$^2$), located in the subtropical Jiangxi Province of China, was selected to assess responses of several flow regime components on both reforestation and fruit tree planting. Three periods of forest changes, including a reference (1961 to 1985), reforestation (1986 to 2000) and fruit tree planting (2001 to 2016) were identified for assessment. Results suggest that the reforestation significantly decreased the average magnitude of high flow by 8.78%, and shortened high flow duration by 2.2 days compared with the reference. In contrast, fruit tree planting significantly increased the average magnitude of high flow by 27.43%. For low flows, reforestation significantly increased the average magnitude by 46.38%, and shortened low flow duration by 8.8 days, while the fruit tree planting had no significant impact on any flow regime components of low flows. We conclude that reforestation had positive impacts on high and low flows, while to our surprise, fruit tree planting had negative effects on high flows, suggesting that large areas of fruit tree planting may potentially become an important driver for some negative hydrological effects in our study area.

**Keywords:** reforestation; fruit tree planting; flow regimes; high flows; low flows

## 1. Introduction

The relationship between forest cover changes and streamflow has long been a heated topic in forested regions [1–3]. Over the past decades, numerous reviews have been made on the effects of forest change on annual mean flow [4–9]. However, research on the impacts of forest change on flow regimes is rather limited [10–13]. Flow regime is composed of five elements: magnitude, duration, timing, variability and frequency [11,14], and the alteration of any element can affect aquatic habitat and biodiversity, as well as ecosystem integrity [15–17]. For example, changes in magnitude and frequency are likely to affect the transport of organic matter and sediments, while changes in flow timing and duration could lead to interference of salmon spawning, and consequently salmon life cycle [11,18]. Therefore, there is a critical need to study how forest cover changes may affect flow regime components where large forest cover change occurs.

Reforestation or afforestation is considered as one of the effective measures to address environmental degradation and climate change impact [19]. To alleviate poverty and environmental degradation, many rural communities in China, particularly in southern China, often grow fruit trees to increase short-term economic benefits and prevent soil erosion. However, forest structures and management strategies resulting from fruit trees and nature forest stands are different. It is still not clear whether fruit tree stands have as similar hydrological functions as natural forests do. Due to more frequent floods and drought events occurring in reforested regions [20], there is a growing concern over the possible negative effects of large-scale fruit tree planting on hydrological functioning. As such, understanding this research topic can support watershed management decisions regarding the relationships between reforestation and water resources.

High and low flows play an important role in the structural composition and function maintenance of riverine aquatic ecosystems by shaping the geomorphologic features of channels and floodplains [10,11,17]. High flow is an indicator of the intensity of floods, and of great significance for public safety. Similarly, low flow regimes are closely related to the functions and structures of riparian plant species [21,22]. Assessing the effectiveness of different forest restoration strategies such as reforestation and fruit trees planting on high and low flows can provide important insights into understanding hydrological processes in forested watersheds.

Jiujushui watershed (261.4 km$^2$) is located in the subtropical region of China. Over the past decades, it has experienced dramatic changes in forest cover, including a forest degradation period in 1960s, reforestation from 1986 to 2000, and fruit tree planting since 2001. In particular, the area of fruit tree planting has been greatly increased over the past 10 years. Such a dramatic forest cover change in the watershed provides a unique opportunity for studying the effects of various reforestation strategies on hydrology. Therefore, the main objective of this study was to the examine whether reforestation and fruit tree planting have led to significant changes in high and low flow regimes in the Jiujushui watershed, and if so, how big the changes have been.

## 2. Materials and Methods

### 2.1. Study Area

The Jiujushui watershed, located in the upper reach of Fu River, is one of the main tributaries of the Poyang Lake basin of Jiangxi Province in the southeast of China (Figure 1). The watershed has a drainage area of 261.4 km$^2$ with the range of slope from 0° to 50°, a main channel drainage length of 41.8 km and an average elevation of 231 m (Figure A1a,b). Red soil, yellow-red soil and mountain yellow soil are the main soil types. Red soil is normally distributed in hilly areas with elevations as low as 500 m below sea level, while yellow-red soil is distributed in areas with an altitude of 500 to 800 m and mountain yellow soil is located in areas with an altitude of more than 800 m. Furthermore, red soil varies with soil depth: A horizon, B horizon and C horizon (Csv). Within the humid subtropical monsoon zone, Jiujushui watershed received an average annual precipitation of 1780 mm between 1961 and 2016, with 855 mm (48.0%) in the wet season from April to June and 225 mm (12.6%) in the dry season from September to November. Annual mean, maximum and minimum temperatures are 18.4 °C, 34.8 °C (in July) and 2.8 °C (in January), respectively (Figure 2). The major land cover types include forest land, farmland and urban. Based on historical land use data, the changes in farmland and urban only accounted for <3.5% (1962–2006) and 0.2% (1996–2005) of the watershed area, respectively, while forest cover change occurred from 36.4% to 77.1% between 1985 and 2016 (reforestation and fruit tree planting).

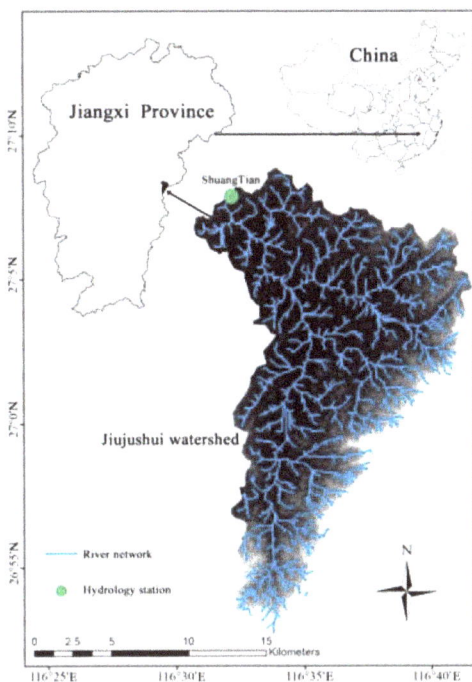

**Figure 1.** Location of the Jiujushui watershed.

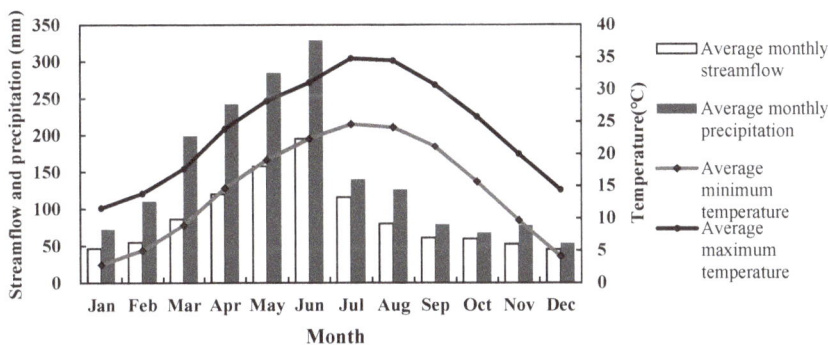

**Figure 2.** Average monthly precipitation and streamflow from 1961 to 2016, with maximum and minimum temperatures from 1961 to 2011 in the Jiujushui watershed.

*2.2. Data Collection*

Stream flow data (1961–2016) used in this study were collected from Shuangtian hydrometric station (Station ID: 62406200) as part of the Chinese National Hydrological Network. Maximum and minimum daily flows were 396 m$^3$s$^{-1}$ (2002) and 0.16 m$^3$s$^{-1}$ (1963), respectively. The annual streamflow varied from 414.2 mm in 1963 to 1820.5 mm in 2016, with an annual mean discharge of 1080.2 mm from 1961 to 2016 (Figure A2). Historical climate data from 1961 to 2011, including daily precipitation and daily maximum, mean and minimum temperatures, were obtained from the Climate Center of Jiangxi Province.

Forest data (forest coverage data and area of fruit tree planting) were obtained from historical forest resources inventory in Nanfeng County. The major forest types included protection forest, timber forest and economic forest, among which *Pinus massoniana, Cunninghamia lanceolata,* citrus and *Phyllostachys heterocycla* (Carr.) *Mitford cv. pubescens* were dominant species in plantation forests. The typical planting density for fruit trees (citrus) in the study watershed is 3.5 m × 4.5 m.

*2.3. Defining the Periods of Forest Changes*

The watershed experienced two distinct forest cover changes over the period of 1961 to 2016. From 1961 to 1985, forest coverage in Jiujushui watershed declined by only 6.3%. Thereafter, a sharp increase of 40.7% from 1986 to 2016 attributed to the Sloping Land Conversion and Mountain-River-Lake Ecological programs in Jiangxi Province, during which fruit tree planting exponentially expanded after 2000 (Figure 3). Therefore, the whole research period was divided into three sub-periods—forest degradation (or the reference period from 1961 to 1985), reforestation (1986–2000) and fruit tree planting (2001–2016)—based on the historical forest changes. It should be noted that the forest degradation period includes 6.3% of forest cover change. Stednick [2] stated that at least of 10~20% change in the watershed area is needed to produce significant hydrological changes. In addition, such a minor change occurred over the 20 years, without significant hydrological effects, that the forest degradation period (1961 to 1985) was treated as the reference or baseline period.

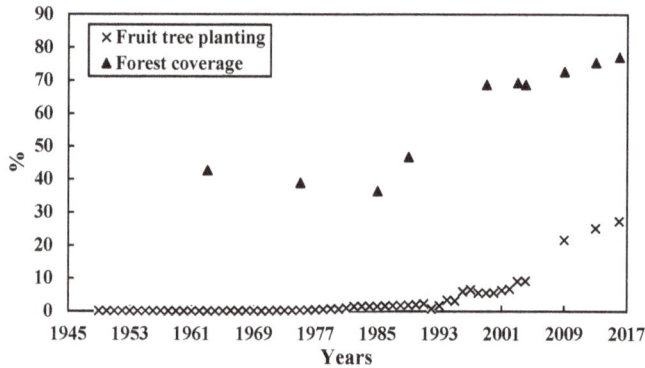

**Figure 3.** Forest changes (forest coverage and fruit tree planting) in the Jiujushui watershed.

*2.4. Defining High and Low Flow Regimes*

In this study, high and low flows are defined as the flows that are equal or more than $Q_{5\%}$, or equal or less than $Q_{95\%}$, in a given year, respectively, through flow duration curves, which represent the relationship between discharges and percentages of discharges below or exceed certain levels for a given time period [10,23,24].

The flow regime includes five elements, which are magnitude, timing, duration, variability and frequency [11]. In this study, magnitude refers to the total amount of water moving through the outlet of the watershed in a given time. The time interval between rainfall peaking and flow peaking represents the timing of high flow, while timing for low flow refers to the average date of low flow occurrence in a year using the paired-wise approach (see the next section) [20]. Duration is defined as the number of days in which daily flow exceeded or was below a given magnitude: high flows refers to the number of days when daily flow exceeded or equaled median value in a given year, while duration of low flows refers to the number of days with daily flow below or equal to median value in a given year. Variability is denoted by the coefficient of variation (CV). Frequency represents how often flow exceeded or was below a given magnitude or return period of high and low flows. Using flood frequency analysis combined with Log-Pearson Type III for analyzing return periods [20,25,26], we divided the return periods of high and low flows into four types: $T_r \leq 1, 1 < T_r \leq 2, 2 < T_r \leq 5$ and $5 < T_r \leq 10$ according to the data, where $T_r$ represents the return period.

## 2.5. Elimination the Effect of Climate Factors on High and Low Flows with The Paired-Year Approach

To minimize the effects of climate variability on high and low flows, the paired-wise approach was used [10,23,24,27], which compares the flow regimes in forest change periods against the reference period under similar climate conditions. As such, the differences between two periods are mainly attributed to forest cover change. In this study, seasonal (wet = April–June, and dry = December–February) and annual precipitation, mean, maximum and minimum temperatures and wind speed were selected as proxies to represent climate conditions over these periods. Firstly, Kendall's Tau and Spearman's Rank were conducted to examine statistical relationships between seasonal and annual climate variables, as well as hydrological variables. As shown in Table A1, annual precipitation (P) and wet-season precipitation ($P_w$), as well as maximum ($T_{maxw}$) and average temperature ($T_{avew}$) in the wet season were significantly correlated with high flows, while P, $P_w$, $T_{maxw}$ and $T_{max}$ were significantly correlated with low flows, where subscript w denotes wet season. Secondly, canonical correlation analysis was used to examine the correlations between two sets of variables, and was elected to determine the highest correlations between sets of climate variables and sets of high and low flows [24]. As a result, $P_w$, $T_{max}$, $T_{maxw}$ and $T_{avew}$ were eventually determined as proxies for similar climate conditions between the reference and reforestation periods during low and high flows (Table A2). Finally, climate variables between the reference and reforestation periods, and between the reference and fruit tree planation periods, were selected (Table A3). It should be noted that high flows are normally associated with storm events. Therefore, the different timings of high flows were selected based on the similarity of storm events in the forest cover change periods (Table A4).

## 3. Results

### 3.1. Responses of High Flows to Reforestation and Fruit Tree Planting

#### 3.1.1. Magnitude

The average magnitude of high flows in the reference period (1961–1985) and reforestation period (1986–2000) was 44.76 $m^3s^{-1}$ and 40.83 $m^3s^{-1}$, respectively. High flows were significantly reduced by 8.78% ($p = 0.018$) when compared to the reference period (Figure 4a). Conversely, the average magnitude of high flows in the fruit tree planting period (2001–2016) was increased by 27.43% ($p = 0.044$) in comparison with the reference period (Figure 4b).

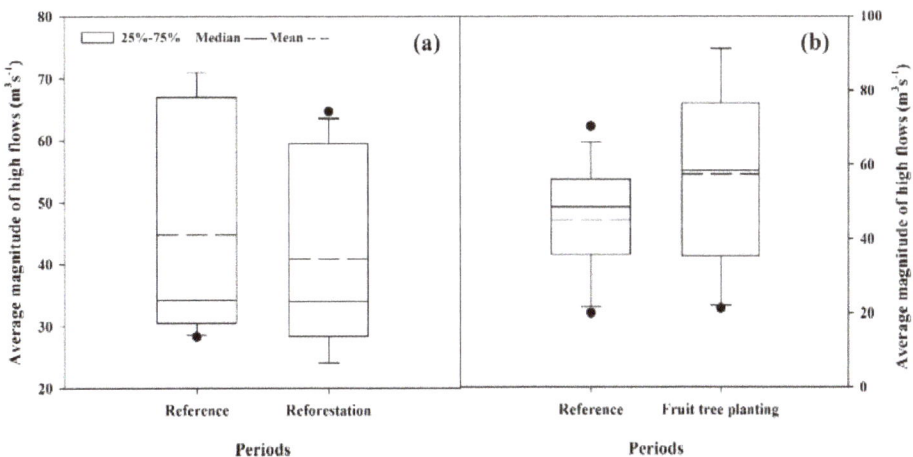

**Figure 4.** Comparison of the magnitudes of high flows (**a**) between the reference and reforestation periods, and (**b**) between the reference and the fruit tree planting periods.

### 3.1.2. Timing

Rainfall events from 25 to 75 mm were selected to find the time intervals between rainfall peaking and flow peaking (Table A4). No statistically significant relationships were detected between forest changes and average timing of high flows in the reforestation period ($p = 0.136$) and the fruit tree planting periods (no delay), respectively.

### 3.1.3. Duration

The analysis from all paired years revealed that the average duration of high flows in the reforestation period was significantly shortened by 2.2 days ($p = 0.033$) than that in the reference period (Figure 5). In contrast, the average duration of high flows was not statistically altered in the fruit tree planting period ($p = 0.235$), but varied with events.

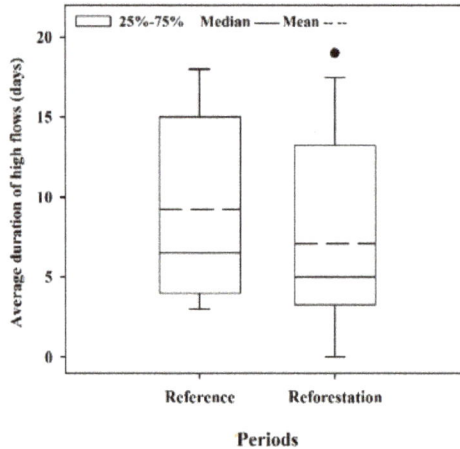

**Figure 5.** Reference period vs. reforestation period: comparison of the duration of high flows.

### 3.1.4. Frequency

Compared with the reference period, the reforestation and fruit tree planting periods had no significant effects on the return periods of $T_r \leq 1$, $1 < T_r \leq 2$, $2 \leq T_r \leq 5$ and $5 \leq T_r \leq 10$ of high flows, respectively ($p = 0.260$ and $p = 0.155$, respectively).

### 3.1.5. Variability

The reforestation and fruit tree planting periods had no statistically significant impacts on the average CV of high flows ($p = 0.911$ and $p = 0.326$, respectively).

### 3.2. Responses of Low Flows to Reforestation and Fruit Tree Planting

### 3.2.1. Magnitude

The average magnitude of low flows in the reforestation period was 46.38% ($p = 0.026$) higher than that in the reference period (Figure 6). In contrast, the magnitude of low flows was not significantly altered in the fruit tree planting period ($p = 0.234$).

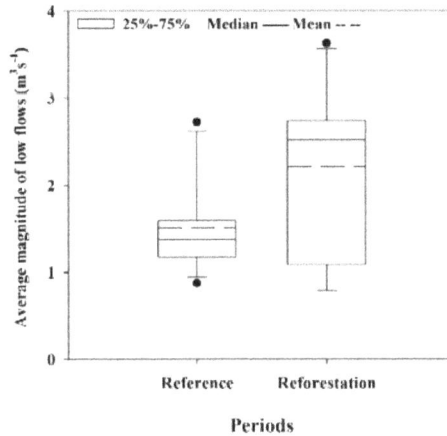

**Figure 6.** Reference period vs. reforestation period: comparison of the average magnitude of low flows.

### 3.2.2. Timing

Reforestation and fruit tree planting had no statistically significant impact on the average timing of low flows ($p = 0.975$ and $p = 0.108$, respectively).

### 3.2.3. Duration

The average duration of low flows in the reforestation period was significantly longer ($p = 0.007$) than that in the reference period. On the contrary, the average duration of low flows was insignificantly related to fruit tree planting ($p = 0.085$). As an example, for the paired years of 1968 and 1994, the daily flows (below or at 1.93 $m^3s^{-1}$) in 1968 (reference year) and in 1994 (reforestation year) were 18 and 0 days, respectively (Figure 7).

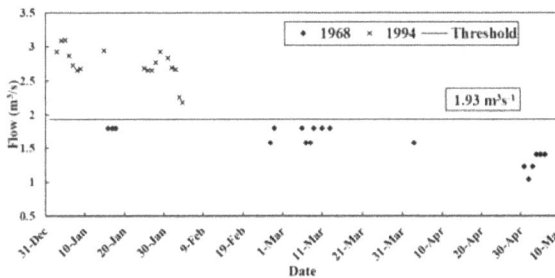

**Figure 7.** Comparison of the duration of low flows between a reference year (1968) and reforestation year (1994) (please refer to the definition of flow duration in the text for further clarification).

### 3.2.4. Frequency

Reforestation and fruit tree planting had insignificant impacts on the return periods ($1 < T_r \leq 2$, $2 < T_r \leq 5$, $5 < T_r \leq 10$) of low flows ($p = 0.231$ and $p = 0.111$, respectively).

### 3.2.5. Variability

The CV of low flows was not significantly related to either reforestation or fruit tree planting, indicating that those two forest practices had similar impacts on the CV of low flows ($p = 0.499$ and $p = 0.689$, respectively).

## 4. Discussion

### 4.1. Effects of Reforestation on High and Low Flows

In the study watershed, reforestation significantly decreased the magnitude and duration of high flows, which is consistent with the general conclusion from other studies [23,28–30]. The reduction of high flows means lower probability of flood occurrence. Reforestation increases leaf area, evapotranspiration and canopy interception of rainfall, and consequently results in reduction of high flows and surface runoff [31]. On the other hand, reforestation enriches the vertical spatial structure of forest ecosystems, abundant understory vegetation and litter layers that can effectively absorb and store water, reducing peak flows [32,33].

Our results also indicated that reforestation had significantly reduced the magnitude and duration of low flows, which are similar to the findings reported in other studies [34–36]. Reforestation can restore the infiltration capacity and water retention ability of soil [23,37,38], increase soil moisture content and groundwater recharge and consequently enhance low flows [39,40]. However, the significant changes in the other flow regime components of low flows were not detected in the reforestation period. This may have been due to the slow forest recovery of soils after severe soil erosion occurred in the reference or forest degradation periods [20].

### 4.2. Effects of Fruit Tree Planting on High and Low Flows

To our surprise, our results indicated that fruit tree planting significantly increased the magnitude of high flows in comparison with the baseline, suggesting that fruit tree planting had negative effects on high flows as the increased high flows produce a higher chance of flooding. Unlike reforestation, fruit tree planting is a distinct planting activity. Fruit tree planting often disturbs the ground surface and reduces surface roughness through intensive land management measures such as soil preparation and removal of understory vegetation and litter layers. As a result, the water-holding capacity of understory vegetation is impaired, which could lead to increased surface runoff [41–44]. Furthermore, removal of weeds and litter can damage the ground surface structure and consequently reduce rainfall infiltration and increase soil and water loss [45].

Our study showed that fruit tree planting had no significant impact on all flow regime components as compared with the reference period. This demonstrates that fruit tree planting did not significantly improve low flows. This is contrary to our expectation, as fruit tree planting is normally expected to play a positive role in soil and water conservation due to the increasing of forest cover. This finding is likely due to intensive land management and resultant reduction of soil infiltration capacity. For instance, some studies demonstrated that the intensive land management in the process of vegetation planting may degrade soil and alter soil infiltration capacity [46], making it impossible for rainfall to infiltrate, and leading to a decrease of soil moisture and groundwater recharge [47].

### 4.3. Contrasted Differences

Both reforestation and fruit tree planting can greatly improve forest coverage. However, their role in flow regimes of high and low flows are different. Our study showed that reforestation had significant and positive effects on the restoration of hydrological processes, while fruit tree planting significantly increased the magnitude of high flows and chance of flooding. Thus, fruit tree planting mainly showed negative effects on flow regimes (Table 1), which is contrary to what we previously anticipated.

The contrasted difference in the responses of flow regimes between the reforestation and the fruit tree planting periods in our study was mainly due to vegetation types and associated management practices [48]. In the study watershed, forest types for reforestation were mainly arbor, while fruit tree species is similar to shrubs due to their similar heights and leaf areas. Lu et al. [49] demonstrated that the runoff magnitude of shrubs was higher, and runoff generation time of shrubs significantly earlier, than arbor. Furthermore, forest stands through reforestation often have well-established understory vegetation and litter layers, while fruit tree planting allows limited understory vegetation and litterfall due to control of weeds from intensive management. Our finding is consistent with Huang et al. [45], who suggested that single-type citrus orchards and farming can significantly increase surface runoff compared with natural vegetation restoration. Thus, vegetation types and associated management practices are critical to flow regimes

**Table 1.** The effects of forest changes on flow regimes in the Jiujushui watershed.

| Period | | Flow Regime Components | | | | |
|---|---|---|---|---|---|---|
| | | Magnitude | Timing | Duration | Frequency | Variability |
| Reference period vs. reforestation period | High flows | ↓ * | - | ↓ * | - | - |
| | Low flows | ↑ * | - | ↓ ** | - | - |
| Reference period vs. fruit tree planting period | High flows | ↑ * | - | - | - | - |
| | Low flows | - | - | - | - | - |

$* p \leq 0.05, ** p \leq 0.01.$

*4.4. Uncertainty Analysis*

There are several uncertainties in this study. Firstly, the paired-wise method has its own strengths and limitations. The current literature indicates that the paired-wise approach is an effective assessment technique for large watersheds (>100 km$^2$). However, its accuracy is largely dependent on the data used to select suitable and comparable pairs. In this study, the combination of Kendall's Tau, Spearman's Rank and canonical correlation over the different seasons ensured that our selected climatic variable was significantly related to both high and low flows. Although annual and seasonal climatic variable were considered, more climate variables in shorter time intervals could be included for better selection in this approach. Secondly, the effects of forest cover change on hydrology are cumulative, which mean that such effects can be prolonged over a long period of time. In this study, averaged hydrological effects between the reference and reforestation or fruit tree planting periods were assessed, which did not differentiate the dynamic and cumulative nature of hydrological effects over the study period. Finally, our study detected significantly negative effect on high flows. This might be related to several mechanisms including site preparation, control of weeds or other human activities. Further studies are needed to understand their relative effects so that the negative hydrological effects caused by fruit tree planting can be minimized. Despite the negative effects on high flow, its effects on the other studied hydrological variables were insignificant in this study, suggesting that the hydrological effects of fruit tree planting may be more complicated than we previously thought, which need more attention.

## 5. Conclusions

Our results show that reforestation has positive effects on high and low flow regimes, while fruit tree planting has negative effects on high flows in our studied watershed. The negative hydrological effects from fruit tree planting suggest that fruit tree planting may not always provide environmental benefits as previously expected, even though it increases forest coverage. To restore soil and water functions in the degradated areas of subtropical regions, caution must be exercised when selecting vegetation types and management practices. It also highlights future studies are needed to fully assess possible contributing mechnisms to increasing of high flows caused by fruit tree planting.

**Author Contributions:** Data curation, H.F.; Investigation, Z.X., Y.G., G.C. and J.X.; Methodology, X.W.; Supervision, W.L.; Writing—original draft, Z.X.; Writing—review & editing, Z.X., W.L. and X.W.

**Acknowledgments:** This work was supported by the National Natural Science Foundation of China (31660234), Jiangxi Provincial Department of Education (GJJ151141) and Scientific Funding by Jiangxi Province (20161BBH80049).

**Conflicts of Interest:** The authors declare no conflict of interest.

# Appendix A

Table A1. Correlation analysis between hydrological variables and climate factors.

| Annual | P | | $T_{max}$ | | $T_{min}$ | | $T_{ave}$ | | Wind Speed | |
|---|---|---|---|---|---|---|---|---|---|---|
| | M-K | S | M-K | S | M-K | S | M-K | S | M-K | S |
| High flow | 0.558 ** | 0.767 ** | −0.173 | −0.267 | 0.074 | 0.12 | −0.081 | −0.119 | 0.048 | 0.07 |
| Low flow | 0.516 ** | 0.704 ** | −0.202 * | −0.282 * | 0.103 | 0.148 | −0.105 | −0.148 | 0.04 | 0.027 |
| **Wet seasons** | **Pw** | | $T_{maxw}$ | | $T_{minw}$ | | $T_{avew}$ | | **Ww** | |
| | M-K | S | M-K | S | M-K | S | M-K | S | M-K | S |
| High flow | 0.640 ** | 0.831 ** | −0.282 ** | −0.352 ** | −0.040 | −0.059 | −0.193 * | −0.246 | −0.051 | −0.054 |
| Low flow | 0.551 ** | 0.739 ** | −0.230 * | −0.287 * | 0.010 | 0.030 | −0.143 | −0.173 | −0.041 | −0.054 |
| **Dry season** | **Pd** | | $T_{maxd}$ | | $T_{mind}$ | | $T_{aved}$ | | **Wd** | |
| | M-K | S | M-K | S | M-K | S | M-K | S | M-K | S |
| High flow | 0.009 | 0.017 | 0.052 | 0.081 | −0.021 | −0.030 | −0.016 | −0.008 | 0.011 | 0.017 |
| Low flow | −0.092 | −0.142 | −0.040 | −0.068 | −0.117 | −0.170 | −0.092 | −0.118 | 0.057 | 0.081 |

Note: P, Tmax, Tave and Tmin are annual mean precipitation, maximum, average, minimum, respectively; Pw and Pd are wet season precipitation (April–June) and dry season precipitation (December–February). M-K and S refer to the methods of Mann–Kendall and Spearman correlation. Significance level set at 0.05, * $p \le 0.05$, ** $p \le 0.01$.

Table A2. Canonical correlation analysis between hydrological variables and climate factors.

| | Canonical R | Hydrological Variable Set (High and Low Flows) |
|---|---|---|
| | Set 1 (P, $T_{max}$,) | 0.774 ** |
| | Set 2 (P, $T_{maxw}$, ) | 0.774 ** |
| | Set 3 (P, $T_{avew}$) | 0.774 ** |
| | Set 4 (P, $T_{max}$, $T_{avew}$) | 0.774 ** |
| | Set 5 (P, $T_{maxw}$, $T_{avew}$) | 0.793 ** |
| | Set 6 ((P, $T_{max}$, $T_{maxw}$) | 0.778 ** |
| Climate variables sets | Set 7 (P, $T_{max}$, $T_{maxw}$, $T_{avew}$) | 0.795 ** |
| | Set 8 ($P_w$, $T_{max}$) | 0.842 ** |
| | Set 9 ($P_w$, $T_{maxw}$) | 0.842 ** |
| | Set 10 ($P_w$, $T_{avew}$) | 0.841 ** |
| | Set 11 ($P_w$, $T_{max}$, $T_{avew}$) | 0.846 ** |
| | Set 12 ($P_w$, $T_{maxd}$, $T_{avew}$) | 0.844 ** |
| | Set 13 ((P, $T_{max}$, $T_{maxw}$) | 0.845 ** |
| | Set 14 ($P_w$, $T_{maxw}$, $T_{maxw}$, $T_{avew}$) | 0.847 ** |

* $p \le 0.05$, ** $p \le 0.01$.

Table A3. Climate and flow variables of paired years in the Jiujushui watershed.

| | Selected Year | Paired Year | $P_w$/mm | | $Q_w$/mm | | $T_{max}$/°C | | $T_{maxw}$/°C | | $T_{avew}$/°C | |
|---|---|---|---|---|---|---|---|---|---|---|---|---|
| | 1968 | 1994 | 1083.1 | 1072.4 | 469.6 | 650.3 | 24.1 | 23.6 | 27.2 | 28.5 | 21.8 | 23.4 |
| | 1979 | 1990 | 609.1 | 614.9 | 278.1 | 284.7 | 24.2 | 23.5 | 26.8 | 27.3 | 21.9 | 22.1 |
| | 1964 | 1986 | 668.3 | 650.8 | 387.8 | 335.4 | 23.7 | 24.0 | 27.7 | 28.2 | 23.1 | 23.1 |
| | 1979 | 1987 | 609.1 | 635.3 | 278.1 | 250.7 | 24.2 | 24.1 | 26.8 | 27.3 | 21.9 | 22.1 |
| | 1968 | 1998 | 1083.1 | 1105.1 | 469.6 | 611.8 | 24.1 | 24.6 | 27.2 | 29.3 | 21.8 | 23.9 |
| Reference period vs. | 1974 | 2000 | 722.2 | 721.6 | 230.7 | 351.7 | 23.6 | 23.3 | 28.5 | 28.0 | 23.0 | 22.8 |
| reforestation period | 1965 | 1993 | 865.6 | 790.4 | 349.2 | 344.1 | 23.6 | 23.6 | 26.4 | 26.9 | 21.4 | 22.1 |
| | 1966 | 1996 | 874.3 | 852.8 | 512.5 | 416.8 | 24.3 | 23.6 | 27.0 | 26.8 | 21.8 | 21.7 |
| | 1982 | 1994 | 978.3 | 1072.4 | 669.0 | 650.3 | 23.3 | 23.6 | 26.8 | 28.5 | 21.6 | 23.4 |
| | 1981 | 1995 | 857.9 | 849.0 | 653.3 | 431.4 | 23.3 | 23.5 | 27.2 | 27.2 | 22.1 | 22.3 |
| | 1972 | 1993 | 819.9 | 790.4 | 283.8 | 344.1 | 23.3 | 23.6 | 26.9 | 26.9 | 21.9 | 22.1 |
| | 1978 | 1987 | 689.4 | 635.3 | 347.9 | 250.7 | 24.2 | 24.1 | 27.1 | 27.3 | 21.9 | 22.1 |

**Table A3.** *Cont.*

| Selected Year | Paired Year | $P_w$/mm | | $Q_w$/mm | | $T_{max}$/°C | | $T_{maxw}$/°C | | $T_{avew}$/°C | |
|---|---|---|---|---|---|---|---|---|---|---|---|
| | 1975 | 2001 | 1090.4 | 1071.2 | 690.4 | 745.3 | 23.1 | 24.0 | 26.6 | 27.2 | 21.8 | 22.3 |
| | 1963 | 2007 | 539.6 | 550.6 | 122.4 | 273.7 | 24.9 | 24.8 | 29.0 | 28.5 | 23.3 | 22.2 |
| | 1980 | 2012 | 1038.3 | 1029.0 | 662.8 | 591.2 | 23.2 | | 27.5 | | 22.2 | |
| | 1980 | 2002 | 1038.3 | 1012.8 | 662.8 | 668.6 | 23.2 | 24.4 | 27.5 | 28.8 | 22.2 | 23.2 |
| | 1964 | 2003 | 668.3 | 643.0 | 387.8 | 371.6 | 23.7 | 25.3 | 27.7 | 28.4 | 23.1 | 23.0 |
| Reference period vs. | 1966 | 2006 | 874.3 | 894.6 | 512.5 | 584.4 | 24.3 | 24.2 | 27.0 | 27.7 | 21.8 | 22.2 |
| fruit tree planting period | 1982 | 2005 | 978.3 | 984.3 | 669.0 | 651.0 | 23.3 | 23.7 | 26.8 | 29.2 | 21.6 | 23.9 |
| | 1976 | 2015 | 852.1 | 816.0 | 543.7 | 348.5 | 22.9 | | 26.3 | | 21.6 | |
| | 1969 | 2002 | 1025.1 | 1012.8 | 359.1 | 668.6 | 23.3 | 24.4 | 28.2 | 28.8 | 22.8 | 23.2 |
| | 1973 | 2005 | 991.1 | 984.3 | 760.3 | 651.0 | 23.5 | 23.7 | 26.9 | 29.2 | 22.4 | 23.9 |
| | 1969 | 2012 | 1025.1 | 1029.0 | 359.1 | 591.2 | 23.3 | | 28.2 | | 22.8 | |
| | 1985 | 2009 | 485.6 | 461.3 | 245.8 | 134.3 | 23.1 | 25.0 | 28.3 | 29.1 | 22.9 | 23.3 |

$P_w$ and $Q_w$ refer to precipitation and streamflow in the wet season, respectively.

**Table A4.** The pairs of rainfall events in the Jiujushui watershed.

| Period | Selected Rainfall Events | Paired Rainfall Events | Peak Rainfall | | Antecedent Rainfall (3-Day Average)/mm | | Time Interval/Day | |
|---|---|---|---|---|---|---|---|---|
| | 1976 20 Apr–26 Apr | 1986 24 Mar–29 Mar | 26.4 | 26.3 | 0 | 0.9 | 0 | 1 |
| | 1964 16 May–22 May | 1987 8 Sep–14 Sep | 33.9 | 33.4 | 0.8 | 1.3 | 0 | 0 |
| | 1964 10 Jan–15 Jan | 1990 19 Feb–25 Feb | 45 | 45.1 | 0.2 | 1.8 | 1 | 1 |
| | 1967 18 May–23 May | 1991 18 Mar–23 Mar | 52.5 | 51.6 | 0.1 | 0.4 | 0 | 1 |
| | 1962 14 Apr–20 Apr | 1992 27 May–2 Jun | 34.7 | 35 | 0.2 | 1.6 | 0 | 2 |
| | 1976 10 Oct–16 Oct | 1993 11 May–17 May | 32.6 | 32.9 | 0.2 | 0 | 1 | 0 |
| Reference period vs. reforestation period | 1976 12 May–18 May | 1994 7 May–13 May | 29.1 | 29 | 0.8 | 0 | 1 | 1 |
| | 1961 17 Nov–23 Nov | 1996 16 Apr–22 Apr | 41 | 40.7 | 0.1 | 0 | 0 | 0 |
| | 1972 30 Oct–5 Nov | 1996 19 Jul–25 Jul | 25.6 | 25.7 | 0.3 | 0.4 | 0 | 0 |
| | 1974 28 Oct–3 Nov | 1996 19 Jul–25 Jul | 25.6 | 25.7 | 0.3 | 0.4 | 0 | 0 |
| | 1968 6 Jun–12 Jun | 1997 28 Apr–4 May | 70.8 | 73.7 | 2.5 | 2.2 | 0 | 1 |
| | 1964 16 May–22 May | 1998 29 Oct–4 Nov | 33.9 | 33.5 | 0.8 | 0.1 | 0 | 1 |
| | 1965 2 Aug–7 Aug | 1997 11 Oct–17 Oct | 41.2 | 41.7 | 0 | 1.2 | 1 | 1 |
| | 1976 27 Sep–3 Oct | 1987 22 Sep–28 Sep | 39.6 | 38.7 | 0.5 | 0.2 | 0 | 1 |
| | 1984 11 Nov–17 Nov | 1990 30 Oct–5 Nov | 44.9 | 44.7 | 1.7 | 0 | 1 | 0 |
| Reference period vs. fruit tree planting period | 1983 20 Aug–26 Aug | 2002 11 May–16 May | 25.7 | 26.1 | 1.3 | 1 | 2 | 1 |
| | 1976 20 Apr–26 Apr | 2014 1 May–7 May | 26.4 | 26.5 | 0 | 0 | 0 | 1 |
| | 1964 16 May–22 May | 2010 11 Feb–17 Feb | 33.9 | 33.4 | 0.8 | 1.2 | 0 | 1 |
| | 1964 11 Oct–17 Oct | 2016 12 May–18 May | 44.4 | 44 | 0.2 | 0 | 0 | 0 |
| | 1966 1 Apr–7 Apr | 2004 9 Apr–14 Apr | 39.7 | 39.3 | 2.0 | 0 | 0 | 0 |
| | 1967 18 May–23 May | 2016 7 Sep–13 Sep | 52.5 | 52 | 0.1 | 0 | 0 | 1 |
| | 1968 6 Jun–12 Jun | 2001 28 Aug–3 Sep | 70.8 | 68.7 | 2.5 | 2.8 | 0 | 0 |
| Reference period vs. fruit tree planting period | 1978 25 Aug–31 Aug | 2005 20 Apr–26 Apr | 30.1 | 30.1 | 0 | 0 | 1 | 0 |
| | 1962 14 Apr–20 Apr | 2006 28 Apr–4 May | 34.7 | 34.6 | 0.2 | 0.7 | 0 | 1 |
| | 1976 10 Oct–16 Oct | 2007 1 Aug–7 Aug | 32.6 | 32.2 | 0.2 | 1.5 | 1 | 1 |
| | 1977 24 Feb–1 Mar | 2006 19 May–25 May | 40.2 | 40.7 | 0 | 0 | 0 | 0 |
| | 1975 3 Jul–10 Jul | 2005 20 Apr–26 Apr | 30.3 | 30.1 | 0.2 | 0 | 1 | 0 |
| | 1972 26 Apr–2 May | 2002 27 May–2 Jun | 26.6 | 26.6 | 0.7 | 0 | 1 | 1 |
| | 1961 17 Nov–23 Nov | 2006 19 May–25 May | 41 | 40.7 | 0.1 | 0 | 0 | 0 |
| | 1965 2 Aug–7 Aug | 2016 14 Mar–19 Mar | 41.2 | 42 | 0 | 0 | 2 | 1 |

Time interval: time between rainfall peaking and flow peaking during a rainfall event.

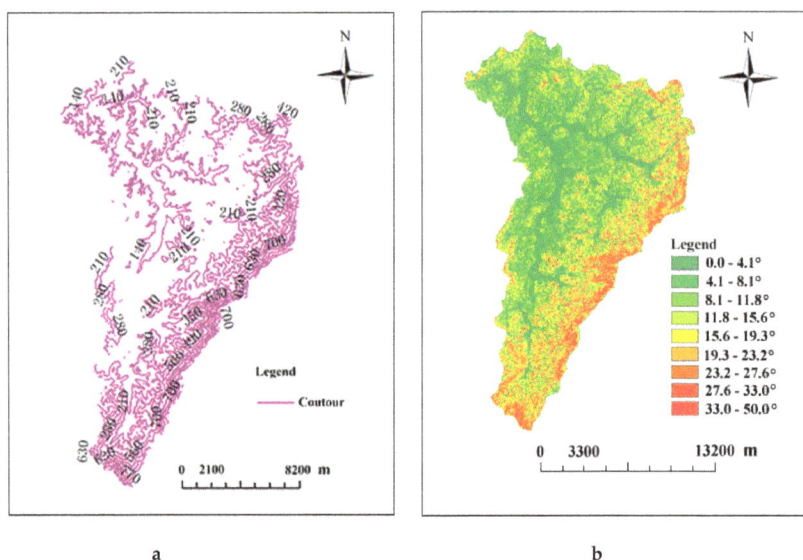

a

b

**Figure A1.** Watershed characteristics: (**a**) watershed elevation, and (**b**) watershed slope.

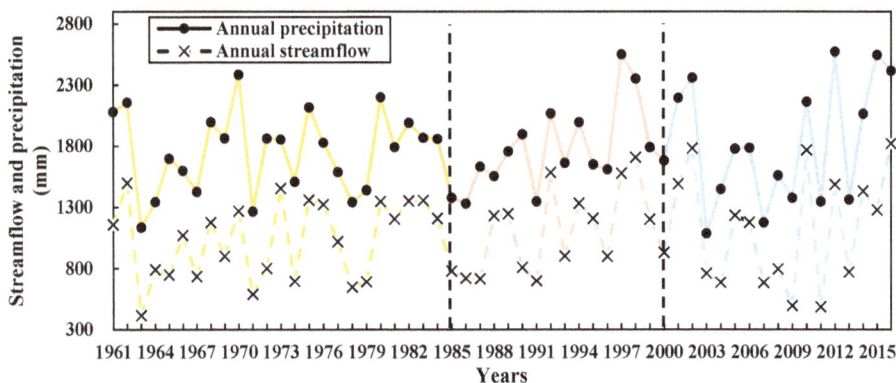

**Figure A2.** Annual precipitation and streamflow from 1961 to 2016 in the Jiujushui watershed (different colors represent three periods: reference, reforestation and fruit tree planting).

## References

1. Wei, X.; Liu, W.; Zhou, P. Quantifying the Relative Contributions of Forest Change and Climatic Variability to Hydrology in Large Watersheds: A Critical Review of Research Methods. *Water* **2013**, *5*, 728–746. [CrossRef]
2. Stednick, J.D. Monitoring the effects of timber harvest on annual water yield. *J. Hydrol.* **1996**, *176*, 79–95. [CrossRef]
3. Farley, K.A.; Jobbágy, E.G.; Jackson, R.B. Effects of afforestation on water yield: A global synthesis with implications for policy. *Glob. Chang. Biol.* **2005**, *11*, 1565–1576. [CrossRef]
4. Andréassian, V. Waters and forests: From historical controversy to scientific debate. *J. Hydrol.* **2004**, *291*, 1–27. [CrossRef]
5. Bruijnzeel, L.A. Hydrological functions of tropical forests: Not seeing the soil for the trees? *Agric. Ecosyst. Environ.* **2004**, *104*, 185–228. [CrossRef]

6.  Li, Q.; Wei, X.; Zhang, M.; Liu, W.; Fan, H.; Zhou, G.; Giles-Hansen, K.; Liu, S.; Wang, Y. Forest cover change and water yield in large forested watersheds: A global synthetic assessment. *Ecohydrology* **2017**, *10*, e1838. [CrossRef]

7.  Zhang, M.; Liu, N.; Harper, R.; Li, Q.; Liu, K.; Wei, X.; Ning, D.; Hou, Y.; Liu, S. A global review on hydrological responses to forest change across multiple spatial scales: Importance of scale, climate, forest type and hydrological regime. *J. Hydrol.* **2017**, *546*, 44–59. [CrossRef]

8.  Wei, X.; Li, Q.; Zhang, M.; Giles-Hansen, K.; Liu, W.; Fan, H.; Wang, Y.; Zhou, G.; Piao, S.; Liu, S. Vegetation cover—Another dominant factor in determining global water resources in forested regions. *Glob. Chang. Biol.* **2018**, *24*, 786–795. [CrossRef] [PubMed]

9.  Li, Q.; Wei, X.; Zhang, M.; Liu, W.; Giles-Hansen, K.; Wang, Y. The cumulative effects of forest disturbance and climate variability on streamflow components in a large forest-dominated watershed. *J. Hydrol.* **2018**, *557*, 448–459. [CrossRef]

10. Zhang, M.; Wei, X. Alteration of flow regimes caused by large-scale forest disturbance: A case study from a large watershed in the interior of British Columbia, Canada. *Ecohydrology* **2014**, *7*, 544–556. [CrossRef]

11. Poff, L.R.; Allan, J.D.; Bain, M.B.; Karr, J.R.; Prestegaard, K.L.; Richter, B.D.; Sparks, R.E.; Stromberg, J.C. The Natural Flow Regime. *Bioscience* **1997**, *47*, 769–784. [CrossRef]

12. Duan, L.; Man, X.; Yu, Z.X.; Liu, Y.; Zhu, B. The effects of forest disturbance on flow regimes of a small forested watershed in northern Daxing'anling, China. *Acta Ecol. Sin.* **2017**, *37*, 1421–1430.

13. Li, Q.; Wei, X.; Yang, X.; Giles-Hansen, K.; Zhang, M.; Liu, W. Topography significantly influencing low flows in snow-dominated watersheds. *Hydrol. Earth Syst. Sci.* **2018**, *22*, 1947–1956. [CrossRef]

14. Schneider, C.; Laizé, C.L.R.; Acreman, M.C.; Flörke, M. How will climate change modify river flow regimes in Europe? *Hydrol. Earth Syst. Sci.* **2012**, *9*, 9193–9238. [CrossRef]

15. Wang, G.X.; Qian, J.; Cheng, G.D. Current situation and prospect of the ecological hydrology. *Adv. Earth Sci.* **2001**, *16*, 314–323.

16. Bunn, S.E.; Arthington, A.H. Basic principles and ecological consequences of altered flow regimes for aquatic biodiversity. *Environ. Manag.* **2002**, *30*, 492–507. [CrossRef]

17. Poff, N.L.; Zimmerman, J.K.H. Ecological responses to altered flow regimes: A literature review to inform the science and management of environmental flows. *Freshw. Biol.* **2010**, *55*, 194–205. [CrossRef]

18. Wharton, C.H.; Lambour, V.W.; Newsom, J. The Fauna of Bottomland Hardwoods in Southeastern United States. *Dev. Agric. Manag. For. Ecol.* **1981**, 87–160.

19. Guariguata, M.R.; Locatelli, B.; Haupt, F. Adapting Tropical Production Forests to Global Climate Change: Risk Perceptions and Actions. *Int. For. Rev.* **2012**, *14*, 27–38. [CrossRef]

20. Liu, W.; Wei, X.; Fan, H.; Guo, X.; Liu, Y.; Zhang, M.; Li, Q. Response of flow regimes to deforestation and reforestation in a rain-dominated large watershed of subtropical China. *Hydrol. Process.* **2015**, *29*, 5003–5015. [CrossRef]

21. Greenberg, L.; Svendsen, P.; Harby, A. Availability of microhabitats and their use by brown trout (*Salmo trutta*) and grayling (*Thymallus thymallus*) in the River Vojman. *Regul. Rivers Reserv. Manag.* **1996**, *12*, 287–303. [CrossRef]

22. Reeves, G.H.; Benda, L.E.; Burnett, K.M.; Bisson, P.A.; Sedell, J.R. A disturbance-based ecosystem approach to maintaining and restoring freshwater habitats of evolutionarily significant units of anadromous salmonids in the Pacific Northwest. *Am. Fish. Soc. Symp. Ser.* **1996**, *17*, 334–349.

23. Liu, W.; Wei, X.; Li, Q.; Fan, H.; Duan, H.; Wu, J.; Krasty, G.H.; Zhang, H. Hydrological recovery in two large forested watersheds of Southeastern China: Importance of watershed property in determining hydrological responses to reforestation. *Hydrol. Earth Syst. Sci.* **2016**, *20*, 4747–4756. [CrossRef]

24. Zhang, M.; Wei, X.; Li, Q. A quantitative assessment on the response of flow regimes to cumulative forest disturbances in large snow-dominated watersheds in the interior of British Columbia, Canada. *Ecohydrology* **2016**, *9*, 843–859. [CrossRef]

25. Mcmahon, T.A.; Srikanthan, R. Log Pearson III distribution—is it applicable to flood frequency analysis of Australian streams? *J. Hydrol.* **1981**, *52*, 139–147. [CrossRef]

26. Sharma, M.; Agarwal, R. Maximum likelihood method for parameter estimation in non-linear models with below detection data. *Environ. Ecol. Stat.* **2003**, *10*, 445–454. [CrossRef]

27. Eastwood, A.; Brooker, R.; Irvine, R.; Artz, R.; Norton, L.; Bullock, J.; Ross, L.; Fielding, D.; Ramsay, S.; Roberts, J. Does nature conservation enhance ecosystem services delivery? *Ecosyst. Serv.* **2016**, *17*, 152–162. [CrossRef]

28. Nadal-Romero, E.; Cammeraat, E.; Serrano-Muela, M.P.; Lana-Renault, N.; Regüés, D. Hydrological response of an afforested catchment in a Mediterranean humid mountain area: A comparative study with a natural forest. *Hydrol. Process.* **2016**, *30*, 2717–2733. [CrossRef]

29. Llorens, P.; Poch, R.; Latron, J.; Gallart, F. Rainfall interception by a *Pinus sylvestris* forest patch overgrown in a Mediterranean moutainous abandoned area I. Monitoring design and results down to the event scale. *J. Hydrol.* **1997**, *199*, 331–345. [CrossRef]

30. Gebrehiwot, S.G.; Taye, A.; Bishop, K. Forest cover and stream flow in a headwater of the Blue Nile: Complementing observational data analysis with community perception. *AMBIO* **2010**, *39*, 284–294. [CrossRef] [PubMed]

31. Dung, B.X.; Gomi, T.; Miyata, S.; Sidle, R. Peak flow responses and recession flow characteristics after thinning of Japanese cypress forest in a headwater catchment. *Hydrol. Res. Lett.* **2012**, *6*, 35–40. [CrossRef]

32. Shen, W.; Peng, S.; Zhou, G.; Lin, Y.; Li, Z. Ecohydrological functions of litter in man-made Acacia Mangium and *Pinus elliotii* plantations. *Acta Ecol. Sin.* **2001**, *21*, 846–850.

33. Robinson, M.; Dupeyrat, A. Effects of commercial timber harvesting on streamflow regimes in the Plynlimon catchments, mid-Wales. *Hydrol. Process.* **2010**, *19*, 1213–1226. [CrossRef]

34. Price, K.; Jackson, C.R.; Parker, A.J.; Reitan, T.; Dowd, G.; Cyterski, M. Effects of watershed land use and geomorphology on stream low flows during severe drought conditions in the southern Blue Ridge Mountains, Georgia and North Carolina, United States. *Water Resour. Res.* **2011**, *47*, 1198–1204. [CrossRef]

35. Ahn, K.Y.; Merwade, V. The effect of land cover change on duration and severity of high and low flows. *Hydrol. Process.* **2017**, *31*, 133–149. [CrossRef]

36. Buttle, J.M. Streamflow response to headwater reforestation in the Ganaraska River basin, southern Ontario, Canada. *Hydrol. Process.* **2011**, *25*, 3030–3041. [CrossRef]

37. Blöschl, G.; Ardoin-Bardin, S.; Bonell, M.; Dorninger, M.; Goodrich, D. At what scales do climate variability and land cover change impact on flooding and low flows? *Hydrol. Process.* **2007**, *21*, 1241–1247. [CrossRef]

38. Zhou, G.; Wei, X.; Chen, X.; Zhou, P.; Liu, X.; Xiao, Y.; Sun, G.; Scott, D.F.; Zhou, S.; Han, L.; et al. Global pattern for the effect of climate and land cover on water yield. *Nat. Commun.* **2015**, *6*, 5918. [CrossRef] [PubMed]

39. Mcvicar, T.R.; Li, L.T.; Niel, T.G.V.; Zhang, L.; Li, R.; Yang, Q.; Zhang, X.; Mu, X.; Wen, Z.; Liu, W.; et al. Developing a decision support tool for China's re-vegetation program: Simulating regional impacts of afforestation on average annual streamflow in the Loess Plateau. *For. Ecol. Manag.* **2007**, *251*, 65–81. [CrossRef]

40. Vertessy, R.; Zhang, L.; Dawes, W.R. Plantations, river flows and river salinity. *Aust. For.* **2003**, *66*, 55–61. [CrossRef]

41. Waterloo, M.J.; Schellekens, J.; Bruijnzeel, L.A.; Rawaqa, T.T. Changes in catchment runoff after harvesting and burning of a *Pinus caribaea* plantation in Viti Levu, Fiji. *For. Ecol. Manag.* **2007**, *251*, 31–44. [CrossRef]

42. Huang, Q.; He, B.H.; Qin, W.; Zuo, C.; Yao, Y.; He, X.; Li, T. The Infiltration Characteristics of Disturbed Soil under Natural Rainfall Conditions. *J. Irrig. Drain. Eng.* **2015**, *34*, 91–95.

43. Pathak, P.; Wani, S.P.; Sudi, R.; Budama, N. Inter-row tillage for improved soil and water conservation and crop yields on crusted Alfisols. *Agric. Sci.* **2017**, *04*, 36–45. [CrossRef]

44. Sileshi, R.; Pitt, R.; Clark, S.; Christan, C. Laboratory and Field Studies of Soil Characteristics of Proposed Stormwater Bioinfiltration Sites. *Proc. Water Environ. Fed.* **2012**, *5*, 241–250. [CrossRef]

45. Huang, H.X.; Xie, X.L.; Wang, K.R. Surface runoff and nutrient loss from red-soil slope-lands under different land use types. *Ecol. Environ. Sci.* **2008**, *17*, 1645–1649.

46. Walker, J.; Reddell, P. Retrogressive Succession and Restoration on Old Landscapes. In *Linking Restoration and Ecological Succession*; Springer: New York, NY, USA, 2007; pp. 69–89.

47. Zhang, M.; Wei, X. The cumulative effects of forest disturbance on streamflow in a large watershed in the central interior of British Columbia, Canada. *Hydrol. Earth Syst. Sci.* **2012**, *9*, 2855–2895. [CrossRef]

48. Lacombe, G.; Ribolzi, O.; De Rouw, A.; Pierret, A. Contradictory hydrological impacts of afforestation in the humid tropics evidenced by long-term field monitoring and simulation modelling. *Hydrol. Earth Syst. Sci.* **2016**, *20*, 2691–2704. [CrossRef]

49. Lu, X.Z.; Kang, L.L.; Zuo, Z.G.; Sun, J.; Ni, Y.X.; Lu, W.X. Effects of different vegetation types on slope runoff process under rainstorm condition. *Guangdong Agric. Sci.* **2016**, *1*, 79–83.

*forests*

MDPI

*Article*

# The Radial Growth of Schrenk Spruce (*Picea schrenkiana* Fisch. et Mey.) Records the Hydroclimatic Changes in the Chu River Basin over the Past 175 Years

Ruibo Zhang [1,2,3,*], Bakytbek Ermenbaev [4], Tongwen Zhang [1], Mamtimin Ali [1], Li Qin [1] and Rysbek Satylkanov [4]

[1] Institute of Desert Meteorology, China Meteorological Administration, Key Laboratory of Tree-ring Physical and Chemical Research of China Meteorological Administration, Key Laboratory of Tree-ring Ecology of Xinjiang Uigur Autonomous Region, Urumqi 830002, China; Zhangtw@idm.cn (T.Z.); Ali@idm.cn (M.A.); Qinhappy@sina.com (L.Q.)
[2] Climate Change Research Center, Institute of Atmospheric Physics, Chinese Academy of Sciences, Beijing 100029, China
[3] Collaborative Innovation Center on Forecast and Evaluation of Meteorological Disasters/Joint International Research Laboratory of Climate and Environment Change/Key Laboratory of Meteorological Disaster, Ministry of Education, Nanjing University of Information Science and Technology, Nanjing 210044, China
[4] Tien-Shan Scientific Center, Institute of Water Problem and Hydropower of National Academy of Sciences of Kyrgyz Republic, Bishkek 720033, Kyrgyzstan; b.ermenbaev@mail.ru (B.E.); r.satylkanov@gmail.com (R.S.)
* Correspondence: river0511@163.com; Tel.: +86-991-2662971

Received: 5 January 2019; Accepted: 26 February 2019; Published: 2 March 2019

**Abstract:** The Chu River is one of the most important rivers in arid Central Asia. Its discharge is affected by climate change. Here, we establish a tree-ring chronology for the upper Chu River Basin and analyze the relationships between radial growth, climate, and discharge. The results show that the radial growth of Schrenk spruce (*Picea schrenkiana* Fisch. et Mey.) is controlled by moisture. We also reconstruct a 175-year standardized precipitation-evapotranspiration index (SPEI) for the Chu River Basin. A comparison of the reconstructed and observed indices reveal that 39.5% of the variance occurred during the calibration period of 1952–2014. The SPEI reconstruction and discharge variability of the Chu River show consistent long-term change. They also show that the Chu River Basin became increasingly dry between the 1840s and the 1960s, with a significant drought during the 1970s. A long and rapid wetting period occurred between the 1970s and the 2000s, and was followed by increasing drought since 2004. The change in the SPEI in the Chu River Basin is consistent with records of long-term precipitation, SPEI and Palmer Drought Severity Indices (PDSI) in other proximate regions of the western Tianshan Mountains. The hydroclimatic change of the Chu River Basin may be associated with westerly wind. This study is helpful for disaster prevention and water resource management in arid central Asia.

**Keywords:** tree rings; Schrenk spruce (*Picea schrenkiana* Fisch. et Mey.); hydroclimatology; Chu River; Tianshan Mountains; climate change; Central Asia

---

## 1. Introduction

It is widely recognized that global warming has occurred since the mid-19th century [1]. However, corresponding hydroclimatic changes demonstrate significant regional variations [2]. Arid Central Asia (ACA) covers $5 \times 10^6$ km², and includes Kazakhstan, Kyrgyzstan, Tajikistan, Turkmenistan, Uzbekistan, and Xinjiang in northwest China. Drought is a major climate disaster in ACA and the cause of considerable agricultural, economic and environmental damage. The Tianshan Mountains, which extend across the region, are known as the "water towers" of Central Asia, and are the largest and most important mountain system in ACA. The region is especially sensitive to climate change [3,4]. Obvious warming in the Tianshan Mountains has been detected at a rate of 0.3 °C/10 years [5], and persistent warming is exacerbating droughts and water shortages. The Chu River, which originates in the Tianshan Mountains, is one of the longest rivers in ACA. Its river basin is shared by Kyrgyzstan (where the river originates) and Kazakhstan, and its waters feed millions of people and support the social development and economic prosperity of both countries. Hence, it is particularly important to understand long-term hydroclimatic changes in the Chu River Basin. However, the sparse and unevenly distributed meteorological and hydrological stations in the region provide limited data for understanding the region's climate and water resource variations [6]. As proxy data is also limited, long-term climate change research for the region is lacking. However, Schrenk spruce (*Picea schrenkiana* Fisch. et Mey.) is distributed throughout the Tianshan Mountains, and its radial growth is an ideal proxy for past climate change [7].

Tree-ring proxies are an important source of high-resolution, absolutely dated information about the hydroclimate of the Common Era (C.E.). They are widespread and well replicated, and they can be statistically calibrated against overlapping instrumental records to produce validated reconstructions and associated estimates of uncertainty in past climate variability at an annual resolution [8]. Recent studies have increasingly shed light on the historical moisture variability of arid Central Asia [7,9,10]. However, because moisture availability is especially variable in mountains, localized hydroclimatic reconstructions are needed. The history of hydroclimatic change in the Chu River Basin is still unclear, despite the river's importance.

In this study, we established a tree-ring-width chronology using tree-ring samples collected in the Chu River Basin in 2014. We analyzed the radial growth of Schrenk spruce and its response to climate, and reconstructed the standardized precipitation-evapotranspiration index (SPEI) to understand past changes in moisture availability. We also examined the relationship between the hydroclimatic changes over the last 175 years and large-scale oscillations in the climate system.

## 2. Data and Methods

### 2.1. Study Area

The Chu River is one of the longest rivers in Central Asia (73°24′–77°04′ E, 41°45′–43°11′ N), with a length of approximately 1067 km and a drainage area of 62,500 km² (Figure 1). The river starts in Kyrgyzstan and runs through the country for 115 km before becoming the border between Kyrgyzstan and Kazakhstan for 221 km. The last 731 km are in Kazakhstan. It is one of the longest rivers in both Kyrgyzstan and in Kazakhstan, and is fed mainly by glaciers and melting snow; rainfall is of secondary importance.

Like most rivers in arid regions, the Chu River is an inland river, originating in the middle ranges of the Tianshan Mountains and disappearing in the desert. After passing through the narrow Boom Gorge, the river enters the comparatively flat Chu Valley, within which the Kyrgyz capital of Bishkek and the Kazakh city of Chu are located. Much of the Chu's water is diverted into a network of canals to irrigate the fertile black soils of the Chu Valley for farming, both on the Kyrgyzstan and Kazakhstan sides of the river. Finally, the Chu River disappears in the Moyynqunm Desert. The study area is located in the headwater region of Chu River Basin [11].

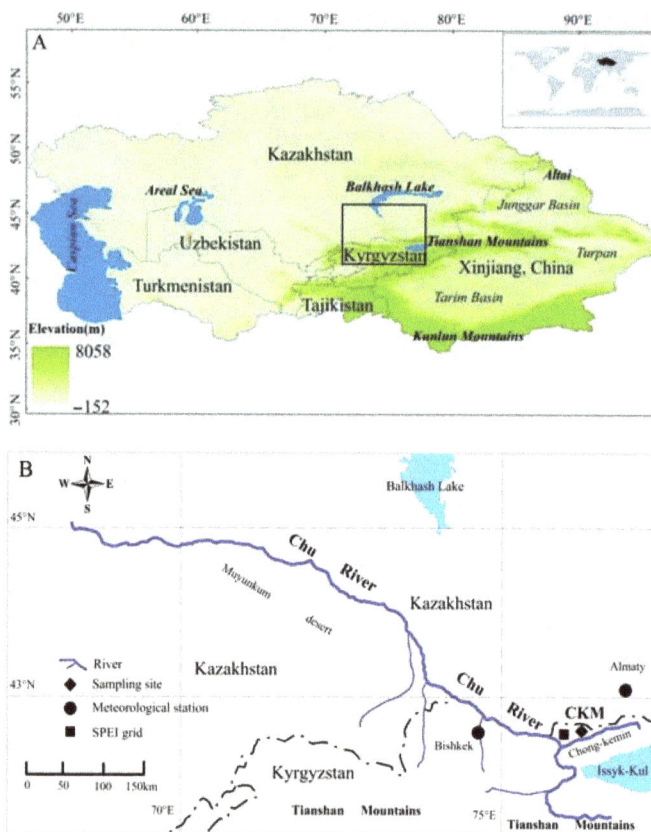

**Figure 1.** Map of tree-ring sample sites information. (**A**) Map of arid Central Asia; (**B**) Sketch map of tree-ring sampling site, meteorological station and SPEI grid.

*2.2. Tree-Ring Data*

We collected tree-ring samples on the southwestern Zailiy Alatau Range in northern Kyrgyzstan. The sampling site was located near the Chong Kemin River (76°23′ E, 42°48′ N, elevation 2400 m, designated the "CKM" group). Forty-one increment cores were collected from 21 trees in virgin forest. The pure Schrenk spruce forests distributed in the shady slopes of mountains. We chose trees with larger slopes, thinner soil layers, less competition and less interference, which radial growth is limited by the climate. The trees with no injury and disease were sampled in order to minimize the signal of non-climatic effects on tree growth.

Following standard dendrochronological methods, all tree-ring cores were brought to the Key Laboratory of Tree-ring Physical and Chemical Research of China Meteorological Administration, dried naturally, mounted, and sanded with progressively finer grains sizes until the ring structures were clear. The samples were then cross-dated with skeleton plots and measured with 0.001 mm precision using a Velmex measuring system [12]. The quality of the cross-dating was checked with the COFECHA program [13] to ensure exact dating. We developed a site chronology using established standardisation techniques with the program ARSTAN [14]. We chose the negative exponential curve (NEC) method to de-trend the growth trend. We used a subsample signal strength (SSS) of 0.85 as an appropriate cut-off criterion for climate reconstructions [15] (Figure 2). The SSS is a measure of decreasing predictive power of transfer functions due to reductions in the sample size of underlying tree-ring series back in time [16].

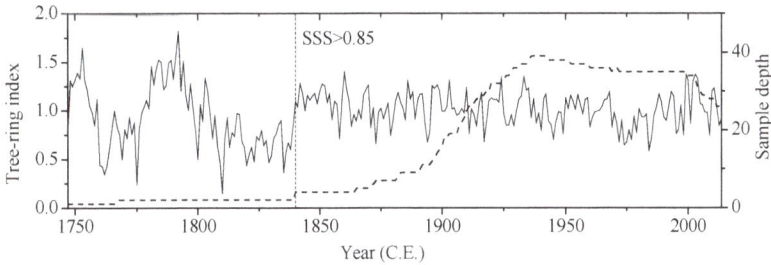

**Figure 2.** Tree-ring-width chronology (STD) in the Chu River Basin.

### 2.3. Meteorological and Hydrological Data

We collected the monthly mean temperature (1896–1988) and precipitation (1895–2004) data from the Frunze meteorological station (43.82 N, 74.58 E, 756.0 m) as a climate background analysis, because it is located in the Chu River Basin. We also used standardized precipitation-evapotranspiration index (SPEI) data (1901–2014) from the nearest grid (42.75° N, 76.25° E). The SPEI is based on monthly precipitation and potential evapotranspiration data from the Climatic Research Unit of the University of East Anglia and can represent drought intensity. The Global SPEI database offers long-term, robust information about drought conditions at the global scale, and has a 0.5 degree spatial resolution. The main advantage of the SPEI Global Drought Monitor is thus its near real-time character, a characteristic best suited for drought monitoring and early warning purposes [17]. The average annual total discharge of the Chu River is derived primarily from runoff from the Chu Valley (not including the Cochkor Valleys). We used the average annual total discharge from 1970–1999 [18].

The mean temperature from 1925 to 1988 and the total precipitation from 1950 to 2000 were analyzed because the meteorological data were discontinuous. The mean temperature was 10.3 °C and average annual precipitation was 429.0 mm over their respective periods. An analysis of the climate data indicates that both the temperature and the precipitation of the Chu River Basin have increased (Figure 3A). The amplitudes are 0.1 °C/10a and 6 mm/10a, respectively. Monthly mean temperature peaks in summer, whereas the average monthly precipitation of the Chu River Basin is bimodal and peaks in spring (March to May) and winter (October to December) (Figure 3B).

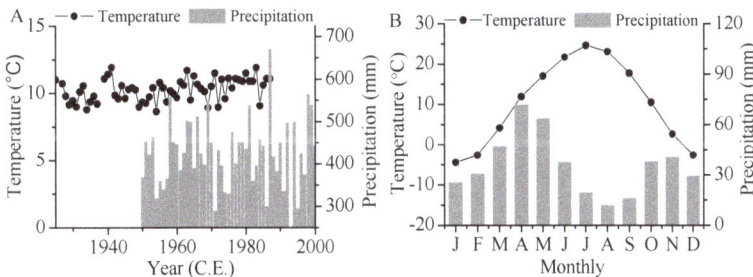

**Figure 3.** Annual (**A**) and monthly (**B**) changes in mean temperature (1925–1988) and total precipitation (1950–2000) at the Frunze meteorological station.

### 2.4. Methods

We used the Pearson correlation coefficient to analyze the relationship between the tree-ring chronologies and SPEI, and a linear regression model to perform the reconstruction [19]. The SPEI reconstruction was conducted on the basis of a split calibration-verification procedure that was designed to test the reliability of the model [20]. The length of the final reconstruction equals the longest nested regression model that still has good calibration and verification results. Spectral properties of both reconstructions were investigated using a multi-taper method (MTM), a powerful tool in spectral estimation that is particularly effective for short time series [21].

## 3. Results

### 3.1. Tree-Ring Response to Climate and the Reconstructed SPEI

Correlation and response analysis revealed significant correlations between the SPEI and tree growth during both the previous and current growing seasons. The greatest single correlation between the CKM tree-ring-width standard chronology (Figure 2) and SPEI from the previous July to current June was 0.629 ($n = 64$, $p < 0.001$). Moisture from the previous July to current June was the dominant climatic factor for tree growth in the Chu River Basin.

Based on the results of the correlation analysis, we reconstructed SPEI from the previous July to current June for the Chu River Basin. The transfer function was

$$SPEI_{p7c6} = 0.911 \times CKM - 0.753 \tag{1}$$

where $SPEI_{p7c6}$ is the SPEI from the previous July to current June, and CKM is the Chro-Kemin chronology detrended by the negative exponential curve fitting with and without application of an adaptive power transformation. For function (1) during the calibration period (1951–2014), the reconstruction explained 39.5% of the variance (38.6% after adjustment for loss of degrees of freedom) in the SPEI data, with $n = 64$, $r = 0.629$, $F_{1, 62} = 40.53$, and $p < 0.0001$ (Figure 4).

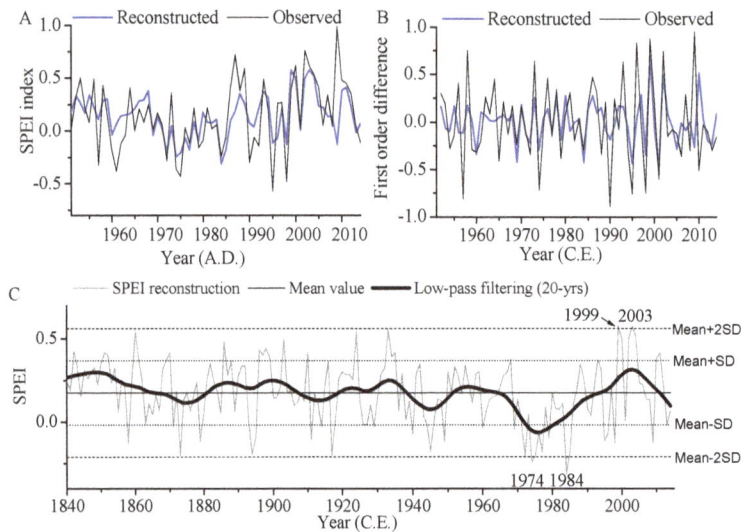

**Figure 4.** Reconstructed SPEI variability of the Chu River Basin. (**A**) Comparison of the reconstructed SPEI (blue line) and the SPEI recorded in grid from CRU (black line) during the common period 1951–2014. (**B**) Comparison of the first differences (year-to-year changes) between the reconstruction (blue line) and the SPEI grid (black line). (**C**) Reconstructed SPEI from the previous July to the current June for the Chu River Basin since 1840 C.E. (gray line). The bold black line shows the data smoothed with a 20-years low-pass filter to emphasize the long-term fluctuations. The solid horizontal line represents the long-term mean for the period 1840–2014; the dashed horizontal lines represent the mean value $\pm 1\sigma$ and the dotted horizontal lines represent the mean value $\pm 2\sigma$.

A leave-one-out cross-validation test indicated that the regression model passed all verification tests (Table 1). We also verified the reliability and stability of the model using split-sample calibration-verification tests. The explained variances are relatively high during the two calibration periods. All of the correlation coefficient (r), first order difference correlation (rd), explained variance ($R^2$), F-test (F), product mean test (t), sign test (ST) and first order difference sign test (ST1) achieved or surpassed the 95% significance level. The reduction of error (RE) and coefficient of efficiency (CE) which are particularly rigorous indicators of reconstructed reliability, were both positive, suggest that the linear regression equation was statistically validated. Finally, we compared the first differences with the SPEI grid and obtained a correlation coefficient of 0.514 ($p < 0.001$, $n = 63$) (Figure 4B). This indicates that there is good consistency in the high frequency changes between the reconstruction and the observed. Equation (1) was therefore used successfully to reconstruct SPEI from the previous July to current June in the Chu River Basin for the period 1840–2014 (Figure 4C).

**Table 1.** Statistics of the leave-one-out cross-validation and the split-sample calibration-verification test model for the SPEI reconstruction.

| | Leave-one-out Cross-validation Test | | | | | |
|---|---|---|---|---|---|---|
| | r | $r_d$ | ST | ST1 | t | RE |
| SPEI | 0.599 ** | 0.470 ** | 47+/17− ** | 42+/21− * | 4.803 ** | 0.357 |

| Calibration | | | | Verification | | | | | |
|---|---|---|---|---|---|---|---|---|---|
| Period | r | $R^2$ | F | Period | r | RE | CE | ST | ST1 |
| 1951–1982 | 0.653 | 0.426 | 22.33 | 1983–2014 | 0.594 | 0.409 | 0.263 | 22+/10− | 22+/9− * |
| 1983–2014 | 0.594 | 0.353 | 16.33 | 1951–1982 | 0.653 | 0.494 | 0.235 | 25+/7− ** | 19+/12− |
| 1951–2014 | 0.629 | 0.395 | 40.53 | | | | | | |

\* indicate significance at the 95% level of confidence. \*\* indicate significance at the 99% level of confidence.

### 3.2. Changes in Moisture over the Past 175 Years

As shown in Figure 4C, the number of drought years and wet years are consistent: 15% (27a) of the years exceeded the mean + 1σ (standard deviation), and 16% (28a) were lower than the mean—1σ. Extreme drought years occurred in 1974 and 1984, when the SPEI was lower than the mean—2σ. Conversely, 1999 and 2003 were extremely moist years with SPEI exceeding the mean + 2σ.

The reconstruction was subjected to 20-year low-pass filtering in order to further understand the low-frequency change in SPEI over the past 175-years (Figure 4C). The results revealed five drying periods and four wetting periods. The drying periods occurred in 1840–1873, 1904–1917, 1934–1945, 1956–1974, and 2004–2014; wetting prevailed in 1874–1903, 1918–1833, 1946–1955, and 1975–2003. Notably, the climate in the Chu river basin over the past 175-years presents a slow process of drought from the 1840s to the 1960s, and a significant drought in the 1970s. Then, the SPEI change exhibited a long period of rapid wetting from the 1970s to 2000s. Since 2004, however, there has again been a strong drying trend, even dropping lower than the average value in the last 3 years.

### 4. Discussion

Many previous studies have confirmed that the radial growth of Schrenk spruce growing at lower elevations is limited by moisture conditions prior to the growing season [7,9,10,22–25]. Zhang et al. [26] analyzed intra-annual radial growth based on data from continuously monitored dendrometers and suggested that moisture between late May to late June is a limiting factor for the radial growth of Schrenk spruce in the Tianshan Mountains. Studies of tree-ring widths and their relation to climate for conifers in arid and semiarid sites iteratively demonstrate that ring-width growth is influenced not only by the climate during the growing season, but also by climatic conditions in the autumn, winter, and spring prior to the growing season [12].

Changes in the SPEI in the Chu River Basin over the past 175-years are consistent with the dry/wet changes in the western Tianshan Mountains (Figure 5). Numerous studies have shown a trend of increasing precipitation from the 1980s [27]. Shi et al. [28] further suggested that the climate in Xinjiang shifted from warm-dry to warm-wet in the 1980s. Several recent studies have also shown that precipitation and the PDSI have decreased since 2004 [9,10]. This study further confirms the moisture fluctuation phenomenon in recent years.

We compared the consistency of the SPEI reconstruction with other studies in the western Tianshan Mountains (Figure 5) and found that past moisture changes in the region are very consistent. The correlation coefficient between the SPEI reconstruction and southern Kazakhstan [9], Dzungarian Alatau [10], and Issyk Lake [7] are 0.596 (Pearson, $n = 175$, $p < 0.0001$), 0.482 (Pearson, $n = 175$, $p < 0.0001$), and 0.399 (Pearson, $n = 131$, $p < 0.0001$), respectively. These strong correlations confirm that the SPEI reconstruction is reliable. To determine the spatial representation of the SPEI reconstruction, we analyzed its spatial correlation with the precipitation, SPEI, and scPDSI data from the CRU-TS grid datasets. The results showed that the SPEI reconstruction successfully represents the changes in climate over the whole of Central Asia during the past century, especially in Kyrgyzstan and southeast Kazakhstan (Figure 6).

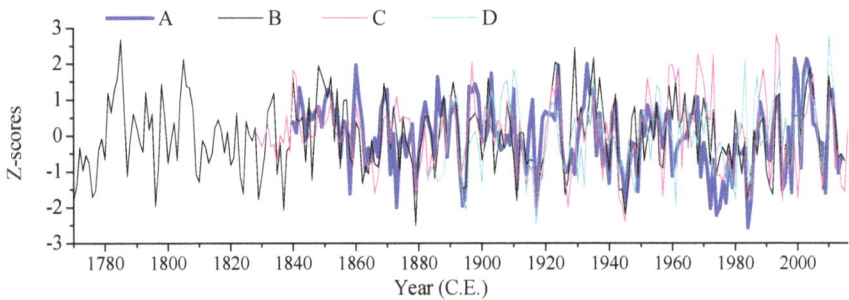

**Figure 5.** Comparison of SPEI reconstruction from this study and other studies in the western Tianshan Mountains. (**A**) Reconstructed SPEI in Chu River Basin (this study); (**B**) Reconstructed precipitation in southern Kazakhstan [9]; (**C**) Reconstructed PDSI in Dzungarian Alatau [10]; (**D**) Reconstructed precipitation in Issyk-Kul, Kyrgyzstan [7].

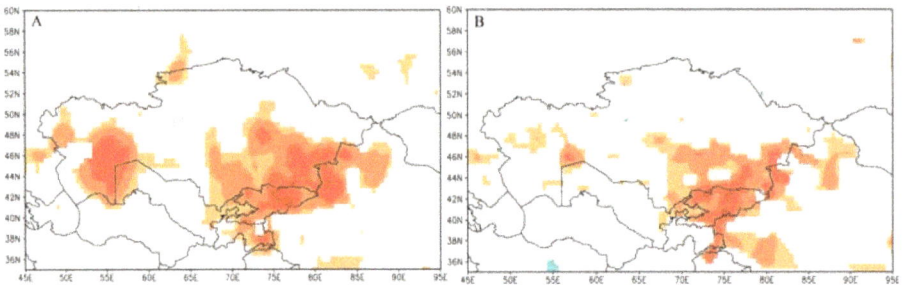

**Figure 6.** *Cont.*

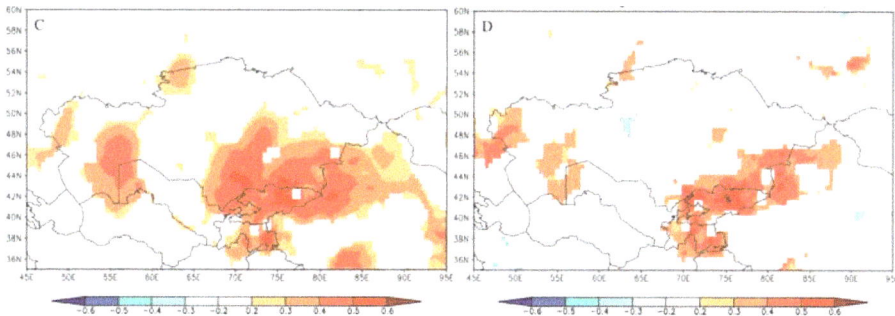

**Figure 6.** Spatial correlation between the reconstruction and previous July to current June grid data (1951–2012). (**A**) Precipitation data from the CRU-TS 3.22; (**B**) Precipitation data from the Global Precipitation Climatology Centre (GPCC) V7; (**C**) SPEI; (**D**) scPDSI.

A comparison of the SPEI reconstruction and discharge variability of the Chu River indicates consistent long-term change (Figure 7). The correlation coefficient between the reconstruction and discharge is 0.540 ($n = 30$, $p < 0.01$) from 1970 to 1999. Discharge out of the Tianshan Mountains is strongly influenced by climate change because moisture mainly is supplied by atmospheric precipitation and snow melt. The SPEI reconstruction is therefore representative of the long-term trend in discharge.

Strong inter-annual SPEI change was identified using the multi-taper method (MTM) [29]. There are significant 2.0-year, 2.6-year, 2.8–2.9-year and 4.4-year periods in the Chu River Basin (Figure 8). These periods suggest that the moisture of the Chu River Basin originates from the westerly circulation. In particular, a study suggested the 2–3-year period is linked to variations in the westerly circulation in the mid troposphere [30]. Other studies have found that the 2–3-year period characterizes precipitation or drought change in arid Central Asia [31–33], and suggest that the periodicity may be related to the tropospheric biennial oscillation (TBO) [34]. Because the upper stream westerly plays an important role in moisture variations in arid Central Asia by influencing the transport of water, the TBO signal is likely related to variations of the westerly.

We also compared the correlation between change in the SPEI and sea surface temperatures (HadISST1). Our results indicate that the SPEI has a significant positive correlation with North Atlantic SSTs (Figure 9). The correlation is strong for both the past 30 years and the past century. These periods have frequently been noted in other dendroclimatology and dendrohydrology studies of the Tianshan Mountains [7,9,10,35] and of other arid and semiarid regions in northern China [36,37]. We therefore posit that changes in the SPEI in the Chu River Basin may be related to westerly circulation.

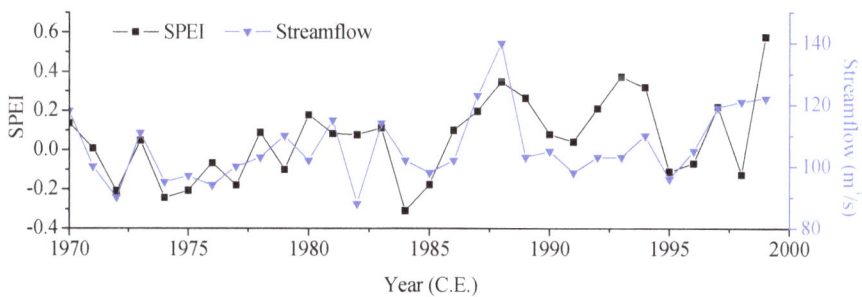

**Figure 7.** Comparison of the SPEI reconstruction and discharge of the Chu River.

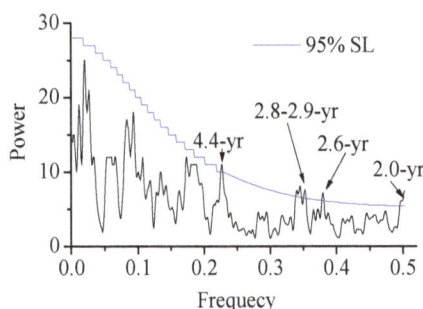

**Figure 8.** Multi-Taper Power spectra for the reconstructed SPEI (AD 1840–2014). 95% SL represents the 95% significance level.

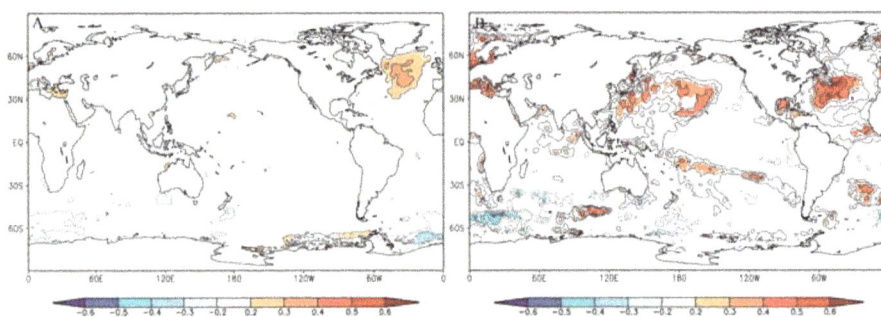

**Figure 9.** Correlation between the SPEI and sea surface temperature (HadISST1). (**A**) Relationship between the reconstruction and July-June HadISST1(1870–2013) $p < 10\%$; (**B**) Relationship between the reconstruction and July-June HadISST1(1981–2010) $p < 10\%$.

## 5. Conclusions

In previous dendroclimatological studies in the Tianshan Mountains, we found that long-term changes in moisture have a strong local character [7,9,10,26,38,39]. The consistency of the change decreases rapidly with increasing distance. It is therefore important to develop a more robust network of reconstruction if we are to fully understand the hydroclimatic changes in the Tianshan Mountains. To this end, we developed the first tree-ring chronology for the Chu River Basin, one of the most important basins in Central Asia. Climate-growth response results suggest that moisture before and early in the growing season is the dominant factor controlling the radial growth of Schrenk spruce in the Chu River Basin. Although moisture varies considerably in mountain areas, our research indicates that the radial growth of Schrenk spruce has a stable response to moisture in the Tianshan mountains.

We also developed a reliable 175-year SPEI reconstruction for the Chu River Basin. We found that past moisture changes in the western Tianshan Mountains are very consistent. Our reconstruction showed a slow, long-term process of drought from the 1840s to the 1960s, followed by a long and rapid wetting from the 1970s to the 2000s. Our reconstruction also provides further evidence for a drought trend since 2004. We posit that long-term changes in the SPEI in the Chu River Basin may be related to large-scale oscillations in the climate system. This study further advances dendroclimatology in the Tianshan Mountains, and is helpful for disaster prevention and water resource management in arid Central Asia.

**Author Contributions:** Conceptualization, R.Z.; Data curation, R.Z. and L.Q.; Formal analysis, R.Z. and L.Q.; Funding acquisition, R.Z.; Investigation, R.Z., B.E., T.Z. and M.A.; Methodology, L.Q.; Resources, R.S.; Validation, L.Q.; Writing—original draft, R.Z.; Writing—review & editing, R.Z.

*Forests* **2019**, *10*, 223

**Funding:** This study was supported by Basic Research Operating Expenses of the Central-level Non-profit Research Institutes (IDM2016006), the Strategic Priority Research Program of Chinese Academy of Sciences (XDA20100306), National Natural Science Foundation of China Projects (41675152, 41405081) and Young Talent Training Plan of Meteorological Departments of China Meteorological Administration.

**Acknowledgments:** We acknowledge the contribution of many people who helped with field and laboratory work: M. Diushen, F. Chen, S. Yu, H. Shang, S. Jiang, W. Liu, F. Yang and Q. He.

**Conflicts of Interest:** The authors declare no conflict of interest.

## References

1. IPCC. *Climate Change 2013: The Physical Science Basis [M/OL]*; Cambridge University Press: Cambridge, UK, 2013. Available online: http://www.ipcc.ch/report/ar5/wg1/#.Uq_tD7KBRR1.2013 (accessed on 5 January 2019).

2. Dai, A.G.; Trenberth, K.E.; Qian, T. A global data set of Palmer Drought Severity Index for 1870–2002: Relationship with soil moisture and effects of surface warming. *J. Hydrometeorol.* **2004**, *5*, 1117–1130. [CrossRef]

3. UNDP. *Central Asia Human Development Report*; UN Development Programme: New York, NY, USA, 2005; 246p.

4. Parry, M.L.; Canziani, O.F.; Palutikof, J.P.; van der Linden, P.J.; Hanson, C.E. (Eds.) *Climate Change 2007: Impacts, Adaptation and Vulnerability*; Cambridge University Press: Cambridge, UK, 2007; 976p.

5. Chen, Y.; Li, W.; Deng, H.; Fang, G.; Li, Z. Changes in Central Asia's WaterTower: Past, Present and Future. *Sci. Rep.* **2016**, *6*, 35458. [CrossRef] [PubMed]

6. Hu, Z.; Zhang, C.; Hu, Q.; Tian, H. Temperature changes in Central Asia from 1979–2011 based on multiple datasets. *J. Clim.* **2014**, *27*, 1143–1167. [CrossRef]

7. Zhang, R.; Yuan, Y.; Gou, X.; He, Q.; Shang, H.; Zhang, T.; Chen, F.; Ermenbaev, B.; Yu, S.; Qin, L.; et al. Tree-ring-based moisture variability in western Tianshan Mountains since A.D. 1882 and its possible driving mechanism. *Agric. For. Meteorol.* **2016**, *218–219*, 267–276. [CrossRef]

8. PAGES Hydro2k Consortium. Comparing proxy and model estimates of hydroclimate variability and change over the Common Era. *Clim. Past.* **2017**, *13*, 1851–1900. [CrossRef]

9. Zhang, R.; Shang, H.; Yu, S.; He, Q.; Yuan, Y.; Bolatov, K.; Mambetov, B.T. Tree-ring-based precipitation reconstruction in southern Kazakhstan, reveals drought variability since A.D. 1770. *Int. J. Climatol.* **2017**, *37*, 741–750. [CrossRef]

10. Zhang, R.; Zhang, T.; Kelgenbayev, N.; He, Q.; Mambetov, B.T.; Dosmanbetov, D.; Shang, H.; Yu, S.; Yuan, Y. A 189-year tree-ring record of drought for the Dzungarian Alatau, arid Central Asia. *J. Asian Earth Sci.* **2017**, *148*, 305–314. [CrossRef]

11. Ma, C.; Sun, L.; Liu, S.; Shao, M.; Luo, Y. Impact of climate change on the streamflow in the glacierized Chu River Basin, Central Asia. *J. Arid Land* **2015**, *7*, 501–513. [CrossRef]

12. Fritts, H.C. *Tree Rings and Climate*; Academic Press, Inc. Ltd.: London, UK; New York, NY, USA, 1976.

13. Holmes, R.L. Computer-assisted quality control in tree-ring dating and measurement. *Tree-Ring Bull.* **1983**, *43*, 69–75.

14. Cook, E.R. A Time-Series Analysis Approach to Tree-Ring Standardization. Ph.D. Dissertation, The University of Arizona Press, Tucson, Arizona, 1985.

15. Wigley, T.M.L.; Briffa, K.R.; Jones, P.D. On the average value of correlated time series, with application in dendroclimatology and hydrometeorology. *J. Appl. Meteorol. Clim.* **1984**, *23*, 201–213. [CrossRef]

16. Buras, A. A comment on the Expressed Population Signa. *Dendrochronologia* **2017**, *44*, 130–132. [CrossRef]

17. Vicente-Serrano, S.M.; Beguería, S.; López-Moreno, J.I. A Multi-scalar drought index sensitive to global warming: The Standardized Precipitation Evapotranspiration Index. *J. Clim.* **2010**, *23*, 1696–1718. [CrossRef]

18. Litvak, R.G.; Morris, B.; Nemaltseva, E.I. Groundwater vulnerability assessment for intermontane valleys using Chu Valley of Kyrghyzstan as an example. In *Groundwater and Ecosystems*; Baba, A., Howard, K.W.F., Gunduz, O., Eds.; Springer: Dordrecht, The Netherlands; pp. 107–120.

19. Cook, E.R.; Kairiukstis, L.A. *Methods of Dendrochronology: Applications in the Environmental Sciences*; Kluwer: Dordrecht, The Netherlands, 1990.

20. Meko, D.M.; Graybill, D.A. Tree-ring reconstruction of Upper Gila River discharge. *Water Res. Bull.* **1995**, *31*, 605–616. [CrossRef]

21. Mann, M.E.; Lees, J.M. Robust estimation of background noise and signal detection in climatic time series. *Clim. Chang.* **1996**, *33*, 409–445. [CrossRef]

22. Yuan, Y.; Jin, L.; Shao, X.; He, Q.; Li, Z.; Li, J. Variations of the spring precipitation day numbers reconstructed from tree rings in the Urumqi River drainage, Tianshan Mts. over the last 370 years. *Chin. Sci. Bull.* **2003**, *48*, 1507–1510. [CrossRef]

23. Yuan, Y.; Li, J.; Zhang, J. 348 year precipitation reconstruction from tree-rings for the North Slope of the middle Tianshan Mountains. *Acta Meteorol. Sin.* **2001**, *15*, 95–104.

24. Solomina, O.N.; Maximova, O.E.; Cook, E.R. Picea Schrenkiana ring width and density at the upper and lower tree limits in the Tien Shan mts Kyrgyz republic as a source of paleoclimatic information. *Geogr. Environ. Sustain.* **2014**, *1*, 66–79. [CrossRef]

25. Li, J.; Gou, X.; Cook, E.R.; Chen, F. Tree-ring based drought reconstruction for the central Tien Shan area in northwest China. *Geophys. Res. Lett.* **2006**, *33*, L07715. [CrossRef]

26. Zhang, R.; Yuan, Y.; Gou, X.; Zhang, T.; Zou, C.; Ji, C.; Fan, Z.; Qin, L.; Shang, H.; Li, X. Intra-annual radial growth of Schrenk spruce (Picea schrenkiana Fisch. et Mey) and its response to climate on the northern slopes of the Tianshan Mountains. *Dendrochronologia* **2016**, *40*, 36–42. [CrossRef]

27. Cheng, H.; Zhang, P.Z.; Spotl, C.; Edwards, R.L.; Cai, Y.J.; Zhang, D.Z.; Sang, W.C.; Tan, M.; An, Z.S. The climatic cyclicity in semiarid-arid central Asia over the past 500,000 years. *Geophys. Res. Lett.* **2012**, *39*, L01705. [CrossRef]

28. Shi, Y.; Shen, Y.; Kang, E.; Li, D.; Ding, Y.; Zhang, G.; Hu, R. Recent and future climate change in Northwest China. *Clim. Chang.* **2007**, *80*, 379–393. [CrossRef]

29. Thomson, D.J. Spectrum estimation and harmonic analysis. *Proc. IEEE* **1982**, *70*, 1055–1096. [CrossRef]

30. Huang, W.; Chen, F.; Feng, S.; Chen, J.; Zhang, X. Interannual precipitation variations in the mid-latitude Asia and their association with large-scale atmospheric circulation. *Chin. Sci. Bull.* **2013**, *58*, 3962–3968. [CrossRef]

31. Chen, F.; Huang, W.; Jin, L.; Chen, J.; Wang, J. Spatiotemporal precipitation variations in the arid Central Asia in the context of global warming. *Sci. China Earth Sci.* **2011**, *54*, 1812–1821. [CrossRef]

32. Fang, K.; Gou, X.; Chen, F.; D'Arrigo, R.; Li, J. Tree-ring based drought reconstruction for the Guiqing Mountain (China): Linkages to the Indian and Pacific Oceans. *Int. J. Climatol.* **2010**, *30*, 1137–1145. [CrossRef]

33. Fang, K.; Gou, X.; Chen, F.; Frank, D.; Liu, C.; Li, J.; Kazmer, M. Precipitation variability during the past 400 years in the Xiaolong Mountain (central China) inferred from tree rings. *Clim. Dyn.* **2012**, *39*, 1697–1707. [CrossRef]

34. Meehl, G.A. The annual cycle and interannual variability in the tropical Pacific and Indian Ocean region. *Mon. Weather Rev.* **1987**, *115*, 27–50. [CrossRef]

35. Wang, T.; Ren, G.; Chen, F.; Yuan, Y. An analysis of precipitation variations in the west-central Tianshan Mountains over the last 300 years. *Quatern. Int.* **2015**, *358*, 48–57. [CrossRef]

36. Liang, E.; Shao, X.; Liu, X. Annual precipitation variation inferred from tree rings since A.D. 1770 for the western Qilian Mts., Northern Tibetan Plateau. *Tree-Ring Res.* **2009**, *65*, 95–103. [CrossRef]

37. Liu, Y.; Sun, J.; Song, H.; Cai, Q.; Bao, G.; Li, X. Tree-ring hydrologic reconstructions for the Heihe River watershed, western China since AD 1430. *Water Res.* **2010**, *44*, 2781–2792. [CrossRef] [PubMed]

38. Zhang, R.; Wei, W.; Shang, H.; Yu, S.; Gou, X.; Qin, L.; Bolatov, K.; Mambetov, B.T. A tree ring-based record of annual mass balance changes for the TS.Tuyuksuyskiy Glacier and its linkages to climate change in the Tianshan Mountains. *Quat. Sci. Rev.* **2019**, *205*, 10–21. [CrossRef]

39. Zhang, R.; Yuan, Y.; Yu, S.; Chen, F.; Zhang, T. Past changes of spring drought in the inner Tianshan Mountains, China, as recorded by tree rings. *Boreas* **2017**, *46*, 688–696. [CrossRef]

![forests](forests logo) **MDPI**

*Article*

# Forest Canopy Can Efficiently Filter Trace Metals in Deposited Precipitation in a Subalpine Spruce Plantation

Siyi Tan, Hairong Zhao, Wanqin Yang, Bo Tan, Kai Yue, Yu Zhang, Fuzhong Wu and Xiangyin Ni *

Long-Term Research Station of Alpine Forest Ecosystems, Provincial Key Laboratory of Ecological Forestry Engineering, Institute of Ecology and Forestry, Sichuan Agricultural University, 211 Huimin Road, Wenjiang District, Chengdu 611130, China; TanSiyii@163.com (S.T.); zhaohrchina@163.com (H.Z.); scyangwq@163.com (W.Y.); bobotan1984@163.com (B.T.); kyleyuechina@163.com (K.Y.); zhangyuchina1996@163.com (Y.Z.); wufzchina@163.com (F.W.)
* Correspondence: nixy@sicau.edu.cn; Tel./Fax: +86-28-86290957

Received: 19 December 2018; Accepted: 2 April 2019; Published: 7 April 2019

**Abstract:** Trace metals can enter natural regions with low human disturbance through atmospheric circulation; however, little information is available regarding the filtering efficiency of trace metals by forest canopies. In this study, a representative subalpine spruce plantation was selected to investigate the net throughfall fluxes of eight trace metals (Fe, Mn, Cu, Zn, Al, Pb, Cd and Cr) under a closed canopy and gap-edge canopy from August 2015 to July 2016. Over the one-year observation, the annual fluxes of Al, Zn, Fe, Mn, Cu, Cd, Cr and Pb in the deposited precipitation were 7.29 kg·ha$^{-1}$, 2.30 kg·ha$^{-1}$, 7.02 kg·ha$^{-1}$, 0.16 kg·ha$^{-1}$, 0.19 kg·ha$^{-1}$, 0.06 kg·ha$^{-1}$, 0.56 kg·ha$^{-1}$ and 0.24 kg·ha$^{-1}$, respectively. The annual net throughfall fluxes of these trace metals were $-1.73$ kg·ha$^{-1}$, $-0.90$ kg·ha$^{-1}$, $-1.68$ kg·ha$^{-1}$, 0.03 kg·ha$^{-1}$, $-0.03$ kg·ha$^{-1}$, $-0.02$ kg·ha$^{-1}$, $-0.09$ kg·ha$^{-1}$ and $-0.08$ kg·ha$^{-1}$, respectively, under the gap-edge canopy and 1.59 kg·ha$^{-1}$, $-1.13$ kg·ha$^{-1}$, $-1.65$ kg·ha$^{-1}$, 0.10 kg·ha$^{-1}$, $-0.04$ kg·ha$^{-1}$, $-0.03$ kg·ha$^{-1}$, $-0.26$ kg·ha$^{-1}$ and $-0.15$ kg·ha$^{-1}$, respectively, under the closed canopy. The closed canopy displayed a greater filtering effect of the trace metals from precipitation than the gap-edge canopy in this subalpine forest. In the rainy season, the net filtering ratio of trace metals ranged from $-66.01\%$ to $89.05\%$ for the closed canopy and from $-52.32\%$ to $33.09\%$ for the gap-edge canopy. In contrast, the net filtering ratio of all trace metals exceeded $50.00\%$ for the closed canopy in the snowy season. The results suggest that most of the trace metals moving through the forest canopy are filtered by canopy in the subalpine forest.

**Keywords:** canopy filtering; closed canopy; forest hydrology; gap-edge canopy; throughfall; trace metal

## 1. Introduction

Trace metals in the environment originate mainly from metal refining, fossil fuel combustion, automotive exhaust emission and other human activities [1]. An increasing number of studies have demonstrated that trace metals mainly exist as particles in the atmosphere and can enter in natural regions with low human disturbance through atmospheric circulation [1,2]. The input of trace metals via atmospheric deposition is a large source of contamination for plants, soil and water, and continuous trace metal input has lasting negative impacts on biogeochemical cycling in ecosystems [3]. Forest ecosystems have often been considered as ecological filters that can efficiently decrease atmospheric pollutants and improve air quality [4,5], but the filtering efficiency of a forest ecosystem is often controlled by precipitation and canopy characteristics.

The term "trace element" is used loosely in the current literature to refer to elements that occur in small concentrations in natural biological systems [6]. Trace metals are introduced into terrestrial ecosystems via two pathways: dissolution in rain and snow (i.e., wet deposition) and direct particulate deposition (i.e., dry deposition) [7]. When precipitation passes through the forest canopy, the deposition precipitation is altered by the wash-off of some particles in the canopy that were deposited in dry periods or by ion exchange, i.e., uptake or leaching [8]. Some trace elements (e.g., Pb and Zn) [9,10] are taken up by the canopy. Lead, Cd and Cr are classified as nonessential trace elements; however, these elements can be highly toxic and can inhibit growth, even cause organismal death [11]. Iron, Mn, Cu, Zn and Al are essential trace elements that participate in plant physiological and biochemical processes, but excessive amounts of these elements can be toxic to plants [6].

In forest ecosystems, canopy gaps are created by dead and fallen trees and by intermediate cuttings, which are the primary modes of forest disturbance and regeneration [12,13]. A gap-edge canopy differs substantially from interior forest zones. Gap-edge and closed canopies represent two different forest canopy conditions, and the coverage of the gap-edge canopy area is less than that of the closed canopy area. The degree of precipitation interception will influence the accumulation of trace metals from precipitation. In addition, the structure of the canopy influences the ability of the canopy to capture suspended particles, and more trace metals may be intercepted by certain canopy structures. Due to the obstruction of the wind profile, which causes local advection and turbulent exchanges, an edge canopy can receive more atmospheric deposition than a closed canopy [9]. Several studies have focused on the effects of closed vs. open canopies on the canopy levels of trace metals received through atmospheric deposition [4,14]. However, no studies have addressed trace metal fluxes or filtration in gap-edge canopy layers [15]. Here, we hypothesized that the fluxes of trace metals are higher in a gap-edge canopy than in a closed canopy, and that the filtration of trace metals is lower in a gap-edge canopy than in a closed canopy.

As important freshwater conservation areas in the Yangtze River basin, subalpine forests in Southwest China play important roles in not only regulating the regional climate and biodiversity but also storing freshwater and conserving water and soil [16]. Since the 1950s, more than 400,000 hectares of pure dragon spruce (*Picea asperata* Mast.) plantations have replaced natural coniferous forests on the Eastern Tibetan Plateau. These plantations are harvested by large-scale industrial logging operations. The forest canopy of these plantations consists of a single canopy level rather than complex, multiple canopy levels as in natural forests. Therefore, these spruce plantations are likely ineffective in intercepting trace metals introduced via direct (dry) deposition.

The migration and transformation of trace metals in forest ecosystems occur through the two external inputs of wet and dry deposition. These processes affect trace metal pollution in various parts of the ecosystem. In spruce plantation ecosystems, certain trace metals play important biogeochemical roles, either as essential trace metals, such as Cu and Zn, or as nonessential trace metals (e.g., Pb, Cd and Cr) [9]. Therefore, quantification of these metals is important. Before precipitation reaches the soil surface, its chemical composition can be modified by contact with vegetation. Throughfall is commonly measured to quantify the load of atmospheric pollutants in forest ecosystems [17,18], where the contents of pollutants in throughfall differ from those in atmospheric precipitation [9]. In the present study, we measured the trace metals in throughfall in an area of an alpine spruce plantation to (1) observe patterns in annual trace metal concentrations and fluxes from the deposited precipitation and (2) compare the trace-metal filtering ability of a gap-edge canopy and a closed canopy in the subalpine spruce plantation. An understanding of these processes can increase our knowledge of the filtering effect of the forest canopy with respect to trace metals, and the main processes that control metal behavior after interaction with the forest canopy. The results of this study would provide insight into the filtration of trace metals deposited through precipitation in different canopy types and provide necessary information on water quality conservation in the upper reaches of the Yangtze River.

## 2. Materials and Methods

### 2.1. Site Description

The experimental site is located at the Long-term Research Station of Alpine Forest Ecosystems, Bipenggou Nature Reserve (102°53′–102°57′ E, 31°14′–31°19′ N; 2458–4619 m a.s.l.), Li County, Sichuan, Southwest China. The site is situated on the eastern edge of the Tibetan Plateau along the upper Yangtze River [19]. The mean annual air temperature is 2~4 °C, and the maximum and minimum temperatures are 23.7 °C and −18.1 °C, respectively. The mean annual precipitation ranges from 801 mm to 850 mm, with most rainfall occurring between May and August. Snowfall mainly occurs from October to April of the following year. The amount of snowfall is approximately 138.56 mm. The canopy forest vegetation is dominated by *Picea asperata* Mast with some understory shrubs (e.g., *Berberis diaphana* Maxin. and *Sorbus rufopilosa* Schneid) and grasses (e.g., *Deyeuxia scab*rescens (Griseb.) Munro ex Duthie). The expanded gap (the canopy gap plus the area that extends to the bases of the surrounding canopy trees) covers 23% of the experimental site [20].

### 2.2. Experimental Design

Three plots with similar topographical and environmental features were selected in a typical spruce forest gap (area: 100 m²) along a gradient from the gap-edge to the closed canopy (closed canopy area: 20 × 20 m) at 3000 m a.s.l. The mean tree age was approximately 60 a. The average diameter at breast height (DBH) and the average tree height in the experimental plots were 19.53 ± 1.99 cm and 7.63 ± 0.45 m, respectively. We selected an open area (20 × 20 m) approximately 50 m from the edge of the spruce plantation forest as the nonforested site to collect precipitation.

### 2.3. Precipitation Observations and Water Sampling

Precipitation: Rainfall was sampled in the nonforested site using 5 custom-made continuous rain gauges (each with a surface collection area of 0.64 m²).

Snowfall: Five cone-shaped collectors (top diameter of 100 cm, bottom diameter of approximately 20 cm) made of PVC and gridding cloth were used to observe and sample snowfall in the open site. Each collector was established 1 m above the ground surface. Each collector drained into a polyethylene (PE) bucket. As the snowfall fell directly into the polyethylene bucket, there was minimal exposure to external conditions and minimal snowfall evaporation.

Throughfall in the rainy season: Throughfall was recorded using 5 PVC rectangular gutters (each with a surface collection area of 400 × 16 cm) that were arranged beneath the closed canopy and gap-edge canopy in each plot. The gutters were established 1 m above the floor to avoid ground-splash effects and at a 5° horizontal angle to promote drainage. The lower end of each gutter was equipped with a plastic bucket.

Throughfall in the snowy season: Five cone-shaped collectors similar to the snowfall collectors were distributed beneath the closed canopy and the gap-edge canopy in each plot, and each collector drained into a PE bucket.

### 2.4. Chemical Analysis

Water samples were collected immediately after each rainfall event during the rainy season from August 2015 to July 2016. Due to the heavy snowfall and cruel natural conditions in winter, snow samples were collected once each month from November 2015 to April 2016. The samples were placed in clean polyethylene bottles. Upon collection, the water was poured from the polyethylene bucket into a graduated cylinder to measure the water volume. Then, the samples were rapidly transported to the laboratory where they were filtered using qualitative filter paper with a diameter of 12.5 cm. The filtered samples were adjusted to a pH of 1~2 with high-purity grade (GR) nitric acid. The concentrations of trace metals (i.e., Fe, Mn, Cu, Zn, Al, Pb, Cd and Cr) were determined using an inductively coupled plasma optical emission spectrometry system (Agilent 7900, US).

*2.5. Calculations*

The throughfall was calculated as follows (Formula (1)):

$$V_j = \frac{V_i'}{S} \times 10,\tag{1}$$

where $V_j$ is the throughfall (mm), $V_i'$ is the volume of water (mL), S is the surface area of collection, and 10 is the unit conversion factor.

The fluxes of trace metals in precipitation and throughfall were calculated using Formula (2) as follows [21]:

$$Flux_j = \frac{VWM_j \times V_j}{100},\tag{2}$$

where $Flux_j$ (kg ha$^{-1}$) is the deposition flux of solute j in different forms of water, $VWM_j$ is the weighted concentration (mg L$^{-1}$) of solute j in different forms of water, $V_j$ is the water of different forms (mm), and 100 is the unit conversion factor.

The net throughfall fluxes (NTFs) and net throughfall ratios (NTRs) were calculated with Equations (3) and (4), respectively [22]:

$$NTF = TF - BP,\tag{3}$$

$$NTR = NTF/BP,\tag{4}$$

where BP and TF represent the bulk precipitation flux (kg ha$^{-1}$) and the throughfall flux (kg ha$^{-1}$), respectively. (Negative and positive N.TFs (NTRs) values represent filtered and leached amounts, respectively.)

*2.6. Statistical Analysis*

All statistical analyses were carried out using IBM SPSS version 20. 0 statistics software (IBM SPSS Statistics Inc, Chicago, IL, USA), Univariate analysis was used to compare the concentrations, fluxes and net throughfall fluxes of trace metals among different canopy types and seasons. The statistical tests were considered significant at the $p < 0.05$ level.

## 3. Results

*3.1. Annual Variations of Trace Metal Concentrations in Precipitation and Throughfall*

After precipitation passed through the canopy, the concentrations of trace metals increased or decreased to different extents between the closed canopy and the gap-edge canopy. In the rainy season, the throughfall concentrations of the essential trace metals Fe, Mn and Cu were higher than those in the precipitation for both the closed canopy and the gap-edge canopy (Figure 1). The concentrations of Mn were 2.63-fold and 1.68-fold higher under the closed canopy and gap-edge canopy, respectively, than in the precipitation. In addition, the concentrations of Fe, Mn and Cu under the closed canopy were higher than those under the gap-edge canopy. Among the nonessential trace metals, only Cr showed a difference between throughfall and precipitation, with a higher gap-edge canopy value than that measured in precipitation (Figure 2). In the snowy season, the throughfall concentrations of all trace metals under the closed canopy and gap-edge canopy were lower than those in the precipitation. The concentrations of trace metals under the closed canopy were higher than those under the gap-edge canopy (except for Al and Pb). In addition, insignificant differences in trace metal concentrations were detected between the gap-edge canopy and the closed canopy, although there were significant seasonal effects on trace metal concentrations, as shown in Table 1 ($p < 0.05$).

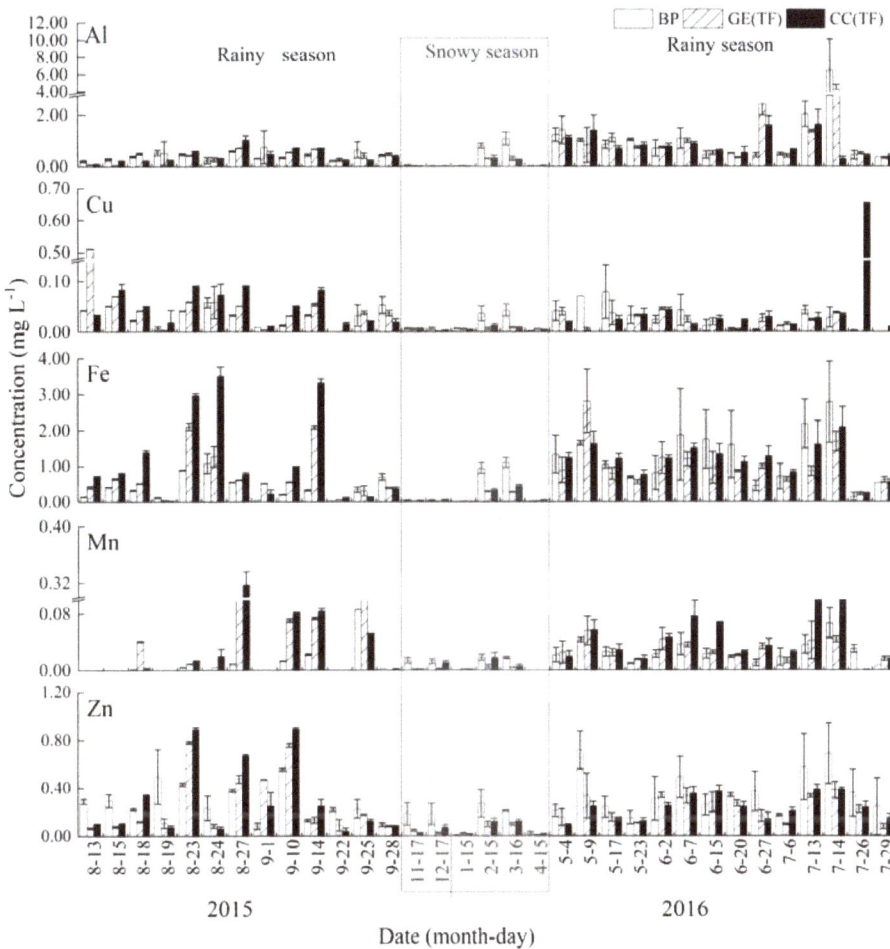

**Figure 1.** Concentrations of essential trace metals in bulk precipitation and throughfall. BP: bulk precipitation, GE: gap-edge canopy, CC: closed canopy. The bars and error bars are the means and 95% confidence intervals, respectively.

## 3.2. Annual Variations of Trace Metal Fluxes in Precipitation and Throughfall

The annual fluxes of Al, Zn, Fe, Mn, Cu, Cd, Cr and Pb were 7.29 kg·ha$^{-1}$, 2.30 kg·ha$^{-1}$, 7.02 kg·ha$^{-1}$, 0.16 kg·ha$^{-1}$, 0.19 kg·ha$^{-1}$, 0.06 kg·ha$^{-1}$, 0.56 kg·ha$^{-1}$ and 0.24 kg·ha$^{-1}$, respectively. The values in the rainy season were higher than those in the snowy season (Figures 3 and 4). The input of all trace metals from the precipitation, gap-edge canopy and closed canopy was 1.15 kg·ha$^{-1}$, 0.29 kg·ha$^{-1}$ and 0.30 kg·ha$^{-1}$, respectively, in the snowy season and 16.68 kg·ha$^{-1}$, 13.02 kg·ha$^{-1}$ and 15.96 kg·ha$^{-1}$, respectively, in the rainy season. Among the trace metal fluxes, the maximum values in the precipitation and under the closed canopy and gap-edge canopy were observed for Fe and Al in both seasons. In the snowy season, the throughfall fluxes of all trace metals under the closed canopy and gap-edge canopy were lower than those in the precipitation. Furthermore, there were no significant differences in trace metal fluxes between the gap-edge canopy and the closed canopy, but seasons had significant effects on the trace metal fluxes, as shown in Table 1 ($p < 0.05$).

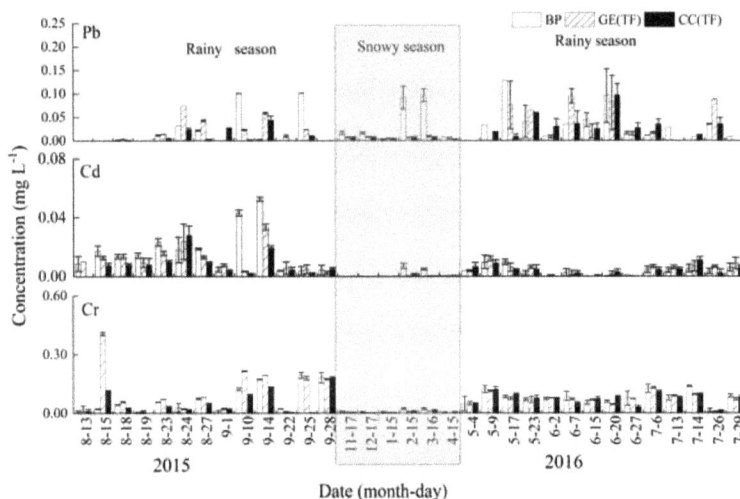

**Figure 2.** Concentrations of nonessential trace metals in bulk precipitation and throughfall. BP: bulk precipitation, GE: gap-edge canopy, CC: closed canopy. The bars and error bars are the means and 95% confidence intervals, respectively.

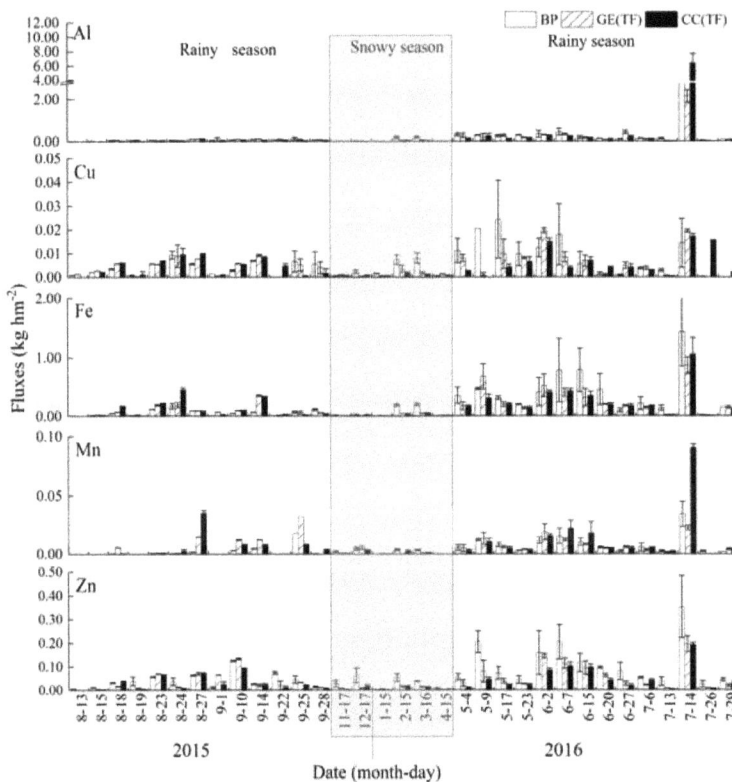

**Figure 3.** The fluxes of essential trace metals in bulk precipitation and throughfall. BP: bulk precipitation, GE: gap-edge canopy, CC: closed canopy. The bars and error bars are the means and 95% confidence intervals, respectively.

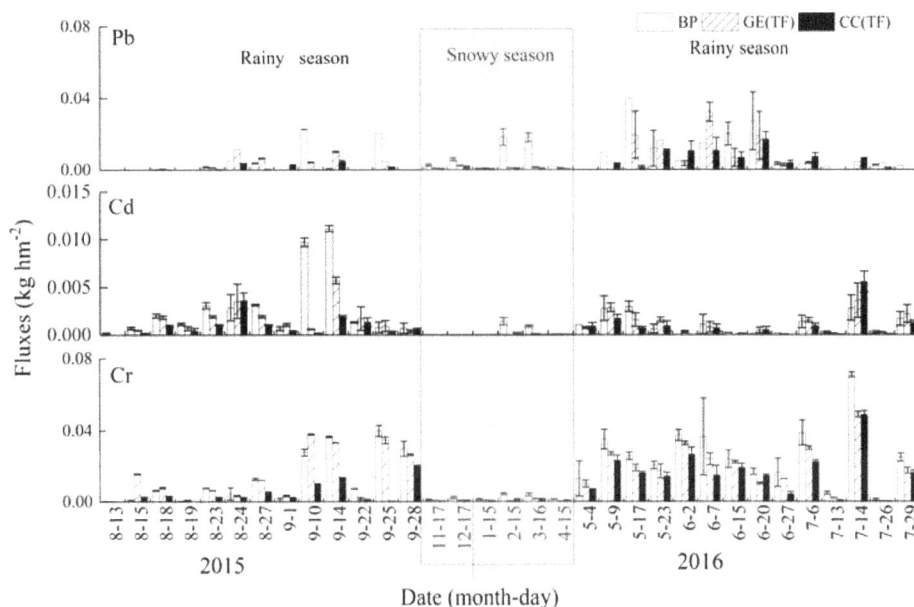

**Figure 4.** The fluxes of essential nonessential trace metals in bulk precipitation and throughfall. BP: bulk precipitation, GE: gap-edge canopy, CC: closed canopy. The bars and error bars are the means and 95% confidence intervals, respectively.

**Table 1.** Univariate analysis results (*F* values) regarding the effects of canopy and season on the concentrations and fluxes of trace metals in throughfall.

| | Item | Al | Zn | Fe | Mn | Cu | Pb | Cd | Cr |
|---|---|---|---|---|---|---|---|---|---|
| Concentration | Canopy | 0.22 | 0.22 | 1.64 | 1.25 | 0.13 | 0.88 | 0.72 | 1.43 |
| | Season | 7.70 ** | 31.83 ** | 44.60 ** | 12.24 ** | 6.89 ** | 11.18 | 37.28 ** | 39.94 ** |
| Fluxes | Canopy | 0.78 | 0.38 | 0.10 | 0.24 | 0.49 | 1.316 | 0.83 | 2.28 |
| | Season | 27.96 ** | 21.93 ** | 52.23 ** | 8.89 ** | 23.58 ** | 9.32 ** | 24.77 ** | 39.59 ** |
| NTF | Canopy | 0.60 | 1.02 | 0.005 | 5.06 * | 0.57 | 5.57 * | 1.32 | 10.59 ** |
| | Season | 0.54 | 0.32 | 0.004 | 0.48 | 0.24 | 0.57 | 0.26 | 4.74 * |

*: $p < 0.05$; **: $p < 0.01$.

### 3.3. Variations of Net Throughfall Fluxes of Trace Metals Under the Closed Canopy and Gap-Edge Canopy

The net throughfall fluxes (NTFs) and net throughfall ratios (NTRs) of Al, Zn, Fe, Mn, Cu, Cd, Cr and Pb are shown in Table 2. The annual NTFs of these respective trace metals under the gap-edge canopy were $-1.41$ kg·ha$^{-1}$, $-0.76$ kg·ha$^{-1}$, $-1.35$ kg·ha$^{-1}$, 0.04 kg·ha$^{-1}$, $-0.02$ kg·ha$^{-1}$, $-0.02$ kg·ha$^{-1}$, $-0.08$ kg·ha$^{-1}$ and $-0.04$ kg·ha$^{-1}$ during the rainy season, and $-0.32$ kg·ha$^{-1}$, $-0.14$ kg·ha$^{-1}$, $-0.33$ kg·ha$^{-1}$, $-0.01$ kg·ha$^{-1}$, $-0.01$ kg·ha$^{-1}$, $-0.00$ kg·ha$^{-1}$, $-0.01$ kg·ha$^{-1}$ and $-0.04$ kg·ha$^{-1}$ during the snowy season. The NTFs of these respective trace metals under the closed canopy were 1.93 kg·ha$^{-1}$, $-0.99$ kg·ha$^{-1}$, $-1.35$ kg·ha$^{-1}$, 0.11 kg·ha$^{-1}$, $-0.03$ kg·ha$^{-1}$, $-0.03$ kg·ha$^{-1}$, $-0.25$ kg·ha$^{-1}$ and $-0.10$ kg·ha$^{-1}$ during the rainy season, and $-0.26$ kg·ha$^{-1}$, $-0.14$ kg·ha$^{-1}$, $-0.30$ kg·ha$^{-1}$, $-0.01$ kg·ha$^{-1}$, $-0.01$ kg·ha$^{-1}$, $-0.00$ kg·ha$^{-1}$, $-0.01$ kg·ha$^{-1}$ and $-0.04$ kg·ha$^{-1}$ during the snowy season. The NTRs of Mn was 33.09% and 89.05% in the rainy season under the two canopies, revealing greater leaching of this trace metal than that of the other trace metals.

**Table 2.** Net throughfall fluxes (NTFs) and net throughfall ratio (NTRs) values and standard deviation (in parentheses) of trace metals for the closed canopy and gap-edge canopy during the rainy and snowy seasons.

| Canopy | Season | Throughfall (mm) | Item | Al | Zn | Fe | Mn | Cu | Cd | Cr | Pb |
|---|---|---|---|---|---|---|---|---|---|---|---|
| Gap-edge canopy | Annual | 645.47 (11.17) | NTF (kg·ha⁻¹) | -1.73 | -0.90 | -1.68 | 0.03 | -0.03 | -0.02 | -0.09 | -0.08 |
| | | | | (0.21) | (0.04) | (0.17) | (0.00) | (0.01) | (0.00) | (0.01) | (0.01) |
| | | | NTR (%) | -23.75 | -39.20 | -23.93 | 19.75 | -20.62 | -32.85 | -26.70 | -35.95 |
| | | | | (12.21) | (21.33) | (11.47) | (6.99) | (13.45) | (10.55) | (7.95) | (14.21) |
| | Rainy season | 530.33 (12.08) | NTF (kg·ha⁻¹) | -1.41 | -0.76 | -1.35 | 0.04 | -0.02 | -0.02 | -0.08 | -0.04 |
| | | | | (1.74) | (0.41) | (1.35) | (0.01) | (0.05) | (0.00) | (0.05) | (0.02) |
| | | | NTR (%) | -36.09 | -52.32 | -32.06 | 33.09 | -25.41 | -39.32 | -18.10 | -27.33 |
| | | | | (44.66) | (28.08) | (32.99) | (8.57) | (50.88) | (1.07) | (11.21) | (13.55) |
| | Snowy season | 115.14 (6.30) | NTF (kg·ha⁻¹) | -0.32 | -0.14 | -0.33 | -0.01 | -0.01 | -0.00 | -0.01 | -0.04 |
| | | | | (0.10) | (0.05) | (0.04) | (0.00) | (0.00) | (0.00) | (0.00) | (0.01) |
| | | | NTR (%) | -100.07 | -101.06 | -82.17 | -49.55 | -61.34 | -95.33 | -79.81 | 101.21 |
| | | | | (32.11) | (38.53) | (4.07) | (19.69) | (16.09) | (12.45) | (22.41) | (17.67) |
| Closed canopy | Annual | 516.18 (10.29) | NTF (kg·ha⁻¹) | 1.59 | -1.13 | -1.65 | 0.10 | -0.04 | -0.03 | -0.26 | -0.15 |
| | | | | (0.54) | (0.04) | (0.16) | (0.01) | (0.01) | (0.00) | (0.01) | (0.01) |
| | | | NTR (%) | 21.83 | -49.23 | -23.47 | 61.11 | -24.22 | -53.50 | -46.81 | -60.33 |
| | | | | (13.41) | (15.76) | (10.91) | (23.41) | (22.13) | (27.99) | (22.31) | (20.34) |
| | Rainy season | 422.26 (11.16) | NTF (kg·ha⁻¹) | 1.93 | -0.99 | -1.35 | 0.11 | -0.03 | -0.03 | -0.25 | -0.10 |
| | | | | (0.83) | (0.50) | (1.57) | (0.02) | (0.06) | (0.00) | (0.05) | (0.01) |
| | | | NTR (%) | 49.70 | -68.45 | -32.61 | 89.05 | -31.88 | -66.01 | -53.55 | -64.55 |
| | | | | (21.51) | (34.67) | (37.10) | (16.87) | (56.43) | (11.33) | (11.44) | (5.41) |
| | Snowy season | 93.91 (5.55) | NTF (kg·ha⁻¹) | -0.26 | -0.14 | -0.30 | -0.01 | -0.01 | -0.00 | -0.01 | -0.04 |
| | | | | (0.22) | (0.05) | (0.04) | (0.00) | (0.00) | (0.00) | (0.00) | (0.01) |
| | | | NTR (%) | -82.04 | -97.21 | -75.45 | -57.47 | 103.21 | -91.41 | -80.56 | 107.21 |
| | | | | (68.77) | (37.51) | (11.90) | (12.21) | (32.04) | (8.78) | (18.97) | (17.69) |

## 4. Discussion

Before reaching the soil surface, the chemical composition of precipitation can be modified by contact with vegetation [22]. In a forest canopy, interactions between precipitation and the canopy lead to ion exchange and changes in element concentrations [23]. In the present study, the concentration differences between the snowy season and rainy season indicated seasonal variation (Figures 1 and 2). Freezing and thawing can damage the cell membranes of plant leaves, and the resulting change in the water content in plant tissues affects the exchange between cells and ions [24–26]. In the present study, seasons had significant effects on the trace metal measurements (Table 1). The extent of enrichment is affected by the solute characteristics [21]. Trace metals with higher concentrations of precipitation are mainly from terrigenous particles, such as Al and Fe, and the Sonja, which supports our findings [27]. Thus, in the present study, the higher concentrations of Al and Fe than of the other trace metals resulted in high fluxes in precipitation. The Al concentration was lower under both the gap-edge canopy and the closed canopy than in the nonforested site during both rainy and snowy seasons, whereas the Fe concentration was higher under both the gap-edge canopy and closed canopy than in the nonforested site in the rainy season. The Al and Fe concentrations accounted for 42.50% and 40.8%, respectively, of the nonforest input in the rainy season and 37.6% and 38.0%, respectively, of the nonforest input in the snowy season.

The brief interactions between vegetation and precipitation create high spatial variability of metal deposition from throughfall, which is very important for elemental cycling in forest ecosystems [28,29]. The concentrations of trace metals in precipitation were higher than those in throughfall in the winter. Snow evaporation and sublimation from the snow sampler during the winter may cause underestimation of total snowfall and overestimation of trace metal concentrations during winter. As such, the trace metal concentration values during winter should be treated with some caution. The annual trace metal fluxes should not be affected for the following two reasons: (1) total amount of trace metals in snow collector would not be affected by snow sublimation or evaporation. (2) concentrations of trace metals in precipitation and throughfall were much lower in snow than in rain, and total fluxes only account for less than 7% of total annual fluxes. Therefore, the fluxes did not differ between precipitation and throughfall in the winter. Consistent with our hypothesis, the results indicated that the annual fluxes of most trace metals (e.g., Zn, Cd, Cr, Cu and Pb) under the gap-edge canopy were higher than those under the closed canopy. Under the gap-edge canopy, the edge affects the wind speed and increases air turbulence, increasing the dry deposition velocities via inflow and advection processes. In addition, the surface of coniferous leaves has a strong capacity to trap dry-deposited particulates [30,31]. As a result, precipitation can wash off more metal particles from a gap-edge canopy than from a closed canopy. In the present study, the throughfall deposition of Zn, Cd, Cr, Cu and Pb under the gap-edge canopy was significantly enhanced relative to that under the closed canopy. The throughfall deposition of Zn, Cd, Cr, Cu and Pb under the gap-edge canopy was 1.20-, 0.14-, 1.57-, 1.05- and 2.68-fold higher, respectively, than that under the closed canopy. Canopy gaps have higher air temperatures and higher levels of solar radiation than do closed forest areas, and evaporation from leaves is another important factor contributing to increased metal concentrations in throughfall [32,33]. Accordingly, the concentrations of Cd, Cr and Pb under the gap-edge canopy were 1.31-, 1.43- and 1.43-fold higher, respectively, than those under the closed canopy.

The net throughfall input is the combined result of leaching and uptake by the canopy [10]. These processes are affected by vegetation type. For example, in pine-oak and oak forests, the elements Fe and Mn in throughfall were primarily derived from leaching. However, in our study, filtering was often the result of trace-metal leaching (except in the case of Mn). This result indicated that the trace metals were filtered (e.g., adsorbed or retained) by the canopy. The concentration of Mn under the closed canopy was 1.11- and 2.38-fold higher than that reported under a pine-oak forest and an oak forest, respectively [34]. In addition, Zn, Fe, Cu, Cd, Cr and Pb were filtered by the gap-edge canopy and the closed canopy, and the net filtering ratios of these metals were higher for the closed canopy than for the gap-edge canopy. In two evergreen oak stands in Spain, the net filtering ratio of Zn was

30.25% and 25.00% lower than that in our study [9]. In addition, the net filtering ratio of Al was 36.09% for the gap-edge canopy, and 49.70% of the Al was leached from the closed canopy during the rainy season. However, in the snowy season, the net filtering ratios of all trace metals exceeded 50%. There are two mechanisms by which foliar structures filter metals: (1) through the absorption and internalization through the cuticle and (2) through the penetration of metals through the stomatal pores [35]. Stomatal openings and cuticle expansion allow high levels of metal penetration from the atmosphere [36]. Canopy retention of Zn and Cd has been reported in previous studies [37,38], and some studies have reported canopy uptake of Zn, Cd, Cu and Pb [39]. Nonessential trace metals, such as Pb [40,41], Cd [42] and Cr [43], can also enter plant leaves via foliar transfer. These metals can penetrate cuticles and accumulate in leaf tissues. In a mid-subtropical forest, the filtration ratios of Pb and Cd by the canopy exceeded 80%, which is higher than that observed in the present study for the gap-edge canopy and closed canopy [44]. Among the essential trace metals, Mn presented the highest levels of leaching for both canopies in the rainy season, and this pattern, observed elsewhere, has been widely attributed to canopy leaching [10,37,45–49]. Additionally, the enrichment factor of Mn demonstrated high Mn enrichment in throughfall. A similar phenomenon was observed by Gandois [50], who attributed the results to internal cycling [37].

## 5. Conclusions

The forest canopy can be regarded as a self-regulating system that filters certain trace metals in deposited precipitation. The annual flux of trace metals in precipitation was 17.83 kg·ha$^{-1}$, and the flux in the rainy season accounted for 93.55% of the total. The trace metals in precipitation were filtered by the closed canopy and gap-edge canopy, with filtration percentages of 4.30% and 21.94%, respectively. Snowfall in the snowy season accounted for 6.45% of the precipitation, and 73.95% and 75.11% of the trace metals in precipitation were filtered by the closed canopy and gap-edge canopy, respectively. Regarding essential trace metals, the closed canopy filtered 49.23%, 23.47%, 24.22%, 60.33%, 53.50% and 46.81% of the Zn, Fe, Cu, Pb, Cd and Cr, respectively, whereas the gap-edge canopy leached 19.75% of the Mn, and it filtered 23.75%, 39.20%, −23.93%, 20.62%, 35.95%, 32.58% and 26.70% of the Al, Zn, Fe, Cu, Pb, Cd and Cr, respectively. However, all of the trace metals demonstrated high net filtering ratios for both the gap-edge canopy and closed canopy in the snowy season. These results provide new insight into the filtration effects of subalpine forest on trace metals deposited via precipitation, and they can inform efforts to protect water quality in the upper reaches of the Yangtze River.

**Author Contributions:** Conceptualization, X.N., W.Y. and F.W.; methodology, K.Y.; software, H.Z.; formal analysis, B.T.; investigation, S.T. and Y.Z.; data curation, S.T. and H.Z.; writing—original draft preparation, S.T.; writing—review and editing, X.N.; supervision, F.W.; project administration, F.W.

**Funding:** This research was funded by the National Key Technologies R & D Program of China (2017YFC0505003), the Key Technologies R & D Program of Sichuan (18ZDYF0307), the Fok Ying-Tong Education Foundation for Young Teachers (161101) and the Sichuan Provincial Science and Technology Project for Youth Innovation Team (2017TD0022).

**Acknowledgments:** We are grateful to Ziyi Liang, Zhuang Wang and Liyan Zhuang for their help with field sampling and laboratory analysis work. This work was supported by the National Key Technologies R & D Program of China (2017YFC0505003), the Key Technologies R & D Program of Sichuan (18ZDYF0307), the Fok Ying-Tong Education Foundation for Young Teachers (161101) and the Sichuan Provincial Science and Technology Project for Youth Innovation Team (2017TD0022).

**Conflicts of Interest:** The authors declare no conflict of interest.

## References

1. Hou, H.; Takamatsu, T.; Koshikawa, M.K.; Hosomi, M. Trace metals in bulk precipitation and throughfall in a suburban area of Japan. *Atmos. Environ.* **2005**, *39*, 3583–3595. [CrossRef]
2. Rauch, J.N.; Pacyna, J.M. Earth's global Ag, Al, Cr, Cu, Fe, Ni, Pb, and Zn cycles. *Glob. Biogeochem. Cycles* **2009**, *23*, GB2001. [CrossRef]

3. Siudek, P.; Frankowski, M. Atmospheric deposition of trace elements at urban and forest sites in central Poland—Insight into seasonal variability and sources. *Atmos. Res.* **2017**, *198*, 123–131. [CrossRef]

4. Mayer, R.; Ulrich, B. Input of atmospheric sulfur by dry and wet deposition to two central European forest ecosystems. *Atmos. Environ.* **1978**, *12*, 375–377. [CrossRef]

5. Wong, C.S.C.; Li, X.D.; Zhang, G.; Qiand, S.H.; Peng, Z. Atmospheric deposition of heavy metals in the Pear Delta, China. *Atmos. Environ.* **2003**, *3*, 767–776. [CrossRef]

6. Nagajyoti, P.C.; Lee, K.D.; Sreekanth, T.V.M. Heavy metals, occurrence and toxicity for plants: A review. *Environ. Chem. Lett.* **2010**, *8*, 199–216. [CrossRef]

7. Lee, C.S.L. Heavy metal concentrations and Pb isotopic composition in urban and suburban aerosols of Hong Kong and Guangzhou, South China—Evidence of the long-range transport of air contaminants. *Acta Geochim.* **2006**, *25*, 123–124. [CrossRef]

8. Draaijers, G.P.J.; Erisman, J.W.; VanLeeuwen, N.F.M.; Romer, F.G.; TEWinkel, B.H.; Veltkamp, A.C.; Vermeulen, A.T.; Wyers, G.P. The impact of canopy exchange on differences observed between atmospheric deposition and throughfall fluxes. *Atmos. Environ.* **1997**, *31*, 387–397. [CrossRef]

9. Avila, A.; Rodrigo, A. Trace metal fluxes in bulk deposition, throughfall and stemflow at two evergreen oak stands in NE Spain subject to different exposure to the industrial environment. *Atmos. Environ.* **2004**, *38*, 171–180. [CrossRef]

10. Zhang, S.; Liang, C. Effect of a native forest canopy on rainfall chemistry in China's Qinling Mountains. *Environ. Earth Sci.* **2012**, *67*, 1503–1513. [CrossRef]

11. Stiller, M.; Sigg, L. Heavy metals in the Dead Sea and their coprecipitation with halite. *Hydrobiologia* **1990**, *197*, 23–33. [CrossRef]

12. Staelens, J.; Houle, D.; An, D.S.; Neirynck, J.; Verheyen, K. Calculating Dry Deposition and Canopy Exchange with the Canopy Budget Model: Review of Assumptions and Application to Two Deciduous Forests. *Water Air Soil Pollut.* **2008**, *191*, 149–169. [CrossRef]

13. Sun, X.Y.; Wang, G.X. The Hydro-chemical Characteristics Study of Forest Ecosystem Precipitation Distribution in Gongga Mountain. *Res. Soil Water Conserv.* **2009**, *16*, 120–124.

14. Guo, J.; Kang, S.; Huang, J.; Zhang, Q.; Tripathee, L.; Sillanpää, M. Seasonal variations of trace elements in precipitation at the largest city in Tibet, Lhasa. *Atmos. Res.* **2015**, *153*, 87–97. [CrossRef]

15. Wuyts, K.; Schrijver, A.D.; Verheyen, K.; Schaub, M. The importance of forest type when incorporating forest edge deposition in the evaluation of critical load exceedance. *iFor. Biogeosci. For.* **2009**, *2*, 385–392. [CrossRef]

16. Yang, W.Q.; Wang, K.Y.; Kellomki, S.; Gong, H.D. Litter dynamics of Three subalpine forests in Western Sichuan. *Pedosphere* **2005**, *15*, 653.

17. Hultberg, H.; Grennfelt, P. Sulphur and seasalt deposition as reflected by throughfall and runoff chemistry in forested catchments. *Environ. Pollut.* **1992**, *75*, 215–222. [CrossRef]

18. Derome, J.; Nieminen, T. Metal and macronutrient fluxes in heavy-metal polluted Scots pine ecosystems in SW Finland. *Environ. Pollut.* **1998**, *103*, 219–228. [CrossRef]

19. Yang, W.Q.; Wang, K.Y.; Kellomki, S.; Jian, Z. Annual and Monthly Variations in Litter Macronutrients of Three Subalpine Forests in Western China. *Pedosphere* **2006**, *16*, 788–798. [CrossRef]

20. Wu, Q.G.; Wu, F.Z.; Yang, W.Q.; Tan, B.; Yang, Y.L.; Ni, X.Y.; He, J. Characteristics of Gaps and Disturbance Regimes of the Alpine Fir Forest in Western Sichuan. *Chin. J. Appl. Environ. Biol.* **2013**, *19*, 922. [CrossRef]

21. Lu, J.; Zhang, S.; Fang, J.; Yan, H.; Li, J. Nutrient Fluxes in Rainfall, Throughfall, and Stemflow in Pinus densata Natural Forest of Tibetan Plateau. *Clean Soil Air Water* **2017**, *45*, 1600008. [CrossRef]

22. Lovett, G.M.; Lindberg, S.E. Dry Deposition and Canopy Exchange in a Mixed Oak Forest as Determined by Analysis of Throughfall. *J. Appl. Ecol.* **1984**, *21*, 1013–1027. [CrossRef]

23. Özsoy, T.; Örnektekin, S. Trace elements in urban and suburban rainfall, Mersin, Northeastern Mediterranean. *Atmos. Res.* **2009**, *94*, 203–219. [CrossRef]

24. Schmidt, R.A.; Gluns, D.R. Snowfall Interception on Branches of Three Conifer Species. *Can. J. For. Res.* **1991**, *21*, 1262–1269. [CrossRef]

25. Bao, W.; Bao, W.; He, B. Redistribution effects of tree canopy of the artificial Pinus tabulaeformis forest on precipitation in the upper stream of Minjiang River. *J. Beijing For. Univ.* **2004**, *26*, 10–16.

26. Xiao, Q.Y.; Yin, C.Y.; Pu, X.Z.; Qiao, M.F.; Liu, Q. Ecophysiological characteristics of leaves and fine roots in dominant tree species in a subalpine coniferous forest of western Sichuan during seasonal frozen soil period. *Chin. J. Plant Ecol.* **2014**, *38*, 343–345.

27. Sonja, G.; Christopher, N.; Alexv, K.; Sergiocgouveia, N.; Helmut, E. Seasonal and within-event dynamics of rainfall and throughfall chemistry in an open tropical rainforest in Rondonia, Brazil. *Biogeochemistry* **2007**, *86*, 155–174.

28. Kimmins, J.P. Some Statistical Aspects of Sampling Throughfall Precipitation in Nutrient Cycling Studies in British Columbian Coastal Forests. *Ecology* **1973**, *54*, 1008–1019. [CrossRef]

29. Zimmermann, A.; Wilcke, W.; Elsenbeer, H. Spatial and temporal patterns of throughfall quantity and quality in a tropical montane forest in Ecuador. *J. Hydrol.* **2007**, *343*, 80–96. [CrossRef]

30. Draaijers, G.P.; Van Ek, R.; Bleuten, W. Atmospheric deposition in complex forest landscapes. *Bound.-Layer Meteorol.* **1994**, *69*, 343–366. [CrossRef]

31. Michopoulos, P.; Bourletsikas, A.; Kaoukis, K.; Daskalakou, E.; Karetsos, G.; Kostakis, M.; Thomaidis, N.S.; Pasias, I.N.; Kaberi, H.; Iliakis, S. The distribution and variability of heavy metals in a mountainous fir forest ecosystem in two hydrological years. *Glob. Nest J.* **2018**, *20*, 188–197.

32. Schliemann, S.A.; Bockheim, J.G. Influence of gap size on carbon and nitrogen biogeochemical cycling in Northern hardwood forests of Upper Peninsula. *Plant Soil* **2014**, *377*, 323–335. [CrossRef]

33. Cornu, S.; Ambrosi, J.P.; Lucas, Y.; Desjardins, T. Origin and behaviour of dissolved chlorine and sodium in Brazilian Rainforest. *Water Res.* **1998**, *32*, 1151–1161. [CrossRef]

34. Silva, I.C.; Rodríguez, H.G. Interception loss, throughfall and stemflow chemistry in pine and oak forests in northeastern Mexico. *Tree Physiol.* **2001**, *21*, 1009. [CrossRef]

35. Säumel, I.; Kotsyuk, I.; Hölscher, M.; Lenkereit, C.; Weber, F.; Kowarik, I. How healthy is urban horticulture in high traffic areas? Trace metal concentrations in vegetable crops from plantings within inner city neighbourhoods in Berlin, Germany. *Environ. Pollut.* **2012**, *165*, 124. [CrossRef]

36. Arvik, J.H.; Zimdahl, R.L. Barriers to the Foliar Uptake of Lead1. *J. Environ. Qual.* **1974**, *3*, 369. [CrossRef]

37. Petty, W.H.; Lindberg, S.E. An intensive 1-month investigation of trace metal deposition and throughfall at a mountain spruce forest. *Water Air Soil Pollut.* **1990**, *53*, 213–226. [CrossRef]

38. Stachurski, A.; Zimka, J.R. Atmospheric input of elements to forest ecosystems: A method of estimation using artificial foliage placed above rain collectors. *Environ. Pollut.* **2000**, *10*, 345–356. [CrossRef]

39. Szarek-Ukaszewska, G. Input of chemical elements to the forest ecosystem on the (Carpathian Foothills, S Poland)—An overview. *Pol. J. Ecol.* **1999**, *47*, 191–213.

40. Chinot, O.; Romain, S.; Martin, P.M. Effects of root and foliar treatments with lead, cadmium, and copper on the uptake distribution and growth of radish plants. *Environ. Int.* **1993**, *19*, 393–404.

41. Schreck, E.; Laplanche, C.; Guédard, M.L.; Bessoule, J.J.; Austruy, A.; Xiong, T.; Foucault, Y.; Dumat, C. Influence of fine process particles enriched with metals and metalloids on *Lactuca sativa* L. leaf fatty acid composition following air and/or soil-plant field exposure. *Environ. Pollut.* **2013**, *179*, 242–249. [CrossRef]

42. Tudoreanu, L.; Phillips, C.J.C. Modeling Cadmium Uptake and Accumulation in Plants. *Adv. Agron.* **2004**, *84*, 121–157.

43. Levi, E.; Dalschaert, X.; Wilmer, J.B.M. Retention and absorption of foliar applied Cr. *Plant Soil* **1973**, *38*, 683–686. [CrossRef]

44. Sun, T.; Ma, M.; Wang, D.Y. Interceptive characteristics of lead and cadmium in a representative forest ecosystem in mid-subtropical area in China. *Acta Ecol. Sin.* **2016**, *36*, 218–225.

45. Heinrichs, H.; Mayer, R. The Role of Forest Vegetation in the Biogeochemical Cycle of Heavy Metals 1. *J. Environ. Qual.* **1980**, *9*, 226–231. [CrossRef]

46. Parker, G.G. Throughfall and Stemflow in the Forest Nutrient Cycle. *Adv. Ecol. Res.* **1983**, *13*, 57–133.

47. Leininger, T.D.; Winner, W.E. Throughfall chemistry beneath Quercusrubra: Atmospheric, foliar, and soil chemistry considerations. *Can. J. For. Res.* **1988**, *18*, 478–482. [CrossRef]

48. Ahmadshah, A.; Rieley, J.O. Influence of tree canopies on the quantity of water and amount of chemical elements reaching the peat surface of a basin mire in the midlands of England. *J. Ecol.* **1989**, *77*, 357–370. [CrossRef]

49. Skřivan, P.; Rusek, J.; Fottova, D.; Burian, M.; Minařík, L. Factors affecting the content of heavy metals in bulk atmospheric precipitation, throughfall and stemflow in central Bohemia, Czech Republic. *Water Air Soil Pollut.* **1995**, *85*, 841–846. [CrossRef]
50. Gandois, L.; Tipping, E.; Dumat, C.; Probst, A. Canopy influence on trace metal atmospheric input on forest ecosystems: Speciation on throughfall. *Atmos. Environ.* **2010**, *44*, 824–833. [CrossRef]

**forests**

*Article*

# Woody Litter Increases Headwater Stream Metal Export Ratio in an Alpine Forest

Ziyi Liang, Fuzhong Wu [ID], Xiangyin Ni [ID], Bo Tan [ID], Li Zhang, Zhenfeng Xu, Junyi Hu and Kai Yue *

Long-Term Research Station of Alpine Forest Ecosystem, Provincial Key Laboratory of Ecological Forestry Engineering, Institute of Ecology and Forestry, Sichuan Agricultural University, Chengdu 611130, China; chn_liangzy@163.com (Z.L.); wufzchina@163.com (F.W.); nixiangyin_922@163.com (X.N.); bobotan1984@163.com (B.T.); zhangli16830116@hotmail.com (L.Z.); sicauxzf@163.com (Z.X.); hujunyi113@163.com (J.H.)
* Correspondence: kkyue@fjnu.edu.cn; Tel.: +86-028-86291112

Received: 8 March 2019; Accepted: 26 April 2019; Published: 30 April 2019

**Abstract:** Headwater streams have low productivity and are closely linked to forest ecosystems, which input a large amount of plant litter into streams. Most current studies have focused on the decomposition process of plant litter in streams, and the effects of non-woody and woody litter on metal transfer, accumulation, and storage in streams are poorly understood. Here, we addressed how non-woody and woody litter affect metals in headwater streams in an alpine forest on the Eastern Tibetan Plateau. This area is the source of many rivers and plays an important regulatory role in the regional climate and water conservation. Through comparisons of five metal concentrations, exports and storage in headwater streams with different input conditions of plant litter, our results showed that the input of woody litter could significantly increase flow discharge and increase the metal export ratio in the water. Similarly, the input of non-woody litter could reduce the metal concentration in the water and facilitate the stable storage of metals in the sediment in the headwater streams. Therefore, allochthonous non-woody and woody litter can affect the concentration of metals in water and sediment, and the transfer and accumulation of metals from upstream to downstream in headwater streams. This study provides basic data and new findings for understanding the effects of allochthonous plant litter on the accumulation and storage of metals in headwater forest streams and may provide new ideas for assessing and managing water quality in headwater streams in alpine forests.

**Keywords:** headwater stream; metals; non-woody litter; woody litter

---

## 1. Introduction

Metals are a critical and complex problem in ecosystems. Metals such as potassium (K), sodium (Na), calcium (Ca), and magnesium (Mg) are essential elements for living organisms, but they pose a threat to organisms when their concentrations are too high [1,2]. The concentrations of some trace metals in headwater streams require close monitoring, such as Fe, Mn, and Cr, because of their toxicity, persistence, tendency to bio-accumulate, and widespread presence in the environment [3]. Riparian vegetation, precipitation, and biogeochemical processes affect stream metal concentrations [4]. In the dry season, riparian plants can influence water quality by intercepting and absorbing nutrients and pollutants [5,6], thereby influencing the dynamics of element concentrations in sediments and streams [7]. Because headwater streams are relatively closely coupled to adjacent forests, they could receive large amounts of organic matter from riparian plants [8]. The forms of organic matter supplied by these riparian forests are divided into non-woody litter (fine litter, e.g., leaves, fruit, small bark fragments, twigs, and flowers) and woody litter (coarse woody ≥ 10 cm and fine woody 1 cm ≤ diameter

≤ 10 cm, e.g., wood, branches, and roots) according to their ecological functions and diameters [9]. Previous studies have demonstrated that leaf and wood inputs are important material for forest stream food webs [10], and contribute to the retention of dissolved nutrients [11]. However, the effects of non-woody and woody litter inputs on the metals in streams are not clear, although studies have suggested that litter decomposition in streams is also an important metal cycling pathway [12].

Seventy-five percent of the global stream length is composed of headwater streams, and forest streams dominate watersheds worldwide [13,14]. Allochthonous plant litter not only represents an important terrestrial subsidy for these aquatic ecosystems, but also plays an essential role in supporting detritus-based food webs in headwater streams [15]. Once allothchonous plant litter is imported into a stream, it is either transported downstream by flow or retained on the streambed, where it is colonized by microorganisms and fragmented by invertebrates, broken down, decomposed, and converted into fine particulate organic matter [16]. Studies have shown that the decomposition rate of leaves in streams is faster than that on the forest floor [17], and broad, soft or senescence leaves can quickly release elements by enhancing microbial activity, thus significantly influencing the water quality in headwater streams [18]. Previous studies have shown that woody litter can affect the morphology and biological functioning of streams by creating new instream habitats [19], and reducing the energy of water during high discharge events [20]. These studies have shown that the input of non-woody and woody litter has an important influence on headwater streams and deserves further study. Moreover, woody litter can intercept and store organic matter and sediments [21], and most metals are transported in association with suspended sediment because of the affinity of metals for the common components of suspended particles [22,23]. Here, we address a hypothesis that there are differences in the contribution of woody litter input and non-woody litter input on the metal concentration and export in the headwater streams of alpine forests.

To test this hypothesis, we conducted a field control experiment by controlling the presence or absence of non-woody or woody litter inputs in headwater streams in an alpine coniferous forest. We compared the variations in the metal concentrations and exports in the headwater streams by removing the inputs of non-woody litter or woody litter in an alpine forest on the Eastern Tibetan Plateau. The objectives were: (1) To evaluate the effects of non-woody and woody litter on the metal concentrations and exports in the headwater streams of an alpine forest and (2) to determine whether non-woody and woody litter are key to controlling the metal concentrations in water and the storage in sediment.

## 2. Material and Methods

### 2.1. Study Site and Experimental Design

The study was conducted at the Long-term Research Station of Alpine Forest Ecosystems, Miyaluo Nature Reserve (102°53'–102°57' E, 31°14"-31°19' N, 2458–4169 m a.s.l. (above sea level)), which is located in Li County, Sichuan, southwestern China. This region is in a transitional zone between the Tibetan Plateau, the Sichuan Basin and the upper Yangtze River. The mean annual air temperature ranges from 2 °C to 4 °C, and the maximum and minimum temperatures are 23 °C and −18 °C, respectively. The mean annual precipitation is approximately 850 mm [24]. Meanwhile, the amount of water in the stream is greater in summer, and smaller or even dry in winter, which is characteristic of a typical seasonal forest stream. The study site is an alpine coniferous forest, and the dominant tree species are *Abies faxoniana* Rehd. and *Picea likiangensis* (Franch) Pritz var. *balfouriana* (Rehd·et Wils) Hillier ex Slavin, and associated species include *Cerasus duclouxii* (Koehne) Yu et Li, *Sabina saltuaria* (Rehd. et Wils.) Cheng et W.T. Wang, and *Betula albosinensis* Burk., interspersed with shrubs are composed of *Salix paraplesia* Schneid., *Rosa omeiensis* Rolfe, and *Rhododendron moupinense* Franch. [25].

To compare the effects of the inputs of non-woody litter and woody litter on streams, it was necessary to ensure that the characteristic conditions of the control streams were consistent and to eliminate the effects of heterogeneity. A common persistent stream that is representative of a typical stream in the alpine coniferous forest was selected, and 6 straight streams with the similar characteristics were artificially excavated near this stream (3 replicates × (1 litter exclusion stream + 1 litter input stream), Figure 1). Water in the selected persistent natural stream was then introduced into the 6 streams. These 6 artificially excavated streams had the same length (50 m), width (0.5 m), and depth (0.15 m), and the streams were at a minimum 2 m apart. In addition, the slope, altitude (3600 m), substrate (mixing of clay, fine sand and gravel), and riparian vegetation of each artificial stream were similar to the original stream.

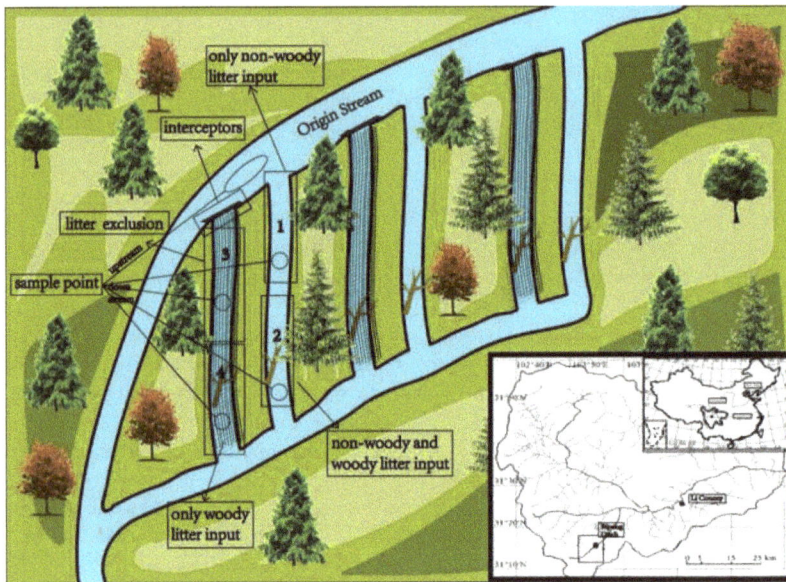

**Figure 1.** Schematic diagram showing the location and experimental design of our study in the alpine forest.

These artificially excavated streams were divided into two sections (25 m in each section). Woody litter was added artificially in the lower part of each stream, and this section of the stream was used as the woody litter input section. The artificially added woody litter consisted of large dead branches from nearby riparian vegetation. The woody litter was equally distributed into each stream to ensure the same control conditions.

Three of these artificial streams were selected as plant litter exclusion streams. A canopy of 1 mm mesh netting that did not affect light transmission was placed above these exclusion streams. This canopy covered the entirety of the exclusion streams and extended over both sides of the stream bank to exclude falling leaves and wood as well as lateral litter input. Meanwhile, interceptors with apertures of 1 mm and 5 mm were erectly installed at the inlet of the exclusion streams. These interceptors were installed to exclude plant litter inputs from the upper stream section, and two apertures were designed to prevent excessive litter that could affect the water flow characteristics from clogging the intercepting net.

After the above treatments, these streams could be divided into four treatments: (1) a stream with only non-woody litter input (the upper half of the reference stream); (2) a stream with non-woody and woody litter inputs (the lower half of the reference stream); (3) a stream with plant litter exclusion (the upper half of the exclusion stream); and (4) a stream with only woody litter input (the lower half of the exclusion stream). These 4 streams with different plant litter input conditions were regarded as downstream streams, and the original streams leading out of these streams were regarded as upstream streams. By comparing the difference in the metal concentration and export between the upstream and downstream, we could determine the influence of the woody litter and non-woody litter inputs on the metals in the forested headwater streams.

## 2.2. Sample Collection and Statistical Analyses

According to previous research and phenological observations, the growing season in this region is from May to October [26]. After entering the snow period, restricted flow and less fallen litter will affect stream characteristics and plant litter inputs. Therefore, the sampling time was chosen from June to October 2017, once a month. Accumulated non-woody and woody litter were removed from the net before each sampling event. The water samples were collected randomly at approximately 1/2 depth of the stream using the pre-cleaned polyethylene bottles, and was carefully taken to avoid disturbing the bottom sediments and collecting surface floats. 2 L water samples were collected at each sample point, and then taken back to the laboratory within 24 hours and stored at 4 °C for chemical analysis. The surface sediments were collected using the pre-cleaned polyethylene bottles at each sample point by a five-point sampling method (use a random number table and tape measure to choose 5 points along the stream reach at random) [27]. After collection, the samples were returned to the laboratory and determined the concentrations of potassium (K), magnesium (Mg), iron (Fe), manganese (Mn), and chromium (Cr), digested using CEM-MARS 5, then tested using inductively coupled plasma spectroscopy (ICP-MS, IRIS Advantage 1000; Thermo Elemental, Waltham, MA, USA) [28].

At each sampling event, the metal exports per unit area of water were calculated as:

$$E = c \times F \tag{1}$$

where $E$ is the metals exports of water in the stream (mg·day$^{-1}$), and $c$ is the metals concentration in the water in the stream (µg·L$^{-1}$), $F$ is the flux of the stream (m$^3$·day$^{-1}$).

The metals storage per unit area of sediment was calculated as:

$$M = \frac{c \times m}{S} \tag{2}$$

where $M$ is the metals storage of sediment in the stream (g m$^{-2}$), $c$ is the metals concentration in the sediment in the stream (g kg$^{-1}$), $m$ is the amount of sediment in the stream (kg), and $S$ is the surface area of the sample at each stream (m$^2$).

The amount of sediment in the stream was calculated as:

$$m = \rho_s \times l \times w \times h_s \tag{3}$$

where $m$ is the amount of sediment in the stream (g), $\rho_s$ is the density of sediment (g m$^{-3}$), $l$ is the length of stream (m), $w$ is the width of stream (m), and $h_s$ is the depth of sediment in the stream (m).

The density of sediment was calculated as:

$$\rho_s = \frac{m_0}{(1 - m_c) \times V} \tag{4}$$

where $\rho_s$ is the density of sediment (g m$^{-3}$), $m_0$ is the drying weight of sediment (g), $m_c$ is the water content of sediment (%), and $V$ is the volume of sediment (m$^3$). The sediments used to calculate density were collected in a volume of 25 cm$^3$ polyethylene bottles at each point.

The cumulative exports rate of metals in the water was calculated as:

$$r = \frac{E_i - E_o}{E_o} \times 100\% \tag{5}$$

where $r$ is the cumulative exports rate of metal in the water (%), $E_i$ is the metals exports from streams with different plant litter input conditions during the study period (g day$^{-1}$), $E_o$ is the metals exports from the origin of streams during the study period (g day$^{-1}$).

Analysis of variance (ANOVA) was used to test the effects of litter input conditions on water characteristics of the study streams. Repeated-measure ANOVA with Tukey's HSD was performed to test the effects of time and litter input conditions on metal concentrations and exports in the water, and metal concentrations and storages in the sediment. Spearman's correlation was selected for test the correlation coefficients between the environmental factors and metal concentrations and exports in the water, and metal concentrations and storages in the sediment [29,30]. All statistical analyses were performed using SPSS 22.0 (IBM SPSS Statistics Inc., Chicago, IL, USA). The water characteristics of the study streams were shown as the average (±SE, $n = 15$) during the study period (Table 1).

Table 1. Water characteristics of the study streams (average values during the study period, mean ± SE, $n = 15$).

| Streams | Temperature (°C) | Dissolved Oxygen (mg/L) | Conductivity (μs/cm) | pH | Illumination (lx) | Discharge (L/s) |
|---|---|---|---|---|---|---|
| origin | 6.87 ± 1.20 a | 7.60 ± 0.23 a | 23.62 ± 5.13 a | 6.42 ± 0.22 a | 12,256 ± 11,061 a | 6.09 ± 0.95 c |
| non-woody litter | 6.87 ± 1.20 a | 7.60 ± 0.23 a | 23.62 ± 5.13 a | 6.42 ± 0.22 a | 13,422 ± 14,124 a | 6.43 ± 1.46 c |
| non-woody and woody litter | 6.71 ± 0.65 a | 7.62 ± 0.27 a | 23.85 ± 4.07 a | 6.54 ± 0.21 a | 7981 ± 6294 b | 9.28 ± 1.79 a |
| litter exclusion | 6.65 ± 0.64 a | 7.51 ± 0.24 a | 24.87 ± 6.37 a | 6.44 ± 0.21 a | 12,405 ± 14,637 a | 6.91 ± 1.83 bc |
| woody litter | 6.71 ± 0.68 a | 7.57 ± 0.28 a | 24.69 ± 3.81 a | 6.40 ± 0.28 a | 8077 ± 5343 b | 8.53 ± 1.81 ab |

Different lowercase letters in the same column denote significant ($p < 0.05$) differences among different litter input conditions based on one-way ANOVA followed by multiple comparisons.

## 3. Results

### 3.1. Dynamics of Metal Concentrations in Water

The order of the concentrations of the metals in the water was K > Mg > Fe > Cr > Mn (Figure 2). Results of the repeated-measures ANOVA showed that time had significant impacts on all metals, while the different plant litter input conditions had significant impacts on K, Mn and Cr concentration (Table 2). While time had a significant effect on the concentrations of the metals in water, the influence pattern of time on the concentrations of these metals was different. Moreover, the concentrations of these metals in the water also had different responses to the input of different plant litter conditions. The exclusion of non-woody litter increased the K concentration in the water downstream. The Mn concentration was reduced in water downstream when woody litter was added to the streams together with non-woody litter. The Mg and Fe concentration of the water was relatively stable, and the allochthonous plant litter input had little effect on the concentration. However the Mg concentration in the water gradually increased during the study period, and the Fe concentration in the water decreased in September and October during the stream with exclusion of non-woody litter. The K concentration was positively correlated with the temperature and dissolved oxygen; the Mg concentration was negatively correlated with the dissolved oxygen and pH; and the Fe concentration was positively correlated with the temperature and conductivity (Table 3).

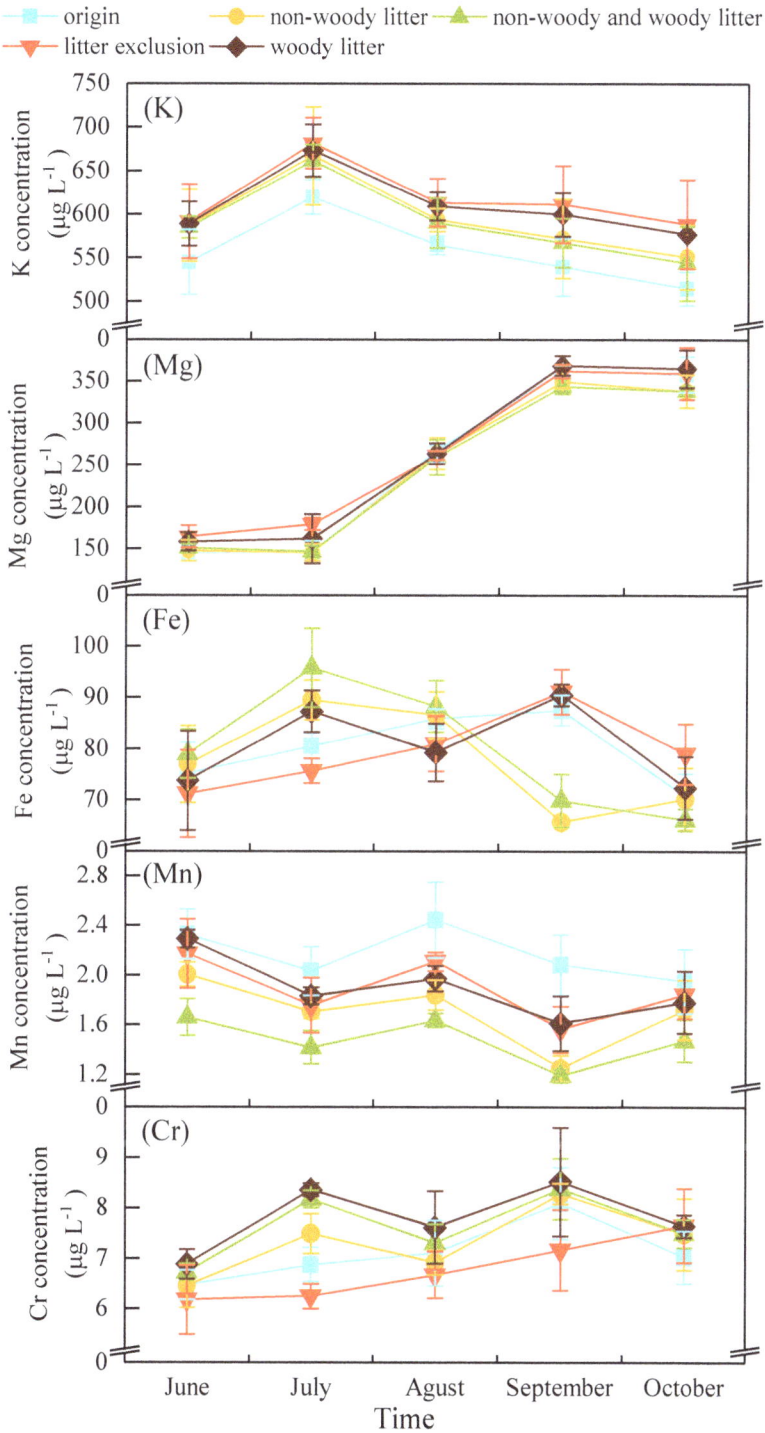

**Figure 2.** The K, Mg, Fe, Mn and Cr concentration of the water in the stream with different plant litter input conditions.

**Table 2.** Effects of time, debris input and their interaction on metals concentration of water tested by repeated-measure ANOVA analyses.

| Factor | df | K | | Mg | | Fe | | Mn | | Cr | |
|---|---|---|---|---|---|---|---|---|---|---|---|
| | | F-Value | p-Value | F-Value | p-Value | F-Value | p-Value | F-Value | p-Value | F-Value | p-Value |
| Time | 4 | 26.743 | <0.001 | 717.856 | <0.001 | 18.730 | <0.001 | 25.690 | <0.001 | 27.040 | <0.001 |
| Input | 4 | 5.042 | <0.05 | 3.186 | 0.062 | 0.740 | 0.586 | 21.203 | <0.001 | 3.778 | <0.05 |
| Time × Input | 16 | 0.194 | 1.000 | 0.829 | 0.647 | 6.582 | <0.001 | 1.194 | 0.314 | 2.268 | <0.05 |
| origin | | 556.67 b | | 255.84 a | | 79.96 a | | 2.17 a | | 7.11 ab | |
| non-woody litter | | 594.20 ab | | 248.72 a | | 77.77 a | | 1.70 bc | | 7.33 ab | |
| non-woody and woody litter | | 589.41 ab | | 247.57 a | | 79.80 a | | 1.47 c | | 7.62 ab | |
| litter exclusion | | 617.30 a | | 265.26 a | | 79.58 a | | 1.89 b | | 6.78 b | |
| woody litter | | 609.62 a | | 263.37 a | | 80.63 a | | 1.90 b | | 7.81 a | |

Different lowercase letters in the same column denote significant ($p < 0.05$) differences among different litter input conditions based on repeated-measure ANOVA followed by multiple comparisons.

**Table 3.** Correlation coefficients ($r$) between the environmental factors and the concentrations and exports of metals in the water.

| Factor | K | | Mg | | Fe | | Mn | | Cr | |
|---|---|---|---|---|---|---|---|---|---|---|
| | Concentration | Export | Concentration | Export | Concentration | Export | Concentration | Export | Concentration | Export |
| Temperature | 0.297 ** | 0.183 | −0.028 | 0.092 | 0.426 ** | 0.291 * | −0.054 | 0.041 | 0.136 | 0.159 |
| Dissolved Oxygen | 0.355 ** | 0.400 ** | −0.483 ** | −0.260 * | 0.054 | 0.281 * | −0.065 | 0.237 * | 0.034 | 0.258 * |
| Conductivity | 0.187 | 0.139 | −0.223 | −0.150 | 0.252 * | 0.208 | 0.188 | 0.226 | −0.199 | 0.016 |
| pH | −0.036 | −0.102 | −0.307 ** | −0.229 * | −0.113 | −0.128 | 0.078 | −0.021 | −0.112 | −0.079 |
| Illumination | −0.104 | −0.264 * | 0.174 | 0.011 | 0.144 | −0.147 | 0.173 | −0.135 | −0.150 | −0.257 * |

*, $p < 0.05$; **, $p < 0.01$, $n = 75$.

## 3.2. Dynamics of Metal Export in Water

The order of the metal export in the water was consistent with the concentration: K > Mg > Fe > Cr > Mn (Figure 3). Results of the repeated-measures ANOVA showed that time had significant impacts on Mg, Fe, Mn and Cr, while the input of non-woody and woody litter had no significant impacts on all metals export (Table 4). Nevertheless, we found that the metal export in the stream with added woody litter was greater than other streams (Figure 4), and the ratio of the export from the treated streams to that from origin stream indicated that woody litter can increase the K, Mg, Fe, Mn and Cr export in the water (Table 5). The K export was positively correlated with dissolved oxygen and negatively correlated with the illumination; the Mg export was negatively correlated with the dissolved oxygen and the pH; the Fe concentration was positively correlated with the temperature and the dissolved oxygen, and the Mn and Cr concentrations were positively correlated with the dissolved oxygen (Table 3).

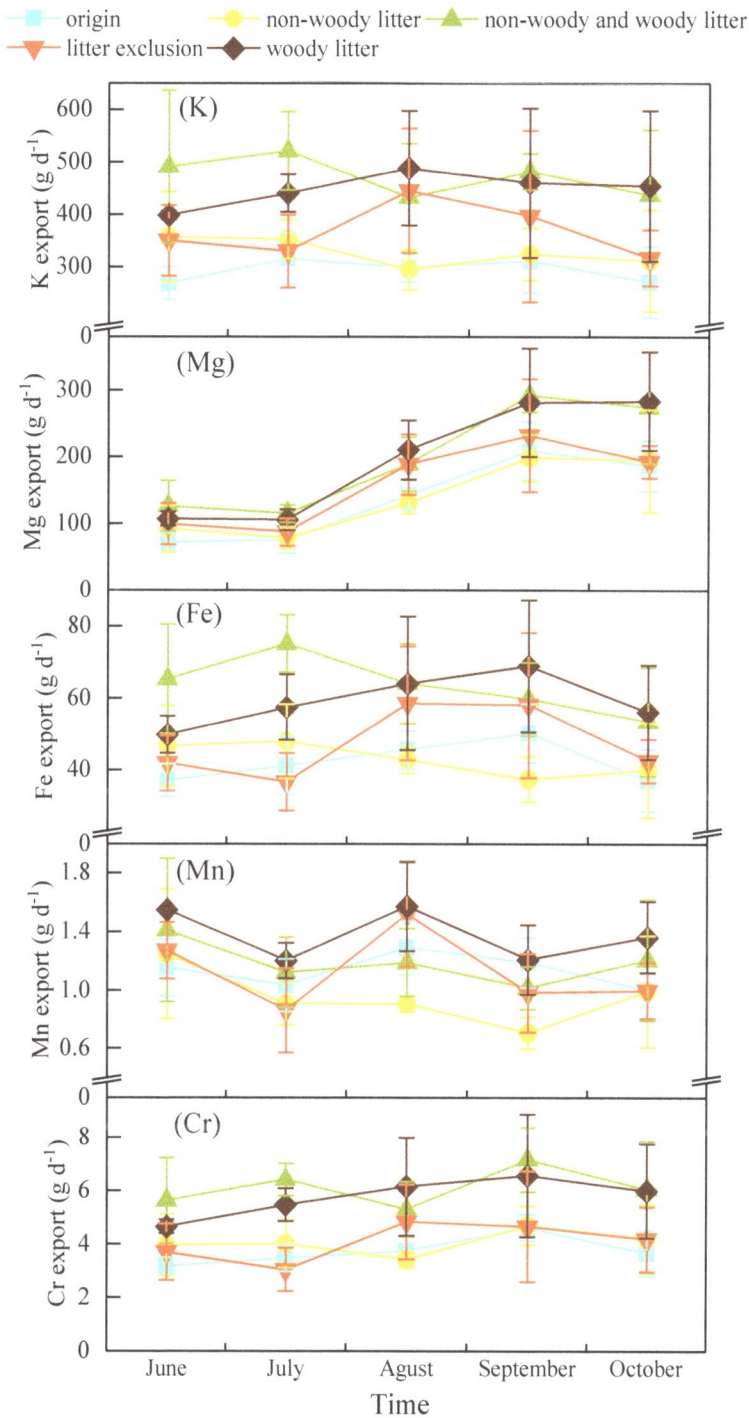

**Figure 3.** The K, Mg, Fe, Mn and Cr export of the water in the stream with different plant litter input conditions.

**Table 4.** Effects of time, debris input and their interaction on metals export of water tested by repeated-measure ANOVA analyses.

| Factor | df | K | | Mg | | Fe | | Mn | | Cr | |
|---|---|---|---|---|---|---|---|---|---|---|---|
| | | F-Value | p-Value | F-Value | p-Value | F-Value | p-Value | F-Value | p-Value | F-Value | p-Value |
| Time | 4 | 1.439 | 0.239 | 90.525 | <0.001 | 5.602 | <0.05 | 14.116 | <0.001 | 8.592 | <0.001 |
| Input | 4 | 3.098 | 0.067 | 2.070 | 0.160 | 3.024 | 0.071 | 1.646 | 0.238 | 3.284 | 0.058 |
| Time × Input | 16 | 1.516 | 0.142 | 1.229 | 0.324 | 3.189 | <0.05 | 2.077 | <0.05 | 1.520 | 0.140 |
| origin | | 292.28 a | | 136.82 a | | 42.13 a | | 1.13 a | | 3.74 a | |
| non-woody litter | | 327.04 a | | 138.41 a | | 42.81 a | | 0.95 a | | 4.05 a | |
| non-woody and woody litter | | 472.06 a | | 199.19 a | | 63.44 a | | 1.19 a | | 6.11 a | |
| litter exclusion | | 367.12 a | | 159.86 a | | 47.46 a | | 1.12 a | | 4.08 a | |
| woody litter | | 448.14 a | | 197.68 a | | 59.23 a | | 1.38 a | | 5.77 a | |

Different lowercase letters in the same column denote significant ($p < 0.05$) differences among different litter input conditions based on repeated-measure ANOVA followed by multiple comparisons.

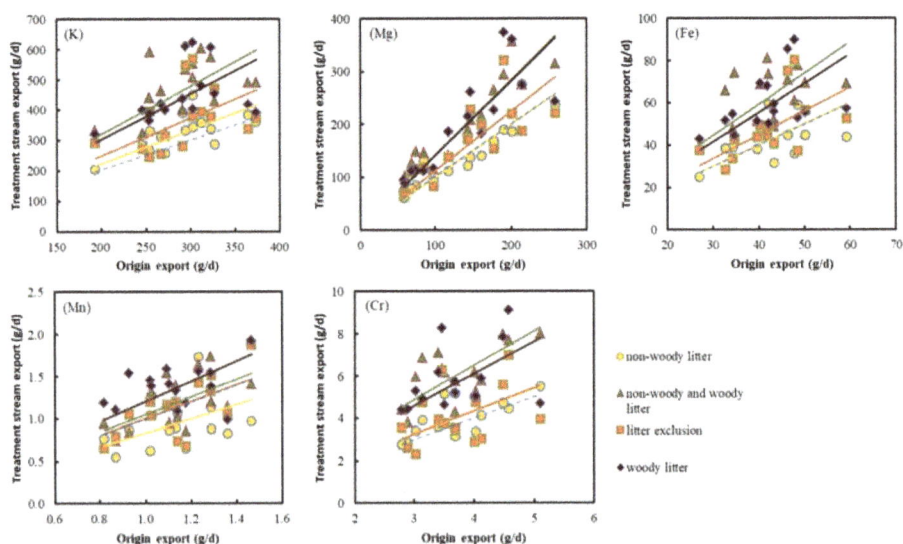

**Figure 4.** Export ratios of the metals (K, Mg, Fe, Mn, Cr) from the origin stream compared to those from the treatment stream (4 input conditions). The lines are the $k$-value trend lines of the treatment/origin streams, and the blue dotted line is the trend line with $k$-value of 1.

**Table 5.** Export ratios of the metals from the origin stream compared to the treatment stream with 4 litter input conditions.

| | K | Mg | Fe | Mn | Cr |
|---|---|---|---|---|---|
| origin | 1.00 c | 1.00 b | 1.00 b | 1.00 b | 1.00 b |
| non-woody litter | 1.12b c | 1.04 b | 1.03 b | 0.84 b | 1.09 b |
| non-woody and woody litter | 1.63 a | 1.50 a | 1.53a | 1.05 a | 1.65 a |
| litter exclusion | 1.27 b | 1.21 b | 1.13 b | 0.99 b | 1.10 b |
| woody litter | 1.55 a | 1.47 a | 1.43 a | 1.24 a | 1.56 a |

Different lowercase letters in the same column denote significant ($p < 0.05$) differences among different litter input conditions based on one-way ANOVA followed by multiple comparisons.

### 3.3. Dynamics of Metals Concentration in Sediment

The order of the concentrations of the metals in the sediment was K > Fe > Mg > Mn > Cr (Figure 5). Results of the repeated-measures ANOVA showed that time had significant impacts on all metals concentration of sediment, and different litter input conditions had significant impacts on Mg,

Fe, Mn and Cr (Table 6). Throughout the study period, the metals concentration of the sediment in the streams gradually decreased from June to October. The input of woody litter increased the Mg, Fe and Mn concentration of the sediment downstream. The variation of Cr concentration was obviously different from that of other metals. Compared with the origin stream, the exclusion of plant litter had no significant effect on the Cr concentration, and the stream with only non-woody litter had a greater concentration than others. The K and Fe concentration of the sediment was positively correlated with the conductivity; the Mg and Fe concentration of the sediment was positively correlated with the dissolved oxygen; and the Mn and Cr concentration of the sediment was positively correlated with the pH (Table 7).

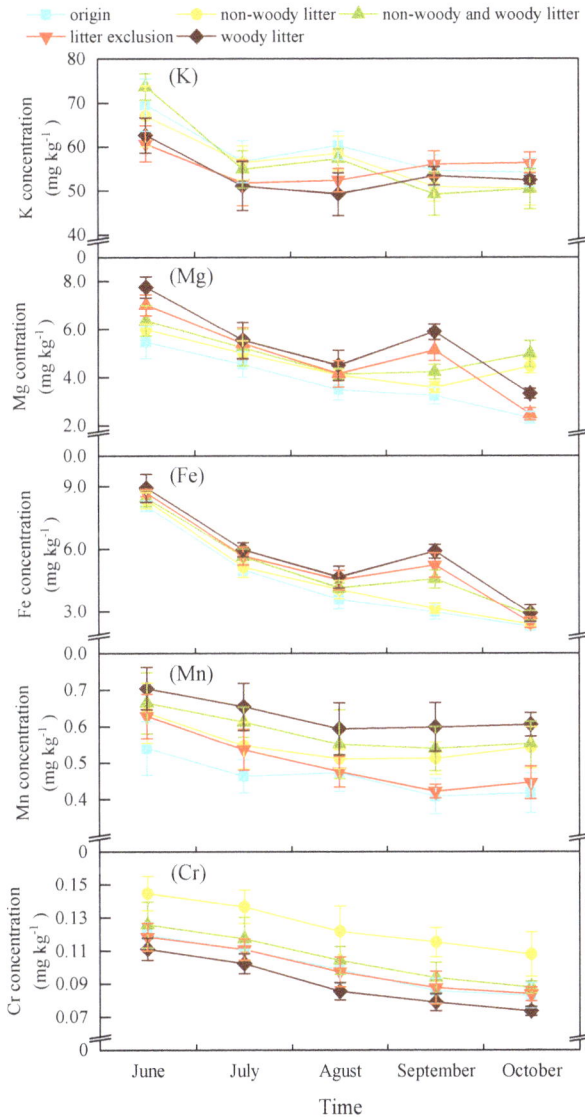

**Figure 5.** The K, Mg, Fe, Mn and Cr concentration of the sediment in the stream with different plant litter input conditions.

**Table 6.** Effects of time, debris input and their interaction on metals concentration of sediment tested by repeated-measure ANOVA analyses.

| Factor | df | K | | Mg | | Fe | | Mn | | Cr | |
|---|---|---|---|---|---|---|---|---|---|---|---|
| | | F-Value | p-Value | F-Value | p-Value | F-Value | p-Value | F-Value | p-Value | F-Value | p-Value |
| Time | 4 | 45.982 | <0.001 | 133.525 | <0.001 | 487.003 | <0.001 | 34.831 | <0.001 | 110.740 | <0.001 |
| Input | 4 | 1.541 | 0.264 | 10.342 | <0.001 | 27.993 | <0.001 | 5.577 | <0.05 | 9.888 | <0.01 |
| Time × Input | 16 | 3.334 | <0.001 | 9.657 | <0.001 | 3.968 | <0.001 | 0.993 | 0.483 | 0.118 | 1.000 |
| origin | | 59.08 a | | 3.80 b | | 4.37 c | | 0.46 b | | 0.10 b | |
| non-woody litter | | 56.71 a | | 4.62 ab | | 4.61 c | | 0.55 ab | | 0.13 a | |
| non-woody and woody litter | | 57.10 a | | 4.98 a | | 5.11 b | | 0.58 ab | | 0.11 b | |
| litter exclusion | | 55.49 a | | 4.83 a | | 5.31 ab | | 0.50 ab | | 0.10 b | |
| woody litter | | 53.78 a | | 5.40 a | | 5.67 a | | 0.63 a | | 0.90 b | |

Different lowercase letters in the same column denote significant ($p < 0.05$) differences among different litter input conditions based on repeated-measure ANOVA followed by multiple comparisons.

**Table 7.** Correlation coefficients ($r$) between the environmental factors and the concentrations and storages of metals in the sediment.

| Factor | K | | Mg | | Fe | | Mn | | Cr | |
|---|---|---|---|---|---|---|---|---|---|---|
| | Content | Storage | Content | Storage | Content | Storage | Content | Storage | Content | Storage |
| Temperature | −0.153 | 0.410 ** | −0.079 | 0.399 ** | 0.035 | 0.363 ** | −0.157 | 0.472 ** | −0.022 | 0.498 ** |
| Dissolved Oxygen | 0.207 | 0.161 | 0.233 ** | 0.255 * | 0.274 * | 0.289 * | −0.018 | 0.082 | 0.195 | 0.192 |
| Conductivity | 0.350 ** | 0.286 * | 0.089 | 0.235 * | 0.276 * | 0.354 ** | 0.095 | 0.226 | 0.197 | 0.273 * |
| pH | −0.025 | −0.256 * | 0.110 | −0.189 | 0.072 | −0.150 | 0.228 * | −0.186 | 0.275 * | −0.128 |
| Illumination | −0.049 | 0.398 ** | −0.192 | 0.284 * | −0.017 | 0.334 ** | −0.138 | 0.414 ** | 0.009 | 0.440 ** |

*, $p < 0.05$; **, $p < 0.01$, $n = 75$.

## 3.4. Dynamics of Metals Storage in Sediment

The order of the storage of the metals in the sediment was K > Fe > Mg > Mn > Cr (Figure 6). Results of the repeated-measures ANOVA showed that time had significant impacts on all metals storage, but different litter input conditions had no significant impacts on all metals storage (Table 8). The amount of metal storage in the sediment was not completely dependent on the metal concentration, as the element concentration in June were the highest during the study period, but their storage were not so. For K, Mn and Cr, the storage in July, August and September was greater than that in June and October. Regarding the time scale, the variation in the metal storage in the sediments after non-woody litter exclusion was greater than that in the streams without non-woody litter exclusion (Figure 6). Temperature and illumination were the dominant environmental factors affecting all metal storage in the sediment in the streams. The K, Mg, Fe, and Cr storage was positively correlated with conductivity; the Mg and Fe storage was positively correlated with dissolved oxygen; and the K storage in the sediment was negatively correlated with pH in the streams (Table 7).

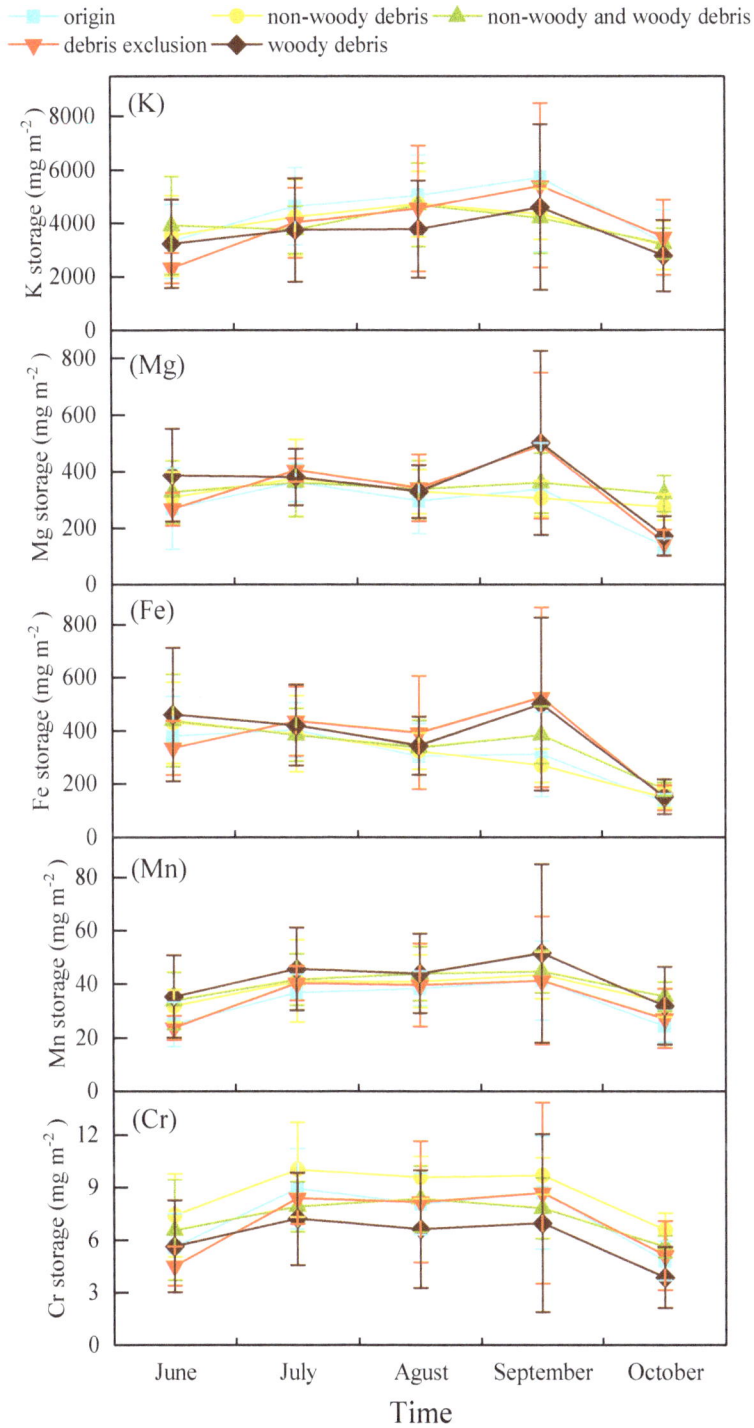

**Figure 6.** The K, Mg, Fe, Mn and Cr storage of the sediment in the stream with different plant litter input conditions.

**Table 8.** Effects of time, debris input and their interaction on metals storage of sediment tested by repeated-measure ANOVA analyses.

| Factor | df | K | | Mg | | Fe | | Mn | | Cr | |
|---|---|---|---|---|---|---|---|---|---|---|---|
| | | F-Value | p-Value | F-Value | p-Value | F-Value | p-Value | F-Value | p-Value | F-Value | p-Value |
| Time | 4 | 10.331 | <0.001 | 11.465 | <0.001 | 20.406 | <0.001 | 12.174 | <0.001 | 16.291 | <0.001 |
| Input | 4 | 0.099 | 0.980 | 0.236 | 0.912 | 0.184 | 0.942 | 0.309 | 0.866 | 0.561 | 0.697 |
| Time × Input | 16 | 0.785 | 0.692 | 1.503 | 0.219 | 1.046 | 0.434 | 0.235 | 0.999 | 0.400 | 0.975 |
| origin | | 4385.02 a | | 278.54 a | | 306.11 a | | 33.12 a | | 7.22 a | |
| non-woody litter | | 3981.85 a | | 316.76 a | | 310.69 a | | 38.00 a | | 8.63 a | |
| non-woody and woody litter | | 3945.64 a | | 339.82 a | | 344.26 a | | 39.80 a | | 7.24 a | |
| litter exclusion | | 3944.64 a | | 329.85 a | | 367.09 a | | 34.39 a | | 6.97 a | |
| woody litter | | 3624.27 a | | 353.65 a | | 375.75 a | | 41.70 a | | 6.06 a | |

Different lowercase letters in the same column denote significant ($p < 0.05$) differences among different litter input conditions based on repeated-measure ANOVA followed by multiple comparisons.

## 4. Discussion

### 4.1. Concentration and Export of Metals in Water

Through comparisons of the concentrations and exports of K, Mg, Fe, Mn, and Cr in headwater streams with different input conditions of plant litter in an alpine forest, we assessed the dynamics of these metals and the effects of non-woody litter and woody litter. In accordance with our hypothesis, our results showed that non-wood debris and wood debris have different contributions to metal in streams. Although there was no significant difference in metal export among streams with different litter input conditions, the input of woody litter significantly increased the metal export ratio in the water. Woody litter formed a major biological pathway for the transfer of elements from riparian vegetation to streams [31], and the woody litter input may affect the export of metals in water by affecting the behavioral factors of metals in headwater streams, such as the hydrology, flow characteristics and properties of the metals themselves [32]. The litter input can change the ionic strength, thus altering metal solubility and mobility [33]. Researchers have stressed the role of woody litter in retaining sediments and organic matter, increasing habitat diversity and providing refuge for aquatic organisms [34], as woody litter decompose more slowly in water than in terrestrial ecosystems. Thus, it is difficult to increase the concentration and export of metals in water by decomposition itself, and to alter metal mobility by changing the water characteristics [12]. The study results showed that the input of woody litter significantly increased the water discharge in the headwater stream (Table 1), and this may be the main reason for the increase in the metal export in the water, because woody litter changed the original flow characteristic of the streams and formed a step [35]. Moreover, our results also showed that there was a significant correlated relationship between the metal export of water and dissolved oxygen. Because of the step formed by the input of woody litter, the process of water flow descending was accompanied by intense water-air exchange, which increases the dissolved oxygen in the downstream, thus affecting ecosystem metabolism and respiration [36].

Many previous experiments on litter decomposition in streams have shown that leaves can decompose rapidly and release elements [18] because submerged leaves experience microbial colonization, are shredded by invertebrates, and are physically abraded [37,38]; thus, non-woody litter may increase the concentration of metals in headwater streams. However, our results showed that the input of non-woody litter reduced the metal concentration in the water, and the streams with plant litter exclusion had higher metal export ratio than the streams with only non-woody litter, which indicates that non-woody litter can reduce the metal export, absorb metals from water and contribute to the self-purification capacity of headwater streams [12]. Direct fresh non-woody litter that falls typically has a lower density than water and moves downstream on the water surface until obstacles protruding above the water surface trap it [39,40]. This litter then becomes waterlogged, sinks, and accumulates on the stream bottom [41,42], and non-woody litter may intercept and absorb the metals from surface runoff during floating. While the input of non-woody litter reduced the metals concentration in the water, the contribution of woody litter input was greater than that of non-woody litter input on the metal export.

### 4.2. Concentration and Storage of Metals in Sediment

By comparing the K, Mg, Fe, Mn, and Cr concentrations and storage of sediment in the headwater streams with different plant litter input conditions, we assessed the effects of non-woody litter and woody litter input for these metals in sediments. The results showed that although non-woody litter and woody litter could affect the metal concentration in the surface sediment, they had no significant effect on the storage of metals in the surface sediment. Compared with the streams with non-woody litter input exclusion, the variation in the metals in the sediment of the streams with non-woody litter input was more stable. Sediment dynamics include the entrainment, transport, deposition, and storage of particulate matter, which includes mineral sediment and particulate organic matter in this study [43]. Allochthonous non-woody and woody litter could influence the sediment dynamics in headwater streams. After non-woody litter is supplied to headwater streams by forest ecosystems, it is initially exposed to the water surface, passively adsorbing and actively taking up suspended sediments in the water [16]. Then, a substantial portion of the non-woody litter entering running waters is buried in the streambed after becoming submerged [44], because of relatively high sediment yields and flow velocities [45]. The input of non-woody litter can reduce the mobility and erodibility of sediments in headwater streams [46], and thus, the variation in the metal storage in the sediment downstream is more stable and is not significantly different from that in the sediment upstream. The exclusion of non-woody litter could cause sediment to move downstream, and thus, non-woody litter could have difficultly contributing to the metal storage in the sediment in headwater streams. The input of woody litter increased the Mg, Fe and Mn concentration and storage in the sediment. As a harmful heavy metal, Cr showed an immobilization pattern in the early stage of litter decomposition [47], and therefore, the concentration and storage of Cr in the sediment increased in the streams with non-woody litter input.

The results showed that the storage of all the metals in the sediment was significantly correlated with temperature and illumination ($p < 0.05$). The addition of woody litter could reduce the water temperature and illumination in the headwater stream. Plant litter input drives ecosystem functioning by promoting periods of intense respiratory activity [48]. High-quality plant litter is preferred by aquatic fungi, bacteria and plankton, but it can be affected when light and temperature are limited [16]. Illumination is a direct mediating factor on ecosystem metabolism [36], and temperature can influence organisms such as bacteria, fungi and microalgae in streams to respond quickly to environmental changes, thus driving a large portion of material cycles [49]. These factors play an important role in regulating primary productivity, decomposition, and disturbance. These factors are also influenced by the input of plant litter, and the exclusion of plant litter could reduce the water temperature in this study, thus affecting the transport, accumulation, and storage of metals in the sediment in the headwater streams.

### 5. Conclusions

Assessing the effects of non-woody litter and woody litter inputs on metals through comparisons of the concentrations, exports and storage of the metals in headwater streams with different plant litter input conditions in an alpine forest is critical to understanding the process controlling the metals in water and sediments by allochthonous plant litter. Through this method of control and comparison, it was found that the input of woody litter can significantly increase the metal export ratio in water. Meanwhile, the input of non-woody litter can reduce the metal concentration in water and facilitate the stable storage of metals in the sediment in the headwater streams. Although the input of woody and non-woody litter had different effects on metal concentration in sediments, they had no significant effect on metal storage in sediments. The dynamics of only five metals in the headwater streams with different input conditions of plant litter were measured in this study, but these metals are representative and contain macroelement, microelement and heavy metal. The input of non-woody and woody litter had contrary effects on the storage of heavy metal Cr in the sediment from other elements. Nevertheless, the study results still showed the responses of the metals to the input of allochthonous organic litter. Knowing this response is critical for understanding the processes controlling the metal concentrations in water and sediments and for assessing and managing the water quality of headwater streams in the alpine forest.

**Author Contributions:** K.Y. and F.W. conceived the idea. Z.L. and J.H. designed the experiment. X.N., Z.X., B.T. and L.Z. provided advice in study design. Z.L. and J.H. conducted the fieldwork. Z.L. collected and analyzed the data. Z.L. wrote the manuscript. All authors contributed to the revision of the manuscript.

**Funding:** This study was financially supported by the National Natural Science Foundation of China (31800373 and 31670526), the National Key Technologies R & D Program of China (2017YFC0505003), the Key Technologies R & D Program of Sichuan (18ZDYF0307) and the Fok Ying-Tong Education Foundation for Young Teachers (161101).

**Acknowledgments:** We would like to thank Junwei Wu, Fan Yang, Zhuang Wang, Liyan Zhuang, Jiao Zhou and Fei Duan for their kind assistance in field work.

**Conflicts of Interest:** The authors declare no conflict of interest.

## References

1. He, Z.L.; Yang, X.E.; Stoffella, P.J. Trace elements in agroecosystems and impacts on the environment. *J. Trace Elem. Med. Biol.* **2005**, *19*, 125–140. [CrossRef]

2. Fu, F.L.; Wang, Q. Removal of heavy metal ions from wastewaters: A review. *J. Environ. Manag.* **2011**, *92*, 407–418. [CrossRef]

3. Audry, S.; Schäfer, J.; Blanc, G.; Jouanneau, J. Fifty-year sedimentary record of heavy metal pollution (Cd, Zn, Cu, Pb) in the lot river reservoirs (France). *Environ. Pollut.* **2004**, *132*, 413–426. [CrossRef]

4. Taka, M.; Aalto, J.; Virkanen, J.; Luoto, M. The direct and indirect effects of watershed land use and soil type on stream water metal concentrations. *Water Resour. Res.* **2016**, *52*, 7711–7725. [CrossRef]

5. Dosskey, M.G.; Vidon, P.; Gurwick, N.P.; Allan, C.J.; Duval, T.P.; Lowrance, R. The role of riparian vegetation in protecting and improving chemical water quality in streams. *J. Am. Water. Resour. Assoc.* **2010**, *46*, 261–277. [CrossRef]

6. Daniels, R.B.; Gilliam, J.W. Sediment and chemical load reduction by grass and riparian filters. *Soil Sci. Soc. Am. J.* **1996**, *60*, 246–251. [CrossRef]

7. Ohta, T.; Shin, K.C.; Saitoh, Y.; Nakano, T.; Hiura, T. The Effects of Differences in Vegetation on Calcium Dynamics in Headwater Streams. *Ecosystems* **2018**, *21*, 1390–1403. [CrossRef]

8. Battin, T.J.; Kaplan, L.A.; Findlay, S.; Hopkinson, C.S.; Marti, E.; Packman, A.I.; Sabater, F. Biophysical controls on organic carbon fluxes in fluvial networks. *Nat. Geosci.* **2008**, *1*, 95–100. [CrossRef]

9. Harmon, M.E.; Nadelhoffer, K.J.; Blair, J.M. Measuring decomposition, nutrient turnover, and stores in plant litter. In *Standard Methods for Long-Term Ecological Research*; Robertson, G.P., Coleman, D.C., Bledsoe, C.S., Sollins, P., Eds.; Oxford University Press: New York, NY, USA, 1999; pp. 202–240.

10. Wallace, J.B.; Eggert, S.L.; Meyer, J.L.; Webster, J.R. Multiple trophic levels of a forest stream linked to terrestrial litter inputs. *Science* **1997**, *277*, 102–104. [CrossRef]

11. Webster, J.R.; Tank, J.L.; Wallace, J.B.; Meyer, J.L.; Eggert, S.L.; Ehrman, T.P.; Ward, B.R.; Bennett, B.L.; Wagner, P.F.; McTammany, M.E. Effects of litter exclusion and wood removal on phosphorus and nitrogen retention in a forest stream. *Verh. Int. Verein. Limnol.* **2000**, *27*, 1337–1340. [CrossRef]

12. Yue, K.; Yang, W.Q.; Peng, Y.; Zhang, C.; Huang, C.P.; Wu, F.Z. Chromium, cadmium, and lead dynamics during winter foliar litter decomposition in an alpine forest river. *Arct. Antarct. Alp. Res.* **2016**, *48*, 79–91. [CrossRef]

13. Downing, J.A.; Cole, J.J.; Duarte, C.M.; Middelburg, J.J.; Melack, J.M.; Prairie, Y.T.; Kortelainen, P.; Striegl, R.G.; McDowell, W.H.; Tranvik, L.J. Global abundance and size distribution of streams and rivers. *Inland Waters* **2012**, *2*, 229–236. [CrossRef]

14. Allan, J.D.; Castillo, M.M. *Stream Ecology: Structure and Function of Running Waters*; Springer: Dordrecht, The Netherlands, 2007.

15. Kobayashi, S.; Kagaya, T. Differences in patches of retention among leaves, woods and small litter particles in a headwater stream: The importance of particle morphology. *Limnology* **2008**, *9*, 47–55. [CrossRef]

16. Gessner, M.O.; Swan, C.M.; Dang, C.K.; Mckie, B.G.; Bardgett, R.D.; Wall, D.H.; Hättenschwiler, S. Diversity meets decomposition. *Trends Ecol. Evol.* **2010**, *25*, 372–380. [CrossRef]

17. Yue, K.; García-Palacios, P.; Parsons, S.A.; Yang, W.Q.; Peng, Y.; Tan, B.; Huang, C.P.; Wu, F.Z. Assessing the temporal dynamics of aquatic and terrestrial litter decomposition in an alpine forest. *Funct. Ecol.* **2018**, *32*, 2464–2475. [CrossRef]

18. Wang, C.Y.; Xie, Y.Z.; Ren, Q.S.; Li, C.X. Leaf decomposition and nutrient release of three tree species in the hydro-fluctuation zone of the Three Gorges Dam Reservoir, China. *Environ. Sci. Pollut. Res.* **2018**, *25*, 23261–23275. [CrossRef]

19. Rinella, D.J.; Booz, M.; Bogan, D.L.; Boggs, K.; Sturdy, M.; Rinella, M.J. Large woody litter and salmonid habitat in the Anchor River basin, Alaska, following an extensive spruce beetle (*Dendroctonus rufipennis*) outbreak. *Northwest Sci.* **2009**, *83*, 57–69. [CrossRef]

20. Tank, J.L.; Rosi-Marshall, E.J.; Griffiths, N.A.; Entrekin, S.A.; Stephen, M.L. A review of allochthonous organic matter dynamics and metabolism in streams. *J. N. Am. Benthol. Soc.* **2010**, *29*, 118–146. [CrossRef]

21. Gomi, T.; Sidle, R.C.; Bryant, M.D.; Woodsmith, R.D. The characteristics of woody litter and sediment distribution in headwater streams, southeastern Alaska. *Can. J. For. Res.* **2001**, *31*, 1386–1399. [CrossRef]

22. Viers, J.; Dupre, B.; Gaillardet, J. Chemical composition of suspended sediments in World Rivers: New insights from a new database. *Sci. Total Environ.* **2009**, *407*, 853–868. [CrossRef]

23. Rodríguez-Blanco, M.L.; Soto-Varela, F.; Taboada-Castro, M.M.; Taboada-Castro, M.T. Using hysteresis analysis to infer controls on sediment-associated and dissolved metals transport in a small humid temperate catchment. *J. Hydrol.* **2018**. [CrossRef]

24. Yue, K.; Yang, W.Q.; Peng, C.H.; Peng, Y.; Zhang, C.; Huang, C.P.; Tan, Y.; Wu, F.Z. Foliar litter decomposition in an alpine forest meta-ecosystem on the eastern Tibetan Plateau. *Sci. Total Environ.* **2016**, *566*, 279–287. [CrossRef]

25. Fu, C.K.; Yang, W.Q.; Tan, B.; Xu, Z.F.; Zhang, Y.; Yang, J.P.; Ni, X.Y.; Wu, F.Z. Seasonal Dynamics of Litterfall in a Sub-Alpine Spruce-Fir Forest on the Eastern Tibetan Plateau: Allometric Scaling Relationships Based on One Year of Observations. *Forests* **2017**, *8*, 314. [CrossRef]

26. Ni, X.Y.; Yang, W.Q.; Tan, B.; He, J.; Xu, L.Y.; Li, H.; Wu, F.Z. Accelerated foliar litter humification in forest gaps: Dual feedbacks of carbon sequestration during winter and the growing season in an alpine forest. *Geoderma* **2015**, *241*, 136–144. [CrossRef]

27. Graça, M.A.S.; Bärlocher, F.; Gessner, M.O. Part 1. Litter dynamics. Coarse benthic organic matter. In *Methods to Study Litter Decomposition a Practical Guide*; Springer: Dordrecht, The Netherlands, 2005.

28. Yue, K.; Yang, W.Q.; Peng, Y.; Zhang, C.; Huang, C.P.; Xu, Z.F.; Tan, B.; Wu, F.Z. Dynamics of multiple metallic elements during foliar litter decomposition in an alpine forest river. *Ann. For. Sci.* **2016**, *73*, 547–557. [CrossRef]

29. Mutema, M.; Chaplot, V.; Jewitt, G.; Chivenge, P.; Blöschl, G. Annual water, sediment, nutrient, and organic carbon fluxes in river basins: A global meta-analysis as a function of scale. *Water Resour. Res.* **2016**, *51*, 8949–8972. [CrossRef]

30. Song, C.L.; Wang, G.X.; Sun, X.Y.; Chang, R.Y.; Mao, T.X. Control factors and scale analysis of annual river water, sediments and carbon transport in China. *Sci. Rep.* **2016**, *6*, 25963. [CrossRef]

31. Richardson, J.S.; Danehy, R.J. A synthesis of the ecology of headwater streams and their riparian zones in temperate forests. *For. Sci.* **2007**, *53*, 131–147. [CrossRef]

32. Huser, B.; Fölster, J.; Köhler, S. Lead, zinc, and chromium concentrations in acidic headwater streams in Sweden explained by chemical, climatic, and land-use variations. *Biogeosciences* **2012**, *9*, 4323–4335. [CrossRef]

33. Landre, A.L.; Watmough, S.A.; Dillon, P.J. The effects of dissolved organic carbon, acidity and seasonality on metal geochemistry within a forested catchment on the Precambrian Shield, central Ontario, Canada. *Biogeochemistry* **2009**, *93*, 271–289. [CrossRef]

34. Elosegi, A.; Díez, J.; Pozo, J. Contribution of dead wood to the carbon flux in forested streams. *Earth Surf. Proc. Landf.* **2007**, *32*, 1219–1228. [CrossRef]

35. Wallace, J.B.; Webster, J.R.; Eggert, S.L.; Meyer, J.L.; Siler, E.R. Large woody litter in a headwater stream: Long-term legacies of forest disturbance. *Int. Rev. Hydrobiol.* **2001**, *86*, 501–513. [CrossRef]

36. Young, R.G.; Matthaei, C.D.; Townsend, C.R. Organic matter breakdown and ecosystem metabolism: Functional indicators for assessing river ecosystem health. *J. N. Am. Benthol. Soc.* **2008**, *27*, 605–625. [CrossRef]

37. Foucreau, N.; Puijalon, S.; Hervant, F.; Piscart, C. Effect of leaf litter characteristics on leaf conditioning and on consumption by *Gammarus pulex*. *Freshw. Biol.* **2013**, *58*, 1672–1681. [CrossRef]

38. Merten, E.C.; Vaz, P.G.; Decker-Fritz, J.A.; Finlay, J.C.; Stefan, H.G. Relative importance of breakage and decay as processes depleting large wood from streams. *Geomorphology* **2013**, *190*, 40–47. [CrossRef]

39. Quinn, J.M.; Phillips, N.R.; Parkyn, S.M. Factors influencing retention of coarse particulate organic matter in streams. *Earth Surf. Proc. Landf.* **2010**, *32*, 1186–1203. [CrossRef]

40. Osei, N.A.; Gurnell, A.M.; Harvey, G.L. The role of large wood in retaining fine sediment, organic matter and plant propagules in a small, single-thread forest river. *Geomorphology* **2015**, *235*, 77–87. [CrossRef]

41. Turowski, J.M.; Hilton, R.G.; Sparkes, R. Decadal carbon discharge by a mountain stream is dominated by coarse organic matter. *Geology* **2016**, *44*, 27–30. [CrossRef]

42. Ruiz-Villanueva, V.; Piégay, H.; Gaertner, V.; Perret, F.; Stoffel, M. Wood density and moisture sorption and its influence on large wood mobility in rivers. *Catena* **2016**, *140*, 182–194. [CrossRef]

43. Wohl, E.; Scott, D.N. Wood and sediment storage and dynamics in river corridors. *Earth Surf. Proc. Landf.* **2017**, *42*, 5–23. [CrossRef]

44. Cornut, J.; Elger, A.; Greugny, A.; Bonnet, M.; Chauvet, E. Coarse particulate organic matter in the interstitial zone of three French headwater streams. *Ann. Limnol.-Int. J. Limnol.* **2012**, *48*, 303–313. [CrossRef]

45. Leithold, E.L.; Blair, N.E.; Perkey, D.W. Geomorphologic controls on the age of particulate organic carbon from small mountainous and upland rivers. *Glob. Biogeochem. Cycles* **2006**, *20*. [CrossRef]

46. Eggert, S.L.; Wallace, J.B.; Meyer, J.L.; Webster, J.R. Storage and export of organic matter in a headwater stream: Responses to long-term detrital manipulations. *Ecosphere* **2012**, *3*, 1–25. [CrossRef]

47. Yue, K.; Yang, W.Q.; Tan, B.; Peng, Y.; Huang, C.P.; Xu, Z.F.; Ni, X.Y.; Yang, Y.; Zhou, W.; Zhang, L.; et al. Immobilization of heavy metals during aquatic and terrestrial litter decomposition in an alpine forest. *Chemosphere* **2019**, *216*, 419–427. [CrossRef]

48. Roberts, B.J.; Mulholland, P.J.; Hill, W.R. Multiple scales of temporal variability in ecosystem metabolism rates: Results from 2 years of continuous monitoring in a forested headwater stream. *Ecosystems* **2007**, *10*, 558–606. [CrossRef]

49. Tockner, K.; Pusch, M.; Borchardt, D.; Lorang, M.S. Multiple stressors in coupled river-floodplain ecosystems. *Freshw. Boil.* **2010**, *55*, 135–151. [CrossRef]

MDPI

St. Alban-Anlage 66

4052 Basel

Switzerland

Tel. +41 61 683 77 34

Fax +41 61 302 89 18

www.mdpi.com

*Forests* Editorial Office

E-mail: forests@mdpi.com

www.mdpi.com/journal/forests